MATHEMATICS
FOR ENGINEERS

Volume 1

Numbers and Equations, Vectors and Matrices,
Functions in one Variable, Differential Calculus

MATHEMATICS FOR ENGINEERS

Volume 1

Numbers and Equations, Vectors and Matrices,
Functions in one Variable, Differential Calculus

Thomas Westermann

University of Applied Sciences Karlsruhe, Germany

World Scientific

NEW JERSEY · LONDON · SINGAPORE · BEIJING · SHANGHAI · HONG KONG · TAIPEI · CHENNAI · TOKYO

Published by

World Scientific Publishing Co. Pte. Ltd.

5 Toh Tuck Link, Singapore 596224

USA office: 27 Warren Street, Suite 401-402, Hackensack, NJ 07601

UK office: 57 Shelton Street, Covent Garden, London WC2H 9HE

British Library Cataloguing-in-Publication Data
A catalogue record for this book is available from the British Library.

Translation from the German language edition:
Mathematik für Ingenieure by Prof Dr Thomas Westermann
Copyright © Springer-Verlag Berlin Heidelberg 2020

MATHEMATICS FOR ENGINEERS
Volume 1: Numbers and Equations, Vectors and Matrices, Functions in
one Variable, Differential Calculus

ISBN 978-981-12-9234-7 (hardcover)
ISBN 978-981-12-9278-1 (paperback)
ISBN 978-981-12-9235-4 (ebook for institutions)
ISBN 978-981-12-9236-1 (ebook for individuals)

For any available supplementary material, please visit
https://www.worldscientific.com/worldscibooks/10.1142/13818#t=suppl

Desk Editor: Tan Rok Ting

Preface

The Karlsruhe University of Applied Sciences started an English-language Bachelor's program in Electrical Engineering in 2021. In order to support our international students with appropriate material, this three-volume textbook was written. The basis for this material is the German textbook "Mathematik für Ingenieure", which has been successfully introduced for years and is now in its 8th edition. Taking into account that different students will attend different mathematics courses, we decided to design three volumes of a series of mathematics books, each part for a single semester.

Mathematical terms are clearly motivated, systematically equated and visualized in many animations. Mathematical proofs are almost completely avoided. Instead, many applications not only support the application of mathematics, but also contribute to a better understanding of mathematics.

The successful acceptance of the German book and the positive response have led us to transfer the presentation and concept of the 8th edition as much as possible to the English lecture notes.

Important formulas and statements are clearly highlighted in order to increase the readability of the books. More than 300 images and sketches support the character of modern textbooks. The color-coded layout provides a clear overview of the presentation of the content, e.g. by adding new terms and definitions in light grey, important statements and sentences in grey.

There are additional styles to make the book easier to read:

— The symbol ⚠ **Caution:** draws your attention to passages that are often misinterpreted, overlooked or ignored.

— Tips and rules help you work through the examples and exercises.

— Definitions and important phrases are highlighted in grey boxes.

— Numerous summaries are highlighted in color.

— Important formulas and results are marked.

— Examples and applications are clearly arranged throughout the text.

— 380 fully worked examples,

— over 360 problems with solutions,

— and more than 200 illustrations and sketches will help you very well for **study and prepare for exams.**

Alongside the topics covered in the book, additional material is available on the website, as well as MAPLE worksheets, which can be downloaded for the current version of MAPLE. The description can be found under the MAPLE tab on the book's website:

$$http://www.imathhome.de/books/mathe/start.htm$$

In the book, the following two symbols explicitly refer to additional information information that can be found on the home page:

① Animations, in gif format are available: By clicking on the appropriate location on the web the animations are played through the Internet browser.

② References indicate the MAPLE descriptions. All MAPLE worksheets are available on the website. An overview of all worksheets can be found in *index.mws.*

I would especially like to thank Mayur Shelke for his valuable and intensive help in translating the German book into an adequate English textbook. I would also like to thank my students Abhijit Karande and Ajeya Raghuveer Simha for fine-tuning and proofreading the English version. Special thanks go to Ms. Andrea Wolf and Ms. Rok Ting of World Scientific Publishing, who made it possible for me to publish these lecture notes in this important and renowned publishing house.

Karlsruhe, February 2024   *Thomas Westermann*

Table of Contents

Chapter 1
Numbers, Equations and Systems of Equations

1
—

Numbers and equations are the most important basic concepts in mathematics, on which all other structures and constructions are based. In this chapter we will cover the basics using both sets and natural numbers. To describe the natural numbers, Peano's axioms are introduced and the principle of mathematical (complete) induction is demonstrated by many examples. The real numbers and elementary laws of arithmetic are presented; the basic rules for powers and logarithms are repeated.

Solving equations is one of the fundamental tasks of mathematics. In this chapter we will also discuss simple equations as well as systems of linear equations that are important for applications, and we will introduce the Gauss algorithm for solving them. Since only a few types of equations are explicitly solvable, we will not systematically cover the solution of general equations, but we will show by example how to work with basic equations.

1

1 Numbers, Equations and Systems of Equations

Numbers and equations are the most important basic concepts in mathematics, on which all other structures and constructions are based. In this chapter we will cover the basics using both sets and natural numbers. To describe the natural numbers, Peano's axioms are introduced and the principle of mathematical (complete) induction is demonstrated by many examples. The real numbers and elementary arithmetic laws are presented; the basic rules for powers and logarithms are repeated.

Solving equations is one of the fundamental tasks of mathematics. In this chapter, we will also discuss simple equations as well as systems of linear equations that are important for applications, and we will introduce the Gauss algorithm for solving them. Since only a few types of equations are explicitly solvable, we will not systematically cover the solution of general equations, but we will show by example how to work with basic equations.

1.1 Sets

"By a *set* M we mean any combination of certain well-defined objects of our view or of our thinking as a whole"; this definition of the set concept was introduced by G. Cantor (1895). This definition of a set is sufficient for our purposes. In the following, we will always use capital letters to describe sets. The objects of a set A are called *elements* of A and are described by lowercase letters.

$a \in A$ means: a is an element of the set A.

$a \notin A$ means: a is not an element of the set A.

Sets are usually introduced with
- listing the elements in curly brackets: $\{a_1, a_2, a_3, a_4, \ldots\}$,
- in the form of a statement: $\{a \in A : a \text{ has the attribute } E\}$.

The *empty set* $\emptyset$ or $\{\}$ contains no elements. B means *subset* of A ($B \subset A$), if every element of B is also an element of A.

Examples 1.1 (Sets):

$\mathbb{N}$	$=$	Set of natural numbers	$= \{1, 2, 3, 4, ...\}.$
$\mathbb{N}_0$	$=$	Set of natural numbers with zero	$= \{0, 1, 2, 3, 4, ...\}.$
$\mathbb{Z}$	$=$	Set of integers	$= \{0, \pm 1, \pm 2, \pm 3, ...\}.$
$\mathbb{Q}$	$=$	Set of rational numbers	$= \{\frac{p}{q} : p \in \mathbb{Z}, q \in \mathbb{N}\}.$
$\mathbb{R}$	$=$	Set of real numbers	

The following applies: $\mathbb{N} \subset \mathbb{N}_0 \subset \mathbb{Z} \subset \mathbb{Q} \subset \mathbb{R}$.   □

Remarks:

(1) The order of the elements in a set is irrelevant. It is therefore
$$\{a, b, c, d\} = \{d, c, a, b\}.$$

(2) Each element of a set is written only once, which means
$$\{a, a, a, b, d, d\} = \{a, b, d\}.$$

⊘ Set Operations

For two sets A and B, the *intersection* $A \cap B$, the *union* $A \cup B$ and the *complement* $A \backslash B$ are defined by

$$A \cap B := \{x : x \in A \text{ and } x \in B\},$$
$$A \cup B := \{x : x \in A \text{ or } x \in B\},$$
$$A \backslash B := \{x : x \in A \text{ and } x \notin B\}.$$

Here ":=" means that the symbol on the left side is defined by the right side of the equation. Similarly, ":⟺" is to be read as a logical equivalence according to the definition of what is on the sides of the colon.

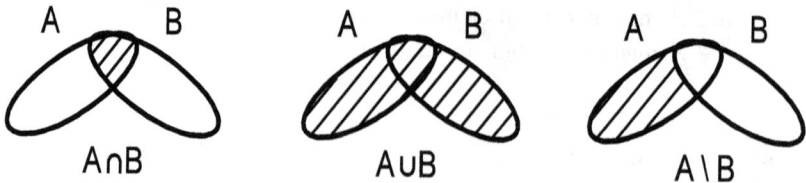

Figure 1.1. Venn Diagrams

The *Venn Diagrams* (see Fig. 1.1) can be used to illustrate sets and set operations. On the left is the intersection of two sets $A \cap B$, in the middle is the union of the sets $A \cup B$, and on the right is the complement of the sets A and B. The Venn diagrams are used to easily illustrate the following calculation rules for sets:

Calculation Rules for Sets

1. $A \cup B = B \cup A$

2. $A \cup A = A$

3. $A \cup (B \cup C) = (A \cup B) \cup C$

4. $A \cup (B \cap C) = (A \cup B) \cap (A \cup C)$

5. $(A \backslash B) \backslash C = A \backslash (B \cup C)$

6. $A \subset B \Leftrightarrow A \cap B = A \Leftrightarrow A \cup B = B \Leftrightarrow A \backslash B = \emptyset$

The *Cartesian product* of two sets M_1 and M_2 is the set consisting of all pairs (x, y), where $x \in M_1$ and $y \in M_2$:

$$M_1 \times M_2 := \{(x, y) : x \in M_1 \text{ and } y \in M_2\}.$$

Example 1.2. $\mathbb{R} \times \mathbb{R}$ consists of all pairs of real numbers. This is nothing more than the plane of numbers; (x, y) is a point on this plane. Instead of $\mathbb{R} \times \mathbb{R}$ we write $\mathbb{R}^2$ for short. □

1.2 Natural Numbers

Natural numbers are the most basic arithmetic objects. They form the basis of our number system. We call all natural numbers the "set of natural numbers" $\mathbb{N}$. The term "natural" numbers for $\mathbb{N} = \{1, 2, 3, \ldots\}$ is certainly well-chosen, because this is how children begin to count, and in all cultures mathematical reasoning begins with these numbers. The zero was invented quite late by the Indians, in 300 BC, and is added to the set of the natural numbers indicated by the index 0: $\mathbb{N}_0$.

It was only with the discovery of zero that Indian mathematicians were able to create a number system that has been adopted throughout the world, using only ten digits (including zero). The number system was later described by A. Ries (1492-1559) in his second book on arithmetic in 1522. There is also an appreciation of the number zero! The basic principle of natural numbers goes back to the mathematician Peano (1858-1939, 1889).

1.2.1 Peano's Axioms

Peano's Axioms

(1) 1 is a natural number.

(2) For every natural number n, there is exactly one successor n' which also belongs to the set of natural numbers.

(3) There is no natural number whose successor is 1.

(4) The successors of two different natural numbers are different from each other.

(5) A subset of the natural numbers contains **all** natural numbers, if two conditions are fulfilled: (1) 1 belongs to the set and (2) the successor n' always belongs to the set for any member n.

Thanks to Peano's axioms we are able to construct the set of natural numbers, because we directly obtain the following conclusions from the axioms:

Conclusions:

(1) The set of natural numbers has an infinite number of distinct elements: Because of $(A1)$ there is at least one natural number: 1. Because of $(A2)$ there is a successor to 1, according to $(A3) \neq 1$: We call it 2. Because of $(A2)$ there is a successor to 2, according to $(A3) \neq 1$ and $(A4) \neq 2$: We call it 3, and so on.

(2) The elements of the natural numbers can be arranged in a certain order, where all the natural numbers are considered step by step:

$$1; 2; 3; 4; 5; ..., n; n+1; ...$$

Here $n + 1$ means the successor of n. This order naturally determines the addition of natural numbers.

(3) Every subset M of the natural numbers $M \subset \mathbb{N}$, containing 1 and with $n \in M$ always having the successor $n + 1 \in M$, is equal to the set of all natural numbers.

Conclusion (3) gives us a principle of proof that can be applied to the most important methods of proof in analysis, namely *Mathematical Induction* or *Complete Induction*.

1.2.2 Mathematical Induction (Complete Induction)

To prove a proposition $A(n)$ for all natural numbers, it is sufficient to apply conclusion (3):

Mathematical Induction

Step 1: **Induction start**: For $n = 1$ the statement is true.

Step 2: **Induction closure** from n to $n + 1$: If the statement is true for any natural number n, then it must also be true for its successor $n + 1$.

If both steps can be performed, the statement is true for all $n \in \mathbb{N}$. After step 1, the statement is true for $n = 1$. After step 2, the statement is also true for the successor 2. Again, after Step 2, the statement is also true for the successor 3, and so on.

Example 1.3. $\quad 1 + 2 + \cdots + n = \dfrac{n(n+1)}{2} \qquad (n \in \mathbb{N})$

The start of the induction is to explicitly check the formula for $n = 1$. We insert $n = 1$ on both the left and the right side of the equation: For $n = 1$, the left side of the equation is 1 and the right side is $\frac{1 \cdot 2}{2} = 1$. So the formula is correct for $n = 1$.

Induction closure from n to $n + 1$: Let $n \in \mathbb{N}$ be arbitrary and let's assume that the formula for this n is correct, i.e. $1 + 2 + \cdots + n = \frac{n(n+1)}{2}$. Under these conditions we need to show that the formula holds for $n + 1$. We start with the left side of the equation; the sum now goes up to $n + 1$. We use the induction assumption and simplify the term until we find the right side for $n + 1$:

$$1 + 2 + \cdots + n + (n + 1) = (1 + 2 + \cdots + n) + (n + 1)$$
$$= \frac{n(n+1)}{2} + (n + 1)$$
$$= \frac{n(n+1) + 2(n+1)}{2} = \frac{1}{2}(n+1)(n+2).$$

This is the formula to be proven for $n + 1$. $\qquad\qquad\qquad\qquad \square$

Note: According to an anecdote, this formula goes back to F. Gauss (1777-1855), who had to calculate the sum of the first 100 numbers. He decided to take the sum of the first and the last, the second and the second last, the third and the third last, and so on:

$$1 + 2 + 3 + \cdots + 98 + 99 + 100 = (1 + 100) + (2 + 99) + (3 + 98) + \cdots .$$

Of the 100 addends, he obtained only $\frac{100}{2}$, each with a value of 101: $1 + 2 + 3 + \cdots + 100 = \frac{100 \cdot 101}{2}$. In fact, both the formula and the method of calculation were already known to Adam Ries (1492-1559; 1522).   □

Abbreviations for sums and products are introduced as follows:

Definition:
(1) *Sum*: For the sum of $a_l, a_{l+1}, \ldots, a_n \in \mathbb{R}$ we write

$$\sum_{k=l}^{n} a_k := a_l + a_{l+1} + \cdots + a_n.$$

(2) *Product*: For the product of $a_l, a_{l+1}, \ldots, a_n \in \mathbb{R}$ we write

$$\prod_{k=l}^{n} a_k := a_l \cdot a_{l+1} \cdot \ldots \cdot a_n.$$

(3) *Factorial*: For each $n \in \mathbb{N}$ we define

$$n! := 1 \cdot 2 \cdot \ldots \cdot n \qquad \textbf{(Factorial of } n \textbf{)} \qquad \text{and} \qquad 0! := 1.$$

Also, $n!$ grows very fast. For example $13! \approx 6 \cdot 10^9$; it would take 100 years to count this number if we could count to 100 in a minute!

Examples 1.4:

① $\displaystyle\sum_{i=5}^{10} i^2 = 5^2 + 6^2 + 7^2 + 8^2 + 9^2 + 10^2 = 355.$

② $\displaystyle\sum_{i=1}^{5} \frac{1}{2i} = \frac{1}{2 \cdot 1} + \frac{1}{2 \cdot 2} + \frac{1}{2 \cdot 3} + \frac{1}{2 \cdot 4} + \frac{1}{2 \cdot 5} = \frac{137}{120}.$

③ $\displaystyle\prod_{i=3}^{6} (2i - 1)^2 = 5^2 \cdot 7^2 \cdot 9^2 \cdot 11^2 = 12006225.$

④ $5! = 1 \cdot 2 \cdot 3 \cdot 4 \cdot 5 = 120.$ □

Example 1.5. $1 + 3 + 5 + \cdots + (2n - 1) = \sum_{k=1}^{n}(2k - 1) = n^2 \quad (n \in \mathbb{N})$

Proof with complete induction. We start the induction with $n = 1$ by inserting $n = 1$ on both sides of the equation: $1 = 1^2$. The formula is therefore correct for $n = 1$.

Induction closure from n to $n + 1$: If n is arbitrary and the formula is correct for this n, then we need to show that the formula also holds for $n + 1$:

$$1 + 3 + 5 + \cdots + (2n - 1) + (2n + 1) =$$
$$= [1 + 3 + \cdots + (2n - 1)] + (2n + 1)$$
$$= n^2 + 2n + 1 = (n + 1)^2.$$

This is the formula for $n + 1$. $\qquad\qquad\square$

Warning: The principle of the **complete** induction refers to both the beginning of the induction and the induction closure. If one of these two parts is missing, the proof is incomplete and the statement may not be correct, as the following two examples show:

⚠ **Example 1.6.** According to L. Euler (1707-1783), the expression

$$p = n^2 - n + 41$$

gives for $n = 1, 2, 3, \ldots, 40$ *prime numbers*: $p = 41, 43, 47, \ldots, 1601$. They can be calculated explicitly by substituting the appropriate n. However, this is not sufficient for a general proof. For $n = 41$ we get $p = 41^2 - 41 + 41 = 41^2$. This is indeed not a prime number. The prime calculation is correct for many single n, but in general it is **not** valid! $\qquad\square$

⚠ **Example 1.7.** We examine the *wrong* formula

$$1 + 2 + 3 + \cdots + n = \frac{n(n + 1)}{2} + 1$$

(see Example 1.3) and show that the induction closure is feasible: We assume that the formula is true for n and show that it is also true for $n + 1$.

$$1 + 2 + 3 + \cdots + n + (n + 1) = \frac{n(n+1)}{2} + 1 + (n + 1)$$
$$= \frac{n(n+1)+2(n+1)}{2} + 1 = \frac{(n+1)(n+2)}{2} + 1.$$

This is the formula for $n + 1$. Although the induction closure is possible, there are **no** natural numbers n for which the formula would be correct. The induction closure therefore loses its meaning if the proof cannot be given for $n = 1$ or for any other fixed integer.   □

1.2.3 Geometric Sum
The geometric sum is important in many applications:

Geometric Sum Formula

For each real number $q \neq 1$: $\displaystyle\sum_{i=0}^{n} q^i = \frac{1 - q^{n+1}}{1 - q}$ $(n \in \mathbb{N}_0)$

Proof by complete induction.

Induction start $n = 0$: $\displaystyle\sum_{i=0}^{0} q^i = q^0 = 1 = \frac{1 - q^1}{1 - q}$.

Induction closure from n to $n + 1$: If n is arbitrary and the formula is correct for n, then we check that it is also valid for $n + 1$:

$$\sum_{i=0}^{n+1} q^i = \sum_{i=0}^{n} q^i + q^{n+1} = \frac{1 - q^{n+1}}{1 - q} + q^{n+1}$$
$$= \frac{1 - q^{n+1} + (1 - q)q^{n+1}}{1 - q} = \frac{1 - q^{n+2}}{1 - q}.$$ □

Alternative Consideration: Often the question arises how we get such a simplification as given by the geometric sum. In particular, if we look at this formula, we obtain the simplification directly by means of the following consideration: We multiply the sum s_n by q

$$s_n = 1 + q + q^2 + q^3 + \cdots + q^n$$
$$q\, s_n = \quad q + q^2 + q^3 + \cdots + q^n + q^{n+1}.$$

Note that both expressions are identical except for the first and last terms. So if we subtract $q\, s_n$ from s_n, we get

$$s_n - q\, s_n = 1 - q^{n+1}.$$

With $s_n - q\, s_n = s_n\,(1 - q)$ and dividing by $1 - q$ for $q \neq 1$ we get the result.

1.2.4 Permutations

A permutation of a set is any possible arrangement of the elements of the set. If $A = \{a_1, a_2, a_3, \ldots, a_n\}$, there is exactly one element for each position in the set. Another arrangement of the set would be for example

$$\{a_2, a_1, a_3, \ldots, a_n\}.$$

The following statement gives the number of different arrangements of a set with n elements:

Permutations: The number of all possible arrangements of a set with n elements $a_1, \ldots, a_n$ is equal to $\quad n! = 1 \cdot 2 \cdot 3 \cdot \ldots \cdot n.$

Proof by complete induction. The induction starts again with $n = 1$: The number of arrangements of the 1-element set $\{a_1\}$ is 1. Since $1! = 1$, the statement is checked for $n = 1$.

Induction closure from n to $n + 1$: We search for the number of all arrangements of an $(n + 1)$-element set $\{a_1, a_2, a_3, \ldots, a_{n+1}\}$. We will consider the element a_1 and all its positions in this set. a_1 can be at the 1st position, then there are $n!$ arrangements for the remaining n elements according to the induction assumption. a_1 can also be at the 2nd position, then there are $n!$ arrangements for the remaining n elements. a_1 can also be at the 3rd position, again there are $n!$ arrangements for the remaining n elements, and so on. So a_1 can be in $n + 1$ different positions, and the remaining n elements still have $n!$ different arrangements. So in total there are $n! \cdot (n + 1) = (n + 1)!$ arrangements. $\qquad\square$

Conclusion: Let $A = \{a_1, \ldots, a_n\}$ be an n-element set. The number of k-element subsets of A is equal to

$$\frac{n!}{k!(n - k)!}.$$

Proof: All k-element subsets of the set $M = \{a_1, \ldots, a_n\}$ can be found by taking only the first k elements from all $n!$ arrangements. Each k-element subset occurs $k!$ times and the remaining $(n - k)!$ times. So the k-element subsets are equal to $\frac{n!}{k!(n-k)!}$. $\qquad\square$

Application: The probability of getting the right combination in the "6 out of 49" lottery game is about 1 in 14 million. Since the number of all 6-element subsets of a 49-element set is according to the last conclusion $\frac{49!}{6!\,43!} = \frac{44\cdot45\cdot46\cdot47\cdot48\cdot49}{1\cdot2\cdot3\cdot4\cdot5\cdot6} = 13.983.816.$

1.2.5 The Binomial Theorem

For two natural numbers n and k with $0 \leq k \leq n$ the following number is introduced

$$\binom{n}{k} := \frac{n!}{k!(n-k)!} = \frac{n(n-1)\ldots(n-k+1)}{k!}.$$

We speak of n over k and call them *binomial coefficients*. The binomial coefficients are determined either by the above formula or by the scheme named after Pascal, called **Pascal's triangle:** Starting from 1, the pyramid given below is extended by 1 to the right and to the left at each step. The numbers in between are the sum of the two numbers above.

$\binom{0}{k}:$ 1

$\binom{1}{k}:$ 1   1

$\binom{2}{k}:$ 1   2   1

$\binom{3}{k}:$ 1   3   3   1

$\binom{4}{k}:$ 1   4   6   4   1

$\binom{5}{k}:$ 1   5   10   10   5   1

$\binom{6}{k}:$ 1   6   15   20   15   6   1

$\phantom{\binom{6}{k}:}\ \vdots$

Note: Using the binomial coefficients, we can briefly formulate the last conclusion: There are $\binom{n}{k}$ ways of choosing k out of n objects. From this statement we derive the binomial theorem:

Binomial Theorem

For any real numbers $a, b \in \mathbb{R}$ and any natural number $n \geq 0$, the following holds:

$$(a+b)^n = \sum_{k=0}^{n} \binom{n}{k} a^{n-k} b^k$$

Proof: If we expand the left side, the term b^k occurs as often as we can choose k factors from n factors, i.e. $\binom{n}{k}$-times (see the note above). The remaining $(n-k)$ factors contribute to a^{n-k}. □

Examples 1.8:

① $(x+y)^0 = 1$
$(x+y)^1 = x + y$
$(x+y)^2 = x^2 + 2xy + y^2$
$(x+y)^3 = x^3 + 3x^2y + 3xy^2 + y^3$

② We use this theorem to calculate the value of the power of $(104)^3$:
$$(104)^3 = (100+4)^3 = 100^3 + 3 \cdot 100^2 \cdot 4 + 3 \cdot 100 \cdot 4^2 + 4^3$$
$$= 1\,000\,000 + 120\,000 + 4\,800 + 64 = 1.124.\,864.$$

③ $\quad 2^n = \sum_{k=0}^{n} \binom{n}{k}; \qquad 0 = \sum_{k=0}^{n} (-1)^k \binom{n}{k}.$ □

1.3 Real Numbers

We assume that the real numbers are available, and we don't deal with their axiomatic structure. For measurements the rational numbers would be sufficient, but for the higher analysis the rational numbers have "too many holes". Only their extension to the real numbers makes calculus possible.

1.3.1 Number Sets and Operations

The natural numbers $\mathbb{N}$ have $+$ and $\cdot$ as their basic arithmetic operations. The equation $x + 1 = 0$ can be formulated within $\mathbb{N}$, but cannot be solved. Therefore, the number range is extended to include all the solutions of the equations,

$$x + n = 0,$$

where $n \in \mathbb{N}_0$. The solutions are $0, -1, -2, -3, \ldots$ and the extended range of numbers is called $\mathbb{Z}$, the integers. In $\mathbb{Z}$, the equation $x + n = 0$ can be solved for each $n \in \mathbb{Z}$.

However, the equation $2 \cdot x = 1$ cannot be solved in $\mathbb{Z}$. To solve these equations, we have to extend $\mathbb{Z}$ by all solutions of equations of the form

$$q \cdot x = p$$

where $p, q \in \mathbb{Z}$ and $q \neq 0$. This results in the set of rational numbers $\mathbb{Q}$. In this set, all equations of the above form are solvable. But the equation

$$x^2 = 2$$

doesn't have a solution in $\mathbb{Q}$. So the rational numbers are extended by all the solutions of equations of the above construction, and the real numbers $\mathbb{R}$ are obtained. Within the real numbers there are also the so-called *transcendent* numbers, such as e and π, which we will examine in more detail in the chapter on Sequences.

Table 1 gives an overview of the number ranges and the corresponding arithmetic operations.

Table 1: Number Sets and Basic Arithmetic Operations

Sets	Basic Operations				Not Solvable
$\mathbb{N}_0$ natural numbers	$+$		$\cdot$		$x + 1 = 0$
$\mathbb{Z}$ integers	$+$	$-$	$\cdot$		$2 \cdot x = 1$
$\mathbb{Q}$ rational numbers	$+$	$-$	$\cdot$	$\backslash$	$x^2 = 2$
$\mathbb{R}$ real numbers	$+$	$-$	$\cdot$	$\backslash$	$x^2 + 1 = 0$

Representation of Real Numbers. The known *number line* from $-\infty$ to $+\infty$ is used to represent the real numbers. Each point on the number line corresponds exactly to one real number.

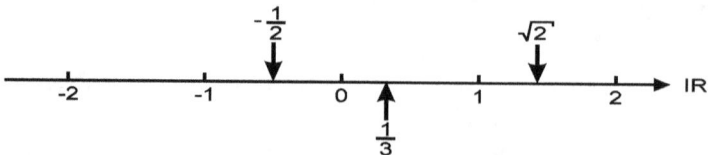

Figure 1.2. Real number line

1.3.2 Calculation Laws for Real Numbers

There are two operators in $\mathbb{R}$, $+$ and $\cdot$. Addition and multiplication of two real numbers yield real numbers. Formally, two mappings $+$ and $\cdot$ are defined:

$$+ : \mathbb{R} \times \mathbb{R} \longrightarrow \mathbb{R} \qquad \text{with} \qquad (x, y) \longmapsto x + y$$

$$\cdot : \mathbb{R} \times \mathbb{R} \longrightarrow \mathbb{R} \qquad \text{with} \qquad (x, y) \longmapsto x \cdot y.$$

The **calculation laws of addition** are applicable

$(A1)$	$x + (y + z) = (x + y) + z$	*Associative Law*
$(A2)$	$x + y = y + x$	*Commutative Law*
$(A3)$	$x + 0 = x$	*Existence of Zero*
$(A4)$	For each x there is $(-x) \in \mathbb{R}$ with	
	$x + (-x) = 0$	*Inverse Element*

The **calculation laws of multiplication** apply

$(M1)$	$x \cdot (y \cdot z) = (x \cdot y) \cdot z$	*Associative Law*
$(M2)$	$x \cdot y = y \cdot x$	*Commutative Law*
$(M3)$	There is a number $1 \in \mathbb{R}$ with $1 \neq 0$,	
	such that $1 \cdot x = x$	*Existence of 1*
$(M4)$	For every $x \in \mathbb{R} \backslash \{0\}$ there is a	
	$x^{-1} \in \mathbb{R}$ with $x \cdot x^{-1} = 1$	*Inverse Element*

The **Distribution law** applies

(D)	$x \cdot (y + z) = x \cdot y + x \cdot z$

All other rules for calculating real numbers can be traced back to these elementary rules. Since these calculation laws are not only valid for the set of real numbers, but also for other constructions, the concept of a *field* is introduced and generalized:

A set K together with two operations $+$ and $\cdot$ defined by

$$+ : K \times K \longrightarrow K \qquad \text{with} \qquad (x, y) \longmapsto x + y$$

$$\cdot : K \times K \longrightarrow K \qquad \text{with} \qquad (x, y) \longmapsto x \cdot y.$$

which satisfy the axioms $(A1)$-$(A4)$, $(M1)$-$(M4)$, (D), is called a **field**.

Examples 1.9:

① Both $(\mathbb{R}, +, \cdot)$ and $(\mathbb{Q}, +, \cdot)$ are fields.

② $(\mathbb{Z}, +, \cdot)$ is not a field because $(M4)$ is violated: For example, $2 \in \mathbb{Z}$ has no inverse element with respect to multiplication, so that $2 \cdot x = 1$.

③ $(\mathbb{N}, +, \cdot)$ is not a field, because $(A4)$ is violated.

④ $(F_2, +, \cdot)$ with $F_2 = \{0, 1\}$ and its operations is a field.

+	0	1
0	0	1
1	1	0

$\cdot$	0	1
0	0	0
1	0	1

The calculation laws are checked directly. F_2 is the smallest field, because each field must contain at least two elements: 0 and 1.   □

1.3.3 Power Calculations

For each real number $a \in \mathbb{R}$ we define the **power** of a by

$$a^0 := 1, \quad a^1 := a, \quad a^n := \underbrace{a \cdot \ldots \cdot a}_{n-times} \qquad (n \in \mathbb{N}).$$

Definition: *The* ***n-th root*** *of a number* $a \geq 0$

$$b := \sqrt[n]{a} := a^{\frac{1}{n}} \qquad (n \in \mathbb{N})$$

is defined as the positive real number b *with the property* $b^n = a$.

With the notation $\sqrt[n]{a} = a^{\frac{1}{n}}$, the n-th roots of a number can be interpreted as powers with rational exponents. For powers of products or quotients of real numbers, the general rules for calculating powers apply, which are summarized in the following table:

Power Calculation Rules

(1) $a^n \cdot b^n = (a \cdot b)^n$ $\qquad\qquad$ (2) $\dfrac{a^n}{b^n} = \left(\dfrac{a}{b}\right)^n$ for $(b \neq 0)$

(3) $\dfrac{a^n}{a^m} = a^{n-m}$ for $(a \neq 0)$ $\qquad\qquad$ (4) $(a^m)^n = a^{n \cdot m}$

(5) $\sqrt[n]{a^m} = a^{m/n}$ for $(a \geq 0)$ $\qquad\qquad$ $(n, m \in \mathbb{N})$

Examples 1.10:

① $\dfrac{a^{5x-2y}}{b^{6m-1}} : \dfrac{a^{4x+y}}{b^{m-2}} = \dfrac{a^{5x-2y-4x-y}}{b^{6m-1-m+2}} = \dfrac{a^{x-3y}}{b^{5m+1}}.$

② $\dfrac{(a^2 b)^2}{2a\sqrt{ab}} = a^4 b^2 \, \frac{1}{2} \, a^{-1} a^{-\frac{1}{2}} b^{-\frac{1}{2}} = \frac{1}{2} \, a^{\frac{5}{2}} b^{\frac{3}{2}}.$

③ $\dfrac{(8a^3 b^{-3})^{-2}}{(12a^{-2}b^{-4})^{-3}} = \dfrac{8^{-2}a^{-6}b^6}{12^{-3}a^6 b^{12}} = \dfrac{3^3}{a^{12}b^6}.$ $\qquad\qquad$ □

1.3.4 Logarithm

Definition: *Given is the equation* $a = b^x$ $(a, b > 0)$. *For given a and b we are looking for the exponent x. We call*

$$x = log_b (a)$$

the **logarithm of** a **to the base** b.

For a fixed base b the **logarithm calculation rules** apply

(1) $log_b(u \cdot v) = log_b(u) + log_b(v)$ $\qquad$ $(u, v > 0)$

(2) $log_b(\frac{u}{v}) = log_b(u) - log_b(v)$ $\qquad$ $(u, v > 0)$

(3) $log_b(u^n) = n \cdot log_b(u)$ $\qquad$ $(u > 0)$

Special logarithms are the logarithm to base 10 $\log a := \log_{10} a$,
the logarithm to base 2 (log dualism) $ld\, a := \log_2 a$,
and the logarithm to base e (natural logarithm) $\ln a := \log_e a$.
There is a relationship between logarithms in different bases

$$\log_b (y) = \frac{\log_c (y)}{\log_c (b)} \qquad (b,\, c,\, y > 0).$$

It is therefore sufficient to be able to calculate the logarithm in one base only (e.g. the natural logarithm). The logarithm to another base is then calculated using the formula above.

Proof of the logarithmic formula: From $b^x = y$ it follows by definition of the logarithm to the base b that $x = \log_b(y)$. On the other hand, for the logarithm to the base c, according to rule (3), we have:

$$\log_c(y) = \log_c(b^x) = x \cdot \log_c(b) \qquad \Rightarrow \qquad x = \frac{\log_c(y)}{\log_c(b)}.$$

This proves the formula above.   □

Examples 1.11:

① $2^x = \frac{1}{8} \Rightarrow x = \log_2 \frac{1}{8} = -\log_2 8 = -3$.

② $10^x = 0.0001 \Rightarrow x = \log_{10} 10^{-4} = -4\,\log_{10} 10 = -4$.

③ $\ln \frac{\sqrt{a}b^{-2}}{\sqrt[3]{c}d^{-3}} = \ln \sqrt{a} + \ln b^{-2} - \ln c^{\frac{1}{3}} - \ln d^{-3} = \frac{1}{2}\ln a - 2\ln b - \frac{1}{3}\ln c + 3\ln d$.

④ $\log\sqrt{\sqrt[3]{a^2 b}\sqrt[4]{a\,c^2}} = \log((a^2 b\,a^{\frac{1}{4}}\,c^{\frac{1}{2}})^{\frac{1}{3}})^{\frac{1}{2}} = \log a^{\frac{1}{3}}b^{\frac{1}{6}}a^{\frac{1}{24}}c^{\frac{1}{12}}$
  $= \frac{9}{24}\log a + \frac{1}{6}\log b + \frac{1}{12}\log c$.   □

1.3.5 Order of the Real Numbers

Among the real numbers there is a certain *order*: Two real numbers $a, b \in \mathbb{R}$ always have one of the following three relations to each other:

$a < b$ (a is to the left of b),   IR

$a = b$ (a is equal to b),   IR

$a > b$ (a is to the right of b).   IR

The **absolute value** of a real number a is the distance from a to the zero point. It is denoted by the symbol $|a|$:

$$|a| := \begin{cases} a & \text{for } a > 0 \\ 0 & \text{for } a = 0 \\ -a & \text{for } a < 0 \end{cases}$$

Examples 1.12: $|3| = 3 \,; |-5| = 5 \,; \left|-\frac{1}{2}\right| = \frac{1}{2} \,; \left|\sqrt{2}\right| = \sqrt{2}.$ □

The **distance** between two numbers x and a on the number line $\mathbb{R}$ is also given using the absolute value $|x - a|$.

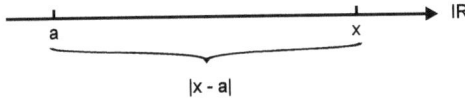

Figure 1.3. Distance between two numbers

According to the definition of the absolute value, we must distinguish between two different cases

$|x - a| = (x - a)$ **if** $x - a \geq 0$ or $x \geq a$,

$|x - a| = -(x - a)$ **if** $x - a < 0$ or $x < a$.

This clause will be important when solving equations and inequalities involving absolute values.

The following rules apply:

(1) $x > 0\,, y > 0 \Rightarrow x + y > 0$.

(2) $x > 0\,, y > 0 \Rightarrow x \cdot y > 0$.

(3) If $x > 0\,, y > 0$, then there is always a natural number $n \in \mathbb{N}$ with

$$n\, x > y \qquad (Archimedes'\ Axiom).$$

Conclusion: Bernoulli's Inequality

$$(1+x)^n \geq 1 + nx \qquad (x \geq -1 \text{ and } n \in \mathbb{N})$$

Proof by complete induction. The equality holds for $n = 1$. Induction closure from n to $n+1$: Since $1 + x > 0$, multiply the induction rule by $(1+x)^n \geq 1 + n\,x$ with $(1+x)$:
$$(1+x)^{n+1} \geq (1+nx)(1+x) = 1 + (n+1)x + nx^2 \geq 1 + (n+1)x. \quad \square$$

Intervals: Subsets of real numbers are called real ranges or intervals. There are finite intervals and infinite intervals. To describe these subsets of $\mathbb{R}$ we introduce the following notations:

(1) **Finite intervals** $(a < b)$

$$
\begin{aligned}
[a, b] &:= \{x : a \leq x \leq b\} && \text{closed interval}\\
[a, b) &:= \{x : a \leq x < b\}\\
(a, b] &:= \{x : a < x \leq b\} && \text{half open intervals}\\
(a, b) &:= \{x : a < x < b\} && \text{open interval}
\end{aligned}
$$

(2) **Infinite intervals**

$$
\begin{aligned}
\mathbb{R}_{\geq a} &:= [a, \infty) &&:= \{x : a \leq x < \infty\}\\
\mathbb{R}_{> a} &:= (a, \infty) &&:= \{x : a < x < \infty\}\\
\mathbb{R}_{\leq a} &:= (-\infty, a] &&:= \{x : -\infty < x \leq a\}\\
\mathbb{R}_{< a} &:= (-\infty, a) &&:= \{x : -\infty < x < a\}
\end{aligned}
$$

1.4 Equations and Inequalities

There are as many ways to solve equations as there are types of equations. We will show examples of how to solve basic equations and inequalities. However, we will not treat with the solution of equations or inequalities systematically, because in many cases they cannot be solved exactly and we therefore have to rely on numerical methods (e.g. see the bisection method §6.4 or Newton's method §7.8).

1.4.1 Equations

Any relationship between (real) quantities in which an equal sign occurs is called an **equation**. If the quantities in the equation occur only as sums and products of powers, they are called **algebraic equations**. The largest exponent indicates the *degree* of the equation.

⊘ 1. Quadratic Equations

The equation

$$x^2 + px + q = 0$$

is called a quadratic equation or equation of second degree. If the equation is in the form of $ax^2 + bx + c = 0$ with $a \neq 0$, we divide by a to get it in the form p/q. For the quadratic equation $x^2 + px + q = 0$, we have the p/q solution formula

---

p/q-Formula

$$x^2 + px + q = 0 \quad \Rightarrow \quad x_{1/2} = -\frac{p}{2} \pm \sqrt{\frac{p^2}{4} - q}.$$

We introduce $D := \frac{p^2}{4} - q$ the (*discriminant*). For $D > 0$ the equation has two different real solutions, for $D = 0$ a double real solution, and for $D < 0$ no real (but two different complex) solutions.

Examples 1.13:

① $x^2 + 2x - 3 = 0$ has two real solutions:
$x_{1/2} = -1 \pm \sqrt{(-1)^2 + 3} = -1 \pm 2$. This means $x_1 = 1$, $x_2 = -3$.

② $x^2 + 4x + 4 = 0$ has a double real solution:
$x_{1/2} = -2 \pm \sqrt{2^2 - 4} = -2$. So $x_1 = -2$ is a double solution.

③ $x^2 - 4x + 13 = 0$ has no real (but two complex) solutions:
$x_{1/2} = 2 \pm \sqrt{4 - 13} = 2 \pm \sqrt{-9} = 2 \pm 3i$. There are no real solutions.
Where i is the imaginary unit (see Chapter 5, Complex Numbers). ☐

⊘ 2. Higher Degree Equations

Higher degree equations are only partially analytically solvable, because in the rarest cases a closed solution exists. Occasionally a substitution leads to a quadratic equation, as the following example shows:

Example 1.14. We are looking for solutions to the equation

$$x^4 - 5x^2 + 4 = 0.$$

We substitute $z = x^2$ and get the quadratic equation

$$z^2 - 5z + 4 = 0.$$

If we use the p/q-formula

$$z_{1/2} = \frac{5}{2} \pm \sqrt{\frac{25}{4} - 4} = \frac{5}{2} \pm \sqrt{\frac{25 - 16}{4}} = \frac{5}{2} \pm \frac{3}{2}.$$

Therefore, $z_1 = 4$, $z_2 = 1$. Because of $z = x^2$ and $x = \pm\sqrt{z}$, we have

$$x_{1/2} = \pm\sqrt{4} = \pm 2$$

$$x_{3/4} = \pm\sqrt{1} = \pm 1.$$

The solutions of the equation are $x = -2, -1, 1, 2$.   □

Sometimes it is possible to guess a zero, and then reduce the problem by polynomial division or by using Horner's scheme. We will describe these techniques in detail in the section on polynomials §4.2.

⊘ 3. Root Equations

Simple root equations are solved by isolating the root, e.g. by placing it on the left side of the equation and the remaining terms on the right side. Square root equations are then squared and solved according to the variable being sought.

⚠ **Caution:** By squaring the equations, we can change the solution set: The equation $x = 1$ has only the number 1 as the solution. If we square the equation, we get $x^2 = 1$. This equation has both $x = 1$ and $x = -1$ as solutions. The squared equations may have more solutions than the original equations. Therefore, it is necessary to check that the resulting numbers are really solutions of the original root equation.

Example 1.15. We look for solutions of the root equation

$$\sqrt{5-x}+3-x=0.$$

We isolate the root

$$\sqrt{5-x}=x-3,$$

square it

$$5-x=x^2-6x+9$$

and rewrite it as

$$x^2-5x+4=0.$$

This is a quadratic equation that can be solved using the p/q-formula. Example 1.14 returns the solutions

$$x_1=4,\qquad x_2=1.$$

Now we have to check whether the two values are solutions of the original root equation. We do this by inserting the values into the root equation:

$x_1=4$: $\sqrt{1}-1=0$. This means that $x_1=4$ satisfies the square root equation.

$x_2=1$: $\sqrt{4}+2=4\neq 0$. This means that $x_1=1$ does not **satisfy** the square root equation.

So the solution set of the square root equation is $\mathbb{L}=\{4\}$.   □

⊘ 4. Absolute Value Equations

The absolute value $|a|$ of a real number a is defined by its distance from zero

$$|a|:=\begin{cases} a & \text{if } a\geq 0 \\ -a & \text{if } a<0 \end{cases}.$$

This definition of the absolute value always leads to a clause distinction in the case of an absolute value equation: Depending on whether there is a positive argument of the absolute value, then the absolute value is replaced by a bracket (), or if there is a negative argument of the absolute value, it is replaced by $-(\ )$.

Example 1.16 (With MAPLE-Worksheet). We look for solutions to the absolute value equation

$$|4x - 1| = -2x + 4.$$

To get an overview of the two functions, we draw the left and right sides of the equation:

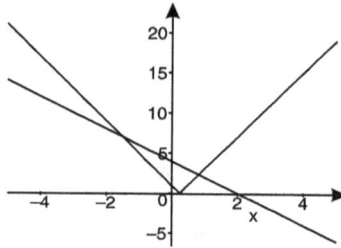

Figure 1.4. Left and right sides of the equation

We identify two intersections of the graphs that need to be determined.

Case 1: $4x - 1 \geq 0$, that is $x \geq \frac{1}{4}$:

The absolute values for $x \geq \frac{1}{4}$ are non-negative numbers. Under these conditions, the absolute value can be replaced by a single bracket:

$$4x - 1 = -2x + 4.$$

Solving for x, we obtain

$$6x = 5 \qquad \text{or} \qquad x = \frac{5}{6}.$$

$x = \frac{5}{6}$ satisfies $x \geq \frac{1}{4}$ under which we calculated the result.

Case 2: $4x - 1 < 0$, that is $x < \frac{1}{4}$:

For $x < \frac{1}{4}$ there are only negative numbers as argument. Under these conditions, the absolute value is replaced by $-(\ )$. In this case

$$-(4x - 1) = -2x + 4 \quad \Rightarrow \quad -4x + 1 = -2x + 4.$$

Solving for x, we get

$$2x = -3 \qquad \text{or} \qquad x = -\frac{2}{3}.$$

$x = -\frac{2}{3}$ satisfies $x < \frac{1}{4}$, under which we calculated the result.

Thus, the solution set is $\mathbb{L} = \{\frac{5}{6}, \frac{-3}{2}\}$.   □

Reference: Due to the widespread use of computers for solving mathematical problems, especially for solving equations, numerical methods are becoming increasingly important. Therefore, two sections of this textbook are devoted to numerical methods.

1.4.2 Inequalities

Equivalent operations acting on inequalities are:

Operations on Inequalities

(1) Add (or subtract) any term on either side sides of the inequality.

(2) Multiply (or divide) both sides by a positive number K$>$0.

(3) Multiply (or divide) both sides by a negative number K$<$0; this changes the sign of the inequality

(from $<$ to $>$), (from $\leq$ to $\geq$), (from $>$ to $<$), (from $\geq$ to $\leq$).

Example 1.17 (With MAPLE-Worksheet). We look for solutions to the absolute value inequality

$$|2x + 2| > 3.$$

As with absolute value equations, when solving inequalities, the absolute value must first be replaced using a case distinction.

Case 1: $2x + 2 \geq 0$, that is $x \geq -1$:
The absolute value for $x \geq -1$ does not contain negative numbers. Then the absolute value bracket can be replaced by a single bracket (...). In this example

$$2x + 2 > 3.$$

Solving for x, we get

$$x > \frac{1}{2}.$$

$x > \frac{1}{2}$ satisfies the condition $x \geq -1$ under which we carried out the calculation.

$$\Rightarrow \mathbb{L}_1 = (\frac{1}{2}, \infty).$$

Case 2: $2x + 2 < 0$, that is $x < -1$:

The argument of the absolute value for $x < -1$ is negative. In this case, the absolute value is replaced by $-(...)$. In this case

$$-(2x + 2) > 3.$$

We multiply by (-1). This changes the sign of the inequality

$$2x + 2 < -3$$

and solving for x, we get

$$x < -\frac{5}{2}.$$

$x = -\frac{5}{2}$ satisfies the condition $x < -1$ under which we carried out the calculation.

$$\Rightarrow \mathbb{L}_2 = (-\infty, -\frac{5}{2}).$$

So the solution set consists of two sub-intervals, the open interval $(-\infty, -\frac{5}{2})$ combined with the open interval $(\frac{1}{2}, \infty)$:

$$\mathbb{L} = \mathbb{L}_1 \cup \mathbb{L}_2 = (-\infty, -\frac{5}{2}) \cup (\frac{1}{2}, \infty). \qquad \square$$

Example 1.18 (With MAPLE-Worksheet). We look for the solutions to the absolute value inequality

$$(x - 2)^2 \le |x|.$$

To get an overview of the two sides of the inequality, we draw the left and right sides:

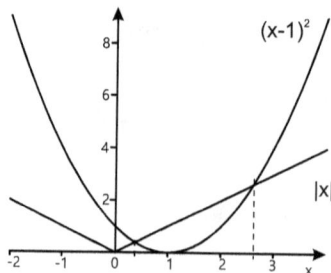

Figure 1.5. Left and right sides of the inequality

We see that there are two points of intersection of the graphs that need to be determined. The solution set then consists of the closed interval in which the quadratic function $(x-1)^2$ is less than or equal to the absolute value function $|x|$.

Intersection points: The graph shows that the intersection points are in the positive x-range. In this range, we can replace the absolute value $|x|$ with a simple bracket, so that we have to solve the equation

$$x = (x-1)^2 \quad \Rightarrow \quad x^2 - 2x + 1 = x \quad \Rightarrow \quad x^2 - 3x + 1 = 0.$$

The p/q-formula gives us

$$x_{1/2} = \frac{3}{2} \pm \sqrt{\frac{9}{4} - 1} = \frac{3}{2} \pm \sqrt{\frac{5}{4}}.$$

So $x_1 = 0.38$ and $x_2 = 2.62$ are the intersection points of the curves. The solution set consists of the closed interval between x_1 and x_2:

$$\mathbb{L} = \left[\frac{3}{2} - \frac{1}{2}\sqrt{5}, \frac{3}{2} + \frac{1}{2}\sqrt{5} \right] = [0.38, 2.62]. \qquad \qquad \square$$

1.5 Systems of Linear Equations

Systems of Linear Equations (**LEq**) play a very important role in theory and applications. In this section we introduce a method for solving arbitrary LEq: the *Gauss algorithm*. For general correlations and statements about LEq we refer to Chapter 3, Matrices and Determinants.

1.5.1 Introduction

Application Example 1.19   (Description of DC Circuits).

Given is the *electrical network* with the resistors $R_1 = 1\Omega$, $R_2 = 5\Omega$, $R_3 = 3\Omega$. Two DC currents $I_A = 1A$ and $I_B = 2A$ are fed into this network. The individual currents I_1, I_2, I_3 are to be found.

To obtain the model equations, we apply **Kirchhoff's laws:** The *node rule* states that the sum of the incoming and outgoing currents in a node is zero. The *mesh rule* states that in a mesh the sum of all voltages is zero.

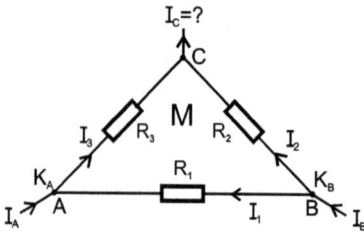

Figure 1.6. Electrical network

In this example, for the node K_A, I_3 flows in and I_A, I_1 flow out

$$(K_A): \quad I_3 = I_A + I_1;$$

for the node K_B, I_B flows in and I_1, I_2 flow out

$$(K_B): \quad I_B = I_1 + I_2.$$

For the mesh with specified current directions, the voltage drop across R_2 is equal to the sum of the voltage drops across R_1 and R_3:

$$(M): \quad R_1I_1 + R_3I_3 = R_2I_2.$$

This gives a *system* of 3 equations for the individual currents I_1, I_2, I_3. □

Method for solving this system of linear equations: We substitute the given values in the equations and introduce a method for solving the system systematically:

$$G_1: \quad 1I_1 - 5I_2 + 3I_3 = 0$$
$$G_2: \quad -1I_1 \qquad + 1I_3 = 1$$
$$G_3: \quad 1I_1 + 1I_2 \qquad = 2$$

I_1	I_2	I_3	$r.S.$
1	−5	3	0
−1	0	1	1
1	1	0	2

This system is solved by eliminating I_1 from equations G_2 and G_3. The sum of equations G_1 and G_2 and the difference between equations G_1 and G_3 is calculated:

$$G_1' = G_1: \qquad 1I_1 - 5I_2 + 3I_3 = \quad 0$$
$$G_2' = G_1 + G_2: \qquad - 5I_2 + 4I_3 = \quad 1$$
$$G_3' = G_1 - G_3: \qquad - 6I_2 + 3I_3 = -2$$

I_1	I_2	I_3	$r.S.$
1	−5	3	0
0	−5	4	1
0	−6	3	−2

Then we take the equation G'_2 and G'_3 and eliminate I_2 from G'_3. To do this, we add 6 times the equation G'_2 to (-5) times the equation G'_3:

$$
\begin{array}{rcr}
-30I_2 + 24I_3 &=& 6 \\
30I_2 - 15I_3 &=& 10 \\
\hline
9I_3 &=& 16
\end{array}
$$

This finally gives

		I_1	I_2	I_3	$r.S.$

$$
\begin{array}{lll}
G''_1 = G_1 : & 1I_1 - 5I_2 + 3I_3 = 0 \\
G''_2 = G'_2 : & -5I_2 + 4I_3 = 1 \\
G''_3 = 6G'_2 - 5G'_3 : & 9I_3 = 16
\end{array}
$$

I_1	I_2	I_3	$r.S.$
1	-5	3	0
0	-5	4	1
0	0	9	16

From the equation G''_3 we get

$$9I_3 = 16 \Rightarrow I_3 = \frac{16}{9}$$

and from the equation G''_2 we get

$$-5I_2 + 4 \cdot \frac{16}{9} = 1 \Rightarrow I_2 = \frac{11}{9}.$$

Substituting these two results into equation G''_1 gives

$$I_1 - 5 \cdot \frac{11}{9} + 3 \cdot \frac{16}{9} = 0 \Rightarrow I_1 = \frac{7}{9}.$$

The currents I_1, I_2, I_3 are thus calculated. ◻

In the right-hand scheme, the variables are not specified; only the coefficients of the variables and the constants of the right-hand side are listed. The coefficient of I_1 always appears first, the coefficient of I_2 second and the coefficient of I_3 third. In principle, this short version of the system of equations is sufficient to solve it. This procedure for solving this system of equations can be generalized ($\rightarrow$ Gauss algorithm) if the variables are linear ($\rightarrow$LEq).

1.5.2 Formulation and Notation

A *linear relationship* between two values x and y exists, if x is proportional to y $(x \sim y)$, so we define a linear equation in x and y if both quantities occur in a linear form, i.e. $a\,x + b\,y = const$. More generally, an equation of the form

$$ax_1 + bx_2 + cx_3 = d$$

is a *linear equation* in x_1, x_2, x_3, because each of the *variables* x_1, x_2 and x_3 is only present in a linear form, i.e. to the power of 1. Any *triple* of real numbers $(x_1, x_2, x_3) \in \mathbb{R}^3 = \mathbb{R} \times \mathbb{R} \times \mathbb{R}$ that satisfies the equation is called a solution.

Examples 1.20:

① $x_1 - x_2 + x_3 = 0$ is a linear equation because the variables x_1, x_2, and x_3 are linear. This equation has e.g. $(0, 1, 1)$, $(1, 1, 0)$, $(1, 2, 1)$ as a solutions.

② The equation $x^2 + 2x - y = 0$ is **not** a linear equation, because the variable x is quadratic.

③ $x_1 - x_2 + x_3 = 0$ and $2x_1 + 3x_2 - x_3 = 0$ form a system of linear equations. □

Definition: *A system of m linear equations in the n unknowns* $x_1, x_2, \ldots, x_n$

$$
\begin{aligned}
a_{11}x_1 + a_{12}x_2 + \cdots + a_{1n}x_n &= b_1 \\
a_{21}x_1 + a_{22}x_2 + \cdots + a_{2n}x_n &= b_2 \\
&\vdots \\
a_{m1}x_1 + a_{m2}x_2 + \cdots + a_{mn}x_n &= b_m
\end{aligned}
$$

is called a **system of linear equations (LEq)**. The real numbers a_{ij} are the **coefficients** and b_i are the right-hand side of the LEq.

The coefficients and the right-hand side of the LEq are written in a compact scheme

$$
\left(
\begin{array}{ccccc|c}
a_{11} & a_{12} & a_{13} & \cdots & a_{1n} & b_1 \\
a_{21} & a_{22} & a_{23} & \cdots & a_{2n} & b_2 \\
\vdots & \vdots & \vdots & & \vdots & \vdots \\
a_{m1} & a_{m2} & a_{m3} & \cdots & a_{mn} & b_m
\end{array}
\right) .
$$

This scheme is called the extended coefficient matrix, or **Matrix** for short. The solid line should remind you that the coefficients are to the left and the constants are to the right of the equals sign.

A system of linear equations is **homogeneous** if all b_i are equal to zero, otherwise **inhomogeneous.**

Each row of the matrix represents an equation; and each column corresponds to an unknown. The solution consists of all n-tuples $(x_1, x_2, ..., x_n)$ that satisfy all m equations.

As we saw in the introductory example, the elimination method only changes the coefficients and the constants, but not the variables. Therefore, when solving LEq, the variables are omitted completely and all calculation steps are performed in matrix notation.

1.5.3 Solving Systems of Linear Equations

Operations that do not change the solution set of a system are called **equivalence manipulations**. The following operations are equivalence operations on a system of linear equations:

(1) The order of the equations can be changed.

(2) An equation can be multiplied by any real number $\lambda \neq 0$.

(3) An equation in the system can be added to another one.

If these three manipulations are applied systematically, as described below, the solution of each LEq can be determined. As shown in the introductory example, at each calculation step one variable is eliminated from the system and thus reduced by one equation, until only one equation remains for one variable. The method, which goes back to Gauss (1777-1855), is called the **Gauss elimination method** or the **Gauss algorithm**. For simplicity, we restrict ourselves to quadratic systems with n equations for the n unknowns. However, the Gauss algorithm can be extended to arbitrary $(n \times m)$ systems.

Elimination Method (Gauss Algorithm)

---

(1) Select an equation with a non-zero coefficient of x_1 as the first equation.

(2) Eliminate the variable x_1 from the remaining $(n-1)$ equations. Multiply the first row by $-\frac{a_{21}}{a_{11}}$ and add it to the second equation. Do the same for the other rows: Add the $-\frac{a_{j1}}{a_{11}}$ multiplication from the first row to the j-th row. This gives us $(n-1)$ equations with $(n-1)$ unknowns $x_2, x_3, \ldots, x_n$.

(3) Step (2) is applied to the reduced system by eliminating the unknown x_2 from rows 3 to n. After a total of $(n-1)$ steps, only one equation remains for the unknown x_n.

(4) The eliminated equations form a *scaled* system of rows from which the unknowns can be calculated in the order $x_n, x_{n-1}, \ldots, x_2, x_1$.

⚠ The above algorithm assumes that none of the coefficients a_{ii} is zero, otherwise the lines must be swapped. If all the remaining coefficients of the variable x_i to be eliminated are equal to zero, this step can be skipped, because the LEq already has the desired form.

However, if the algorithm is performed numerically, there will be inaccuracies in the calculation, even if these coefficients are very small. To keep these errors as small as possible, it is advisable to swap the lines at each step, so that the line with the largest coefficient a'_{ii} is selected as the top equation. This is called *pivoting*.

Examples 1.21:
① **A system with exactly one solution:** The solution set of the LEq is searched for

$$
\begin{aligned}
2x_1 + x_2 - x_3 &= 3 \\
3x_1 + 5x_2 - 4x_3 &= 1 \\
4x_1 - 3x_2 + 2x_3 &= 2 .
\end{aligned}
$$

In matrix notation, this LEq is as follows

$$
\begin{array}{cc}
G_1: \\
G_2: \\
G_3:
\end{array}
\left(
\begin{array}{ccc|c}
2 & 1 & -1 & 3 \\
3 & 5 & -4 & 1 \\
4 & -3 & 2 & 2
\end{array}
\right)
$$

To solve the system we use the Gaussian algorithm. We copy the first line, multiply G_1 by (-3) and add the result to 2 times the second line. We also multiply the first line by (-2) and add the result to the third line:

$$
\begin{array}{cc}
G_1': \\
G_2': \\
G_3':
\end{array}
\left(
\begin{array}{ccc|c}
2 & 1 & -1 & 3 \\
0 & 7 & -5 & -7 \\
0 & -5 & 4 & -4
\end{array}
\right)
\quad
\begin{array}{l}
(G_1) \\
(2G_2 - 3G_1) \\
(G_3 - 2G_1)
\end{array}
$$

We leave the first two equations unchanged and modify the last one so that the coefficient of x_2 becomes zero.

$$
\begin{array}{cc}
G_1'': \\
G_2'': \\
G_3'':
\end{array}
\left(
\begin{array}{ccc|c}
2 & 1 & -1 & 3 \\
0 & 7 & -5 & -7 \\
0 & 0 & 3 & -63
\end{array}
\right)
\quad
\begin{array}{l}
(G_1') \\
(G_2') \\
(7G_3' + 5G_2')
\end{array}
$$

Now we can easily calculate the solutions: The last equation gives

$$
3x_3 = -63 \Rightarrow x_3 = -21.
$$

Substituted into G_2'': $7x_2 - 5 \cdot (-21) = -7 \Rightarrow x_2 = -16$.
Both used in G_1'': $2x_1 + (-16) - (-21) = 3 \Rightarrow x_1 = -1$.
So the system has **a unique solution** $(-1; -16; -21)$ and the solution set is given by

$$
\mathbb{L} = \left\{ (x_1, x_2, x_3) \in \mathbb{R}^3 : \begin{pmatrix} x_1 \\ x_2 \\ x_3 \end{pmatrix} = \begin{pmatrix} -1 \\ -16 \\ -21 \end{pmatrix} \right\}.
$$

The system $(\prime\prime)$ is called a system with an *upper triangular matrix* because the entries below the *main diagonal* $(a_{11}^{\prime\prime}, a_{22}^{\prime\prime}, a_{33}^{\prime\prime})$ are zero. If the system has an upper triangular shape, the elimination procedure is complete. The unknowns x_1, x_2, x_3 are then determined by *reverse resolution*.

② **The solution has one free parameter:** To solve the system

$$\begin{aligned} x_1 - 3x_2 + 2x_3 &= 4 \\ -2x_1 + x_2 + 3x_3 &= 2 \\ 2x_1 - 16x_2 + 18x_3 &= 28 \end{aligned}$$

we modify the coefficient matrix in two steps to obtain the *triangular form*

$$\begin{array}{c} G_1 \\ G_2 \\ G_3 \\ G_1' \\ G_2' \\ G_3' \\ G_1'' \\ G_2'' \\ G_3'' \end{array} \quad \begin{array}{c} \left(\begin{array}{ccc|c} 1 & -3 & 2 & 4 \\ -2 & 1 & 3 & 2 \\ 2 & -16 & 18 & 28 \end{array} \right) \\ \left(\begin{array}{ccc|c} 1 & -3 & 2 & 4 \\ 0 & -5 & 7 & 10 \\ 0 & -15 & 21 & 30 \end{array} \right) \\ \left(\begin{array}{ccc|c} 1 & -3 & 2 & 4 \\ 0 & -5 & 7 & 10 \\ 0 & 0 & 0 & 0 \end{array} \right) \end{array} \qquad \begin{array}{c} \\ \\ \\ \\ (G_2 + 2G_1) \\ (G_3 + G_2) \\ \\ \\ (G_3' - 3G_2') \end{array}$$

It follows from the last row that $\boxed{0 \cdot x_3 = 0}$, which is true for every x_3. Therefore, we set $x_3 = \lambda$ (arbitrarily). Inserting this into G_2'' we get

$$-5x_2 + 7\lambda = 10 \Rightarrow x_2 = -2 + \frac{7}{5}\lambda.$$

Inserting both into G_1'' gives

$$x_1 = 4 + 3\left(-2 + \frac{7}{5}\lambda\right) - 2\lambda = -2 + \frac{11}{5}\lambda.$$

To get a better representation, we choose $\lambda = 5k$ so that the solution is

$$\mathbb{L} = \left\{ (x_1, x_2, x_3) \in \mathbb{R}^3 : \begin{pmatrix} x_1 \\ x_2 \\ x_3 \end{pmatrix} = \begin{pmatrix} -2 \\ -2 \\ 0 \end{pmatrix} + k \begin{pmatrix} 11 \\ 7 \\ 5 \end{pmatrix} \text{ with } k \in \mathbb{R} \right\}.$$

③ **The system has no solution:** We look at the system of ② by modifying the last equation: We replace the constant 28 with 27. By elementary manipulations we obtain

$$\begin{pmatrix} 1 & -3 & 2 & | & 4 \\ 0 & -5 & 7 & | & 10 \\ 0 & 0 & 0 & | & -1 \end{pmatrix}.$$

The last row results in $\boxed{0 \cdot x_3 = -1}$. This equation cannot be fulfilled because the left side always returns zero. Therefore, $\mathbb{L} = \{\}$.

④ **Homogeneous LEq:** From the example ② we can immediately identify the solution set of the homogeneous LEq

$$\begin{array}{rrrl} x_1 - & 3x_2 + & 2x_3 & = 0 \\ -2x_1 + & x_2 + & 3x_3 & = 0. \\ 2x_1 - & 16x_2 + & 18x_3 & = 0 \end{array}$$

Since the elementary line manipulations give

$$\begin{pmatrix} 1 & -3 & 2 & | & 0 \\ 0 & -5 & 7 & | & 0 \\ 0 & 0 & 0 & | & 0 \end{pmatrix}.$$

By resolving backwards we get $\boxed{0 \cdot x_3 = 0}$ from row 3. So x_3 is arbitrary. We set $x_3 = 5\,k$ and insert in line 2 as follows

$$-5x_2 + 7 \cdot 5k = 0 \Rightarrow x_2 = 7k$$

and insert both into row 1:

$$x_1 = +3 \cdot 7k - 2 \cdot 5k = 11k.$$

Therefore

$$\mathbb{L} = \left\{ (x_1, x_2, x_3) \in \mathbb{R}^3 : \begin{pmatrix} x_1 \\ x_2 \\ x_3 \end{pmatrix} = k \begin{pmatrix} 11 \\ 7 \\ 5 \end{pmatrix} \text{ and } k \in \mathbb{R} \right\}. \qquad \square$$

The Examples ① - ④ suggest the following general conclusion, which we will examine in more detail in Chapter 3: Matrices and Determinants.

Solution Behavior of Systems of Linear Equations

(1) An **inhomogeneous** LEq has either exactly one solution or infinitely many solutions or no solution at all.

(2) A **homogeneous** LEq has either exactly one solution, namely the trivial zero solution $\vec{x} = \begin{pmatrix} 0 \\ \vdots \\ 0 \end{pmatrix}$, or infinitely many solutions.

(3) If the inhomogeneous LEq is solvable, the general solution consists of all the homogeneous solutions plus one solution of the inhomogeneous system:

$$\mathbb{L}_i = \mathbb{L}_h + \vec{x}_s \,,$$

where $\mathbb{L}_i = $ *solution set of the inhomogeneous LEq*, $\mathbb{L}_h = $ *solution set of the associated homogeneous LEq and x_s is a special solution of the inhomogeneous system.*

Application Example 1.22 (Chemical Reaction).

From quartz (SiO_2) and caustic soda $(NaOH)$ we obtain sodium silicate $(Na_2\,SiO_3)$ and water (H_2O):

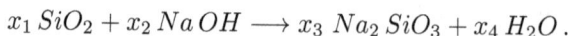

$$x_1\,SiO_2 + x_2\,NaOH \longrightarrow x_3\,Na_2\,SiO_3 + x_4\,H_2O\,.$$

We are looking for the proportions of the substances x_1, x_2, x_3, x_4 for which the reaction takes place. Since only integer multiples are possible, the natural numbers x_1, x_2, x_3, x_4 have to be determined in such a way that each of the chemical elements Si, O, Na, H occurs equally often on both sides of the reaction equation. This results in the following homogeneous system of linear equations:

$Si:\quad x_1 = x_3$
$Na:\quad x_2 = 2x_3$
$O:\quad 2x_1 + x_2 = 3x_3 + x_4$
$H:\quad x_2 = 2x_4.$

In matrix form, the LEq is as follows

$$\left(\begin{array}{ccc|cc} 1 & 0 & -1 & 0 & 0 \\ 0 & 1 & -2 & 0 & 0 \\ 2 & 1 & -3 & -1 & 0 \\ 0 & 1 & 0 & -2 & 0 \end{array}\right) \hookrightarrow \dots \hookrightarrow \left(\begin{array}{ccc|cc} 1 & 0 & -1 & 0 & 0 \\ 0 & 1 & -2 & 0 & 0 \\ 0 & 0 & 1 & -1 & 0 \\ 0 & 0 & 0 & 0 & 0 \end{array}\right).$$

The last row gives $0 \cdot x_4 = 0$. So x_4 is arbitrary. We choose $x_4 = k$ and insert it into row 3, followed by $x_3 = k$. Both results inserted into row 2 and row 1 give $x_2 = 2k$ and $x_1 = k$. The solution with the smallest numbers of the substances is therefore (for $k = 1$)

$$SiO_2 + 2\,Na\,OH \longrightarrow Na_2\,SiO_3 + H_2O. \qquad \square$$

Application Example 1.23   (Mixing of Alloys).

Stainless steel is an alloy of iron, chromium and nickel. For example, V2A steel consists of 74% iron, 18% chromium and 8% nickel. The table below shows the existing alloys (I - IV) with which 1000 kg of V2A steel is to be mixed.

	I	II	III	IV
Iron	70%	72%	80%	85%
Chrome	22%	20%	10%	12%
Nickel	8%	8%	10%	3%

If x_1, x_2, x_3, x_4 are the proportions of alloys I - IV in units of kg, then the sum of all alloy proportions in kg is given by

$$x_1 + x_2 + x_3 + x_4 = 1000.$$

The conservation equations for the individual constituents iron, chromium and nickel are as follows

$$0.7\,x_1 + 0.72\,x_2 + 0.8\,x_3 + 0.85\,x_4 = 740$$
$$0.22\,x_1 + 0.2\,x_2 + 0.1\,x_3 + 0.12\,x_4 = 180$$
$$0.08\,x_1 + 0.08\,x_2 + 0.1\,x_3 + 0.03\,x_4 = 80.$$

Note that for 1000 kg of 74% iron alloy, the weight of iron is 740 kg. The same applies to chromium and nickel.

These four equations define an inhomogeneous system of linear equations for the four unknowns x_1, x_2, x_3, x_4:

$$\begin{pmatrix} 1 & 1 & 1 & 1 & 1000 \\ 70 & 72 & 80 & 85 & 74000 \\ 22 & 20 & 10 & 12 & 18000 \\ 8 & 8 & 10 & 3 & 8000 \end{pmatrix} \hookrightarrow \begin{pmatrix} 1 & 1 & 1 & 1 & 1000 \\ 0 & 2 & 10 & 15 & 4000 \\ 0 & 0 & -2 & 5 & 0 \\ 0 & 0 & 0 & 0 & 0 \end{pmatrix}.$$

We arbitrarily choose $x_4 = k$. Row 3 gives $x_3 = \frac{5}{2}k$. The two results in row 2 and row 1 give $x_2 = 2000 - 20k$ and $x_1 = -1000 + 16.5k$. For the solution to be feasible, all proportions must be x_1, x_2, x_3, $x_4 \geq 0$. Because of the condition for x_2, $100 \geq k$. And because of the condition for x_1, it follows that $k \geq 60.6$. This means that for $100 \geq k \geq 60.6$ the problem can be solved physically.   □

1.6 Proofs

Mathematics can partly be seen as a set of statements that are derived (=*proved*) from basic statements in a purely logical way. These statements are then generally valid and cannot be disproved under the given conditions (*axioms*). This is the principle of mathematics going back to Euclid (around 300 BC). In his "Elements", Euclid was the first to list proven mathematical laws (=*propositions*) rather than observed laws of nature. Since then, this approach has been a fundamental difference between mathematics and the natural sciences. In the sciences, a law of nature is considered to be confirmed when several independent experiments repeatedly confirm the same statement. A law of nature is valid until it is disproved by another experiment.

Although the focus of this textbook is on the application of mathematics rather than on rigorous mathematical proofs, it is important to introduce the main methods of proof.

1.6.1 Mathematical Induction

Mathematical induction is one of the most important elementary proof methods in mathematics. A proposition $A(n)$ is true for all natural numbers if it is checked explicitly for $n = 1$ and $A(n + 1)$ can be proved under the assumption of $A(n)$. This method is discussed in detail in Section 1.2.2.

1.6.2 Direct Proof

A direct proof is the method of deriving a statement directly from preconditions or valid formulas. Examples are the proof of the binomial theorem (1.2.5) or the following proof of the geometric sum formula:

Theorem (Geometric Sum):

For any real number $q \neq 1$ holds: $\displaystyle\sum_{i=0}^{n} q^i = \frac{1 - q^{n+1}}{1 - q}$ $(n \in \mathbb{N}_0)$.

Proof: We define

$$s_n := \sum_{i=0}^{n} q^i = q^0 + q^1 + \cdots + q^n \tag{1}$$

and multiply this equation by q

$$q \cdot s_n = q^1 + q^2 + \cdots + q^{n+1} . \tag{2}$$

Subtracting equation (2) from equation (1) gives

$$s_n - q \cdot s_n = q^0 - q^{n+1} .$$

Therefore,

$$(1 - q) \cdot s_n = 1 - q^{n+1} \Rightarrow s_n = \sum_{i=0}^{n} q^i = \frac{1 - q^{n+1}}{1 - q} .$$

This is a direct proof of the geometric sum formula.   □

1.6.3 Proof by Contradiction

Another common method is to prove by contradiction. To prove a statement, we assume its opposite, which leads to a contradiction. An example is the Euclidean proof of the theorem that there are an infinite number of prime numbers.

Definition: *A natural number $p > 1$ is called* **Prime** *if it is divisible only by 1 and by itself .*

Theorem: There are an infinite number of prime numbers.

Proof by Contradiction: Suppose there are only finitely many prime numbers, namely $p_1, p_2, \ldots, p_n > 1$. Then, we consider the natural number

$$m := p_1 \cdot p_2 \cdot \ldots \cdot p_n + 1.$$

This number m is greater than 1, because the prime number 2 appears as a factor. The number m cannot be another prime number, since we have assumed that $p_1, p_2, \ldots, p_n$ are all prime numbers. Therefore, m is divisible by at least one $p_i \in \{p_1, p_2, \ldots, p_n\}$. So p_i divides both $p_1 \cdot p_2 \cdot \ldots \cdot p_n$ and 1. But this is a contradiction, since 1 has no divisors greater than 1. So we have turned the assumption (that there is only a finite number of primes) into a contradiction. If there is not a finite number of primes, then there must be an infinite number of primes.   □

1.6.4 Counterexample
Giving a counterexample to a statement is also a possible form of proof.
- All prime numbers are odd. (The counterexample is the number 2.)
- The formula $p = n^2 - n + 41$ gives prime numbers. (The counterexample is $n = 41$; see Example 1.6.)

1.7 Problems on Numbers, Equations and Systems of Equations

1.1   Specify the following sets by enumerating their elements:
   a) $\{x : x \text{ is prime and } x < 20\}$
   b) $\{x : x \text{ is real and } x^2 + 1 = 0\}$

1.2   The sets $A = \{x \in \mathbb{R} : 0 < x < 2\}$ and $B = \{x \in \mathbb{R} : 1 \leq x \leq 3\}$ are given. Determine graphically and computationally
   (i) $A \cap B$,   (ii) $A \cup B$,   (iii) $A \times B$,   (iv) $A \backslash B$.

1.3   Form the union, intersection and both complements of the following sets
   a) $M_1 = \{2, 4, 6, \ldots\}$, $M_2 = \{3, 6, 9, \ldots\}$
   b) $M_1 = \{x : x^2 + x - 2 = 0\}$, $M_2 = \{x : x^2 - 3x + 2 = 0\}$

1.4   There are three sets M_1, M_2, and M_3. Use Venn diagrams to show
   a) $M_1 \cap (M_2 \cup M_3) = (M_1 \cap M_2) \cup (M_1 \cap M_3)$
   b) $M_1 \cup (M_2 \cap M_3) = (M_1 \cup M_2) \cap (M_1 \cup M_3)$

1.5   Show by mathematical induction that all $n \in \mathbb{N}$
   a) $1^2 + 2^2 + 3^2 + \ldots + n^2 = \sum_{k=1}^{n} k^2 = \dfrac{n\,(n+1)\,(2n+1)}{6}$

   b) $2^0 + 2^1 + 2^2 + \ldots + 2^n = \sum_{k=0}^{n} 2^k = 2^{n+1} - 1$

   c) $\dfrac{1}{1 \cdot 2} + \dfrac{1}{2 \cdot 3} + \dfrac{1}{3 \cdot 4} + \ldots + \dfrac{1}{n\,(n+1)} = \dfrac{n}{n+1}$

1.6   Show by mathematical induction
   a) $2^n \leq n!$ for each $n \geq 4$
   b) $2n + 1 \leq 2^n$ for each $n \geq 3$
   c) $n^2 \leq 2^n$ for each $n \neq 3$

1.7   Calculate
   a) $\dbinom{n}{0}$, $\dbinom{n}{n}$, $\dbinom{3}{1}$, $\dbinom{3}{2}$, $\dbinom{4}{0}$, $\dbinom{4}{1}$, $\dbinom{4}{2}$, $\dbinom{4}{3}$, $\dbinom{4}{4}$
   b) 102^4

1.8   Check this
   $$\sum_{k=1}^{n} a_{k-1} = \sum_{k=0}^{n-1} a_k \quad ; \quad \sum_{k=0}^{n-1} a_{k+1} = \sum_{k=1}^{n} a_k \ .$$

1.9   Show $\dbinom{n}{k} \dfrac{1}{n^k} \leq \dfrac{1}{k!}$ for each $n \in \mathbb{N}$.

1.10 Expand the binomials
   a) $(x + 4)^5$ b) $(1 - 5y)^4$ c) $(a^2 - 2b)^3$.

1.11   Use Problem 1.5 to determine the total value of $\sum\limits_{k=71}^{125} (k^2 + 1)$.

1.12   Find a formula for the sum
   a) $1^3 + 2^3 + 3^3 + \ldots + n^3$ b) $1^4 + 2^4 + 3^4 + \ldots + n^4$

1.13   Simplify the expressions as much as possible
   a) $\dfrac{18\,x^{a+4}}{2\,y^{5a+7}} : \dfrac{4\,x^{7-3a}}{9\,y^{8+5a}}$ b) $(a^{n+1}b^{x-1} + a^n b^x + a^{n-1}b^{x+1}) : (a^{n-2}b^{x-1})$

1.14   Simplify the root terms symbolically as much as possible
   a) $\dfrac{x(2r^2 - 4x^2)}{\sqrt{r^2 - x^2}} - 8x\sqrt{r^2 - x^2}$ b) $2\sqrt{(x-k)^2 + x^2} - \dfrac{(2x-k)^2}{\sqrt{2x^2 - 2kx + k^2}}$
   c) $\sqrt{6x^2 - 6}\sqrt{\dfrac{3x-3}{2x+2}}$

1.15   Calculate
   a) $\sqrt{\sqrt[3]{a^6 b^{12}}}$ b) $\sqrt[4]{a^2 \sqrt[3]{a^2}}$ c) $\sqrt[3]{\sqrt{a^6 b^8}}$ d) $\sqrt[3]{a^3\sqrt{a^2\sqrt[5]{a^8\sqrt[4]{a^3}}}}$
   e) $\dfrac{\sqrt[6]{a^5 \sqrt[3]{a^2}}}{\sqrt[3]{a^2 \sqrt[6]{a^4}}} : \dfrac{\sqrt{a^3 \sqrt[9]{a^7}}}{\sqrt[9]{a^7\sqrt{a}}}$

1.16   Calculate
   a) $\operatorname{ld} 2^4$, $\log\sqrt{10}$, $\ln e^3$ b) $\ln(\sqrt{e})^3$, $\ln\sqrt{\frac{1}{\sqrt[3]{e^2}}}$, $\ln\sqrt{e^{3(\ln e^2 + \ln e^6)}}$
   c) $\log {}^{n+1}\!\sqrt{a^n \sqrt[m]{b^{-1}}}$

1.17   Show that the two sets, together with the arithmetic operations $+$ and $\cdot$
       fulfil the field axioms.
   a) $(\{a + b\sqrt{2} \text{ with } a, b \in \mathbb{Q}\}, +, \cdot)$ with the arithmetic operations in $\mathbb{R}$.
   b) $(F_2, +, \cdot)$ with link tables from the Example 1.9 ④.

1.18   Specify the real solutions of the following quadratic equations:
   a) $4\,x^2 + 8\,x - 60 = 0$ b) $x^2 - 4\,x + 13 = 0$ c) $-1 = -9\,(x - 2)^2$
   d) $5\,x^2 + 20\,x + 20 = 0$ e) $(x - 1)\,(x + 3) = -4$

1.19   Set the parameter c so that the equation $2\,x^2 + 4\,x = c$ has exactly one real
       solution.

1.20   What are the real solutions of the equations?
   a) $-2\,x^3 + 8\,x^2 = 8\,x$
   b) $t^4 - 13\,t^2 + 36 = 0$
   c) $\frac{1}{2}\,(3x^2 - 6)\,(x^2 - 25)\,(x + 3) = 0$

1.21   Solve the following root equations:
   a) $\sqrt{-3 + 2\,x} = 2$ b) $\sqrt{x^2 + 4} = x - 2$
   c) $\sqrt{x - 1} = \sqrt{x + 1}$ d) $\sqrt{2\,x^2 - 1} + x = 0$

1.22   What real solutions do the absolute value equations have?
   a) $\left|x^2 - x\right| = 24$ b) $|2\,x + 4| = -\,(x^2 - x - 6)$

1.23 Find the real solutions of the following inequalities:

a) $2x - 8 > |x|$ b) $x^2 + x + 1 \geq 0$ c) $|x| \leq x - 2$ d) $|x - 4| > x^2$

1.24 Solve the following systems of equations:

a)
$$\begin{aligned} 4x_1 &+& 2x_2 &+& 4x_3 &=& 10 \\ x_1 &+& x_2 &+& x_3 &=& 3 \\ 2x_1 &+& 3x_2 &+& 3x_3 &=& 8 \end{aligned}$$

b)
$$\begin{aligned} 2x_1 &+& x_2 &+& x_3 &=& 7 \\ 2x_1 &+& 2x_2 &+& x_3 &=& 10 \\ 3x_1 & & &+& x_3 &=& 5 \end{aligned}$$

c)
$$\begin{aligned} 2x_1 &+& x_2 &+& x_3 &=& 7 \\ 2x_1 &+& x_2 &+& x_3 &=& 0 \\ 3x_1 & & &+& x_3 &=& 5 \end{aligned}$$

1.25 Find the solution set of the following systems:

a)
$$\begin{aligned} x_1 &-& 3x_2 &+& x_3 &=& -3 \\ -3x_1 &+& x_2 &+& x_3 &=& 5 \end{aligned}$$

b)
$$\begin{aligned} x_1 &+& x_2 &+& x_3 &=& 6 \\ x_1 &+& 2x_2 &+& x_3 &=& 7 \\ 2x_1 &+& x_2 &+& 2x_3 &=& 11 \end{aligned}$$

c)
$$\begin{aligned} x_1 &+& x_2 &+& x_3 &=& 7 \\ x_1 &+& 2x_2 &+& x_3 &=& 7 \\ 2x_1 &+& x_2 &+& 2x_3 &=& 11 \end{aligned}$$

1.26 Find the solution set of the systems of linear equations:

a)
$$2x_1 + 3x_2 + 4x_3 = 4$$

b)
$$\begin{aligned} x_1 &-& x_2 &+& x_3 &=& 1 \\ -3x_1 &+& 3x_2 &-& 3x_3 &=& -3 \\ 5x_1 &-& 5x_2 &+& 5x_3 &=& 5 \end{aligned}$$

c)
$$\begin{aligned} x_1 &-& x_2 &+& x_3 &=& 1 \\ -3x_1 &+& 3x_2 &-& 3x_3 &=& -1 \\ 5x_1 &-& 5x_2 &+& 5x_3 &=& 5 \end{aligned}$$

1.27 Find the solution of the corresponding homogeneous systems 1.24 to 1.26. What general statements can be made with respect to these homogeneous systems?

1.28 In the following chemical reactions, the variables $x_1, x_2, \ldots$ represent natural numbers that are as small as possible to balance the reactions:

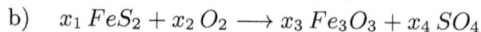

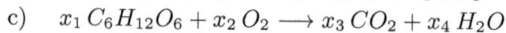

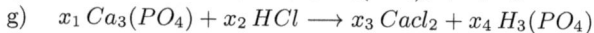

a) $x_1\, Fe + x_2\, O_2 \longrightarrow x_3\, Fe_2O_3$

b) $x_1\, FeS_2 + x_2\, O_2 \longrightarrow x_3\, Fe_3O_3 + x_4\, SO_4$

c) $x_1\, C_6H_{12}O_6 + x_2\, O_2 \longrightarrow x_3\, CO_2 + x_4\, H_2O$

d) $x_1\, C_3H_5N_3O_9 \longrightarrow x_2\, CO_2 + x_3\, H_2O + x_4\, N_2 + x_5\, O_2$

e) $x_1\, NH_3 + x_2\, CuO_2 \longrightarrow x_3\, N_2 + x_4\, Cu + x_5\, H_2O$

f) $x_1\, Al + x_2\, H_2SO_4 \longrightarrow x_3\, Al_2(SO_4)_3 + x_4\, H_2$

g) $x_1\, Ca_3(PO_4) + x_2\, HCl \longrightarrow x_3\, Cacl_2 + x_4\, H_3(PO_4)$

Chapter 2
Vectors and Vector Calculus

2

Vectors are an indispensable tool for describing physical quantities. While the temperature of a body, the density of a homogeneous medium, the ohmic resistance of an electrical element are characterized by a real number (together with a unit), this is not possible for the following physical quantities:

The velocity of a point mass is determined by the magnitude of the velocity and its direction. The force acting on a point of mass is described by the magnitude of the force and the direction in which the force acts. In physics and engineering, whenever quantities occur that can be described by magnitude (size, length) and direction, we speak of vectors.

In the section of straight lines and plans some applications of vectors and vector calculus are given: The description of straight lines and planes in $\mathbb{R}^3$ as well as distance calculations and the position of points, straight lines and planes in relation to each other.

The definition of vectors is transferred from $\mathbb{R}^3$ to $\mathbb{R}^n$ and some elementary arithmetic operations (addition and scalar multiplication) are extended to these vectors. The concept of a vector space is defined with general considerations. Central to the characterization of vector spaces are the concepts of linear combination, linear independence and the basis as a minimal generating system.

2

2 Vectors and Vector Calculus

Vectors are an indispensable tool for describing physical quantities. While the temperature of a body, the density of a homogeneous medium, the ohmic resistance of an electrical element are characterized by a real number (together with a unit), this is not possible for the following physical quantities:

The velocity of a point mass is determined by the magnitude of the velocity and its direction. The force acting on a point of mass is described by the magnitude of the force and the direction in which the force acts. In physics and engineering, vectors are used to describe quantities that can be described by magnitude (size, length) and direction.

First we discuss vectors in $\mathbb{R}^2$ in Section 2.1 and extend the concept of vectors and vector operations to $\mathbb{R}^3$ in Section 2.2. The descriptions of lines and planes in $\mathbb{R}^3$ are introduced in Section 2.3. Extensions to $\mathbb{R}^n$ are made in Section 2.4.

We define a vector in general terms:

> **Definition:** A **Vector** $\vec{a}$ is a class of directed lines (arrows) that match in direction and magnitude.

Two directed distances $\overrightarrow{AB}$ (start point A, end point B) and $\overrightarrow{CD}$ (start point C, end point D) represent the same vector if they have the same direction and the same magnitude. This is why they are often called *direction vectors*. Vectors created by parallel translation are therefore equal.

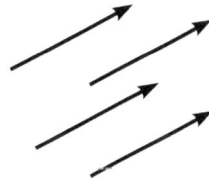

Historically, vector calculus is a relatively new discipline compared to differential calculus, for example. The foundations of modern vector calculus go back to Großmann (1809 - 1877; 1844). The formalism of vectors and vector calculus thus came much later than, for example, the complex numbers.

In the following we will use a Cartesian *coordinate system* to describe the vectors and their arithmetic operations.

2.1 Vectors in $\mathbb{R}^2$

The *two-dimensional space* $\mathbb{R}^2$ is defined by two perpendicular coordinate axes (**Cartesian Coordinate System**). In such a coordinate system a vector $\vec{a}$ from point $P_1 = (x_1, y_1)$ to point $P_2 = (x_2, y_2)$ is defined by its **components**, by its projections onto the coordinate axes:

$$\vec{a} := \begin{pmatrix} x_2 - x_1 \\ y_2 - y_1 \end{pmatrix}.$$

Figure 2.1. Direction vectors (left) and position vector (right)

This **Direction Vector** does not depend on the particular position in $\mathbb{R}^2$. $\vec{a}$ and $\vec{a}_1$ represent the same vector. In contrast to direction vectors, we define **Position Vectors** when the vector goes from the origin O to the point P:

$$\vec{r}(P) := \begin{pmatrix} x \\ y \end{pmatrix}.$$

2.1.1 Multiplying a Vector by a Scalar

$$\lambda \vec{a} = \lambda \begin{pmatrix} a_x \\ a_y \end{pmatrix} := \begin{pmatrix} \lambda a_x \\ \lambda a_y \end{pmatrix}; \quad \lambda \in \mathbb{R}.$$

The product of a scalar $\lambda \in \mathbb{R}$ with a vector $\vec{a}$ is again a vector. The multiplication is done **component by component**. Geometrically, this corresponds to the *stretching* of the vector $\vec{a}$ by the factor λ. If $\lambda = -1$ then $-\vec{a}$ has the same length as $\vec{a}$, but in the opposite direction. Sometimes it is convenient to write the factor to the right of the vector. So we set $\vec{a}\lambda := \lambda \vec{a}$.

2.1.2 Addition of two Vectors

$$\vec{a} + \vec{b} = \begin{pmatrix} a_x \\ a_y \end{pmatrix} + \begin{pmatrix} b_x \\ b_y \end{pmatrix} := \begin{pmatrix} a_x + b_x \\ a_y + b_y \end{pmatrix}.$$

The sum of two vectors is a vector. The addition is done component by component. The subtraction is done similarly. Geometrically, the addition of two vectors is equivalent to the addition of forces in a *parallelogram* (see Fig. 2.2).

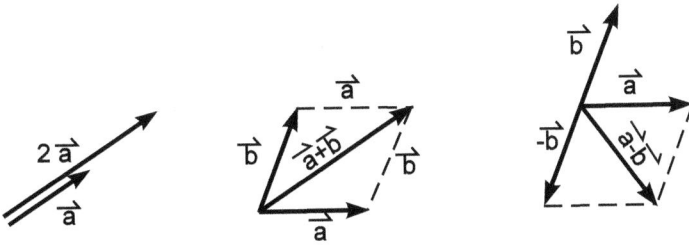

Figure 2.2. Addition and subtraction through the parallelogram of forces

2.1.3 The Magnitude (Size) of a Vector

The *magnitude* (= *length*) of a vector $\vec{a}$ is calculated using the Pythagorean Theorem:

$$|\vec{a}| = \sqrt{a_x^2 + a_y^2}.$$

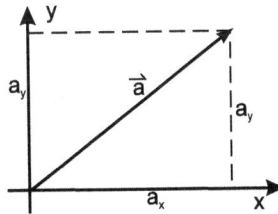

Figure 2.3. Magnitude of a vector

The magnitude of a vector is sometimes also written as $\|\vec{a}\|$.

Examples 2.1:

① Given are the vectors $\vec{a} = \begin{pmatrix} 5 \\ 3 \end{pmatrix}$, $\vec{b} = \begin{pmatrix} 4 \\ 1 \end{pmatrix}$, $\vec{c} = \begin{pmatrix} -3 \\ -2 \end{pmatrix}$. Then

$$\vec{d} = -\vec{a} + 3\vec{b} + 2\vec{c} = \begin{pmatrix} -5 \\ -3 \end{pmatrix} + \begin{pmatrix} 12 \\ 3 \end{pmatrix} + \begin{pmatrix} -6 \\ -4 \end{pmatrix} = \begin{pmatrix} 1 \\ -4 \end{pmatrix}.$$

② Three forces acting on a point mass

$$\vec{F_1} = \begin{pmatrix} 2N \\ 1N \end{pmatrix}, \vec{F_2} = \begin{pmatrix} -1N \\ 5N \end{pmatrix}, \vec{F_3} = \begin{pmatrix} -4N \\ 2N \end{pmatrix}.$$

The magnitude of the resulting force F_R is searched for:

$$\vec{F_R} = \vec{F_1} + \vec{F_2} + \vec{F_3}$$

$$= \begin{pmatrix} 2N \\ 1N \end{pmatrix} + \begin{pmatrix} -1N \\ 5N \end{pmatrix} + \begin{pmatrix} -4N \\ 2N \end{pmatrix} = \begin{pmatrix} -3N \\ 8N \end{pmatrix}$$

$$F_R = \left| \vec{F_R} \right| = \sqrt{9N^2 + 64N^2} = \sqrt{73}N.$$

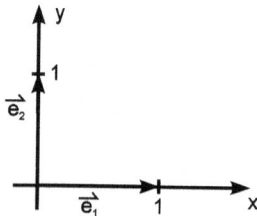

Figure 2.4. Unit vectors

③ A vector of magnitude 1 is called a **unit vector**. Special unit vectors are the *coordinate unit vectors*

$$\vec{e}_1 := \begin{pmatrix} 1 \\ 0 \end{pmatrix} \quad \text{and} \quad \vec{e}_2 := \begin{pmatrix} 0 \\ 1 \end{pmatrix}.$$

These vectors have the direction of the corresponding coordinate axes and the length 1. With these unit vectors, each vector $\vec{a}$ can be written as

$$\vec{a} = a_x \vec{e}_1 + a_y \vec{e}_2$$

(*Linear combination* of $\vec{e}_1$ and $\vec{e}_2$).

④ Given is a vector $\vec{a} = \begin{pmatrix} a_x \\ a_y \end{pmatrix}$. We look for the **unit vector in direction** $\vec{a}$. Because of $|\vec{a}| = \sqrt{a_x^2 + a_y^2}$

$$\vec{e}_a = \frac{1}{|\vec{a}|} \vec{a} = \frac{1}{\sqrt{a_x^2 + a_y^2}} \begin{pmatrix} a_x \\ a_y \end{pmatrix}$$

is the unit vector we are looking for because $|\vec{e}_a| = 1$. The direction of $\vec{e}_a$ and $\vec{a}$ is the same.   □

2.1.4 The Scalar Product of two Vectors

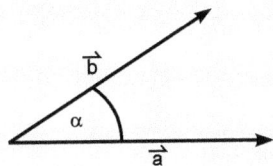

Figure 2.5. Angle between $\vec{a}$ and $\vec{b}$

Definition: *The* **Scalar Product** *(Dot Product) of two vectors* $\vec{a} = \begin{pmatrix} a_x \\ a_y \end{pmatrix}$ *and* $\vec{b} = \begin{pmatrix} b_x \\ b_y \end{pmatrix}$ *is defined as the real number*

$$\vec{a} \cdot \vec{b} := |\vec{a}| \cdot |\vec{b}| \cdot \cos \alpha, \qquad (1)$$

where α is the angle between the vectors $\vec{a}$ and $\vec{b}$. Common notations for the scalar product are also $< \vec{a}, \vec{b} >$ *or* $(\vec{a}, \vec{b})$.

By definition, the **calculation rules** apply to the scalar product.

(S$_1$)	$\vec{a} \cdot \vec{b}$	$= \vec{b} \cdot \vec{a}$	Symmetry
(S$_2$)	$\lambda \cdot (\vec{a} \cdot \vec{b})$	$= (\lambda \cdot \vec{a}) \cdot \vec{b} = \vec{a} \cdot (\lambda \cdot \vec{b})$	Associative
(S$_3$)	$\vec{a} \cdot (\vec{b} + \vec{c})$	$= (\vec{a} \cdot \vec{b}) + (\vec{a} \cdot \vec{c})$	Distributive

Instead of proving these statements, we use the rules to get a very simple representation of the scalar product: Obviously, we have

$$\vec{e}_1 \cdot \vec{e}_1 = 1, \quad \vec{e}_1 \cdot \vec{e}_2 = 0, \quad \vec{e}_2 \cdot \vec{e}_2 = 1.$$

So for the two vectors

$$\vec{a} = \begin{pmatrix} a_x \\ a_y \end{pmatrix} = a_x \vec{e}_1 + a_y \vec{e}_2 \text{ and } \vec{b} = \begin{pmatrix} b_x \\ b_y \end{pmatrix} = b_x \vec{e}_1 + b_y \vec{e}_2 :$$

$$\begin{aligned} \vec{a} \cdot \vec{b} &= (a_x \vec{e}_1 + a_y \vec{e}_2) \cdot (b_x \vec{e}_1 + b_y \vec{e}_2) \\ &= a_x b_x \vec{e}_1 \cdot \vec{e}_1 + a_x \vec{e}_1 \cdot b_y \vec{e}_2 + a_y \vec{e}_2 \cdot b_x \vec{e}_1 + a_y b_y \vec{e}_2 \cdot \vec{e}_2. \end{aligned}$$

Scalar Product

$$\Rightarrow \vec{a} \cdot \vec{b} = a_x b_x + a_y b_y. \qquad (2)$$

Note: The scalar product of two vectors can easily be computed without knowing the angle α between the vectors $\vec{a}$ and $\vec{b}$ just by **adding the products of the first components and the second components.**

Application Example 2.2 (Work at constant force).

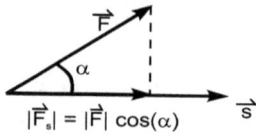

Physically, the scalar product is the work done under the influence of a constant force. If force and motion are in the same direction, then the work is $W = F \cdot s$.

$$|\vec{F}_s| = |\vec{F}| \cos(\alpha)$$

Figure 2.6. Force in direction $\vec{s}$

But if a mass can only move along a direction $\vec{e}_s$ that does not match to the direction of the force, then the work is the component of the force $|\vec{F}_s|$ in the direction $\vec{e}_s$ multiplied by the distance moved $|\vec{s}| = |s\,\vec{e}_s|$:

$$W = \left|\vec{F}_s\right| \cdot |\vec{s}| = \left|\vec{F}\right| \cdot \cos\alpha \cdot |\vec{s}|$$

$$W = \vec{F} \cdot \vec{s}. \qquad \qquad \Box$$

Application Example 2.3 (Angle between two vectors).

Identities (1) and (2) give an important formula for calculating the angle between two vectors:

$$\cos\alpha = \frac{\vec{a} \cdot \vec{b}}{|\vec{a}| \cdot |\vec{b}|} = \frac{a_x\, b_x + a_y\, b_y}{\sqrt{a_x^2 + a_y^2}\sqrt{b_x^2 + b_y^2}}.$$

Using the inverse function of the cosine (see Section 4.7 Arc functions), the angle α enclosed by the vectors is obtained between 0 and 180 °. $\qquad \Box$

Important Application: $\vec{a} \perp \vec{b}$

If $\vec{a}$ and $\vec{b}$ are perpendicular to each other, then $\alpha = 90°$ and $\cos \alpha = 0$. Therefore, the following applies

$$\vec{a} \cdot \vec{b} = 0 \;\; \Leftrightarrow \;\; \vec{a} \perp \vec{b}.$$

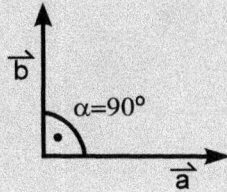

⚠ **Caution:** Unlike the product of two real numbers, the scalar product is zero not only if at least one of the two factors is the zero vector, but also if the two vectors are perpendicular to each other.

Examples 2.4:

① Determine the scalar product of $\vec{a} = \begin{pmatrix} 4 \\ 2 \end{pmatrix}$ and $\vec{b} = \begin{pmatrix} -1 \\ 3 \end{pmatrix}$.

$$\vec{a} \cdot \vec{b} = \begin{pmatrix} 4 \\ 2 \end{pmatrix} \cdot \begin{pmatrix} -1 \\ 3 \end{pmatrix} = 4 \cdot (-1) + 2 \cdot 3 = 2.$$

② The vectors $\vec{a} = \begin{pmatrix} 1 \\ 2 \end{pmatrix}$ and $\vec{b} = \begin{pmatrix} -2 \\ 1 \end{pmatrix}$ are *orthogonal*, i.e. they are perpendicular to each other.

We show that the scalar product is zero:

$$\vec{a} \cdot \vec{b} = \begin{pmatrix} 1 \\ 2 \end{pmatrix} \cdot \begin{pmatrix} -2 \\ 1 \end{pmatrix} = 1 \cdot (-2) + 2 \cdot 1 = 0.$$

③ The magnitude of a vector can be calculated from the scalar product:

For $\vec{a} = \begin{pmatrix} a_x \\ a_y \end{pmatrix}$ we get $\vec{a} \cdot \vec{a} = \begin{pmatrix} a_x \\ a_y \end{pmatrix} \cdot \begin{pmatrix} a_x \\ a_y \end{pmatrix} = a_x^2 + a_y^2 = |\vec{a}|^2$

$$\Rightarrow \;\; |\vec{a}| = a = \sqrt{\vec{a} \cdot \vec{a}} = \sqrt{a_x^2 + a_y^2}.$$

④   Given is the vector $\vec{a} = \begin{pmatrix} 2 \\ 1 \end{pmatrix}$. We are looking for the angles α and β which the vector encloses with the coordinate axes. From the scalar product of $\vec{a}$ with $\vec{e}_1$ and $\vec{e}_2$ we can calculate these angles:

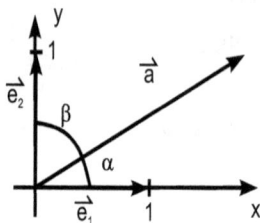

Figure 2.7. Vector and its angles with respect to axes

$$\cos\alpha = \frac{\vec{a} \cdot \vec{e}_1}{|\vec{a}| \cdot |\vec{e}_1|} = \frac{a_x}{|\vec{a}|} = \frac{2}{\sqrt{5}} \Rightarrow \alpha = 26,6°.$$

$$\cos\beta = \frac{\vec{a} \cdot \vec{e}_2}{|\vec{a}| \cdot |\vec{e}_2|} = \frac{a_y}{|\vec{a}|} = \frac{1}{\sqrt{5}} \Rightarrow \beta = 63,4°.$$

⑤   Given is a vector $\vec{a} = \begin{pmatrix} a_x \\ a_y \end{pmatrix}$. We are looking for a perpendicular vector $\vec{n}$ with magnitude 1:

A vector perpendicular to $\vec{a}$ is obtained by swapping the x component with the y component of the vector $\vec{a}$ and then changing the sign of one component. $\vec{N} = \begin{pmatrix} -a_y \\ a_x \end{pmatrix}$ is therefore perpendicular to $\vec{a}$, because

$$\vec{a} \cdot \vec{N} = \begin{pmatrix} a_x \\ a_y \end{pmatrix} \cdot \begin{pmatrix} -a_y \\ a_x \end{pmatrix} = 0.$$

If we divide $\vec{N}$ by its magnitude, we obtain the corresponding **normal unit vector**

$$\vec{n} := \vec{e}_n = \frac{1}{N}\vec{N} = \frac{1}{\sqrt{a_x^2 + a_y^2}} \begin{pmatrix} -a_y \\ a_x \end{pmatrix}.$$

□

2.1.5 Geometric Application

Each point $P = (x, y)$ in $\mathbb{R}^2$ corresponds to exactly one position vector $\vec{r}(P) = \begin{pmatrix} x \\ y \end{pmatrix}$. So a straight line g through the two points $P_1 = (x_1, y_1)$ and $P_2 = (x_2, y_2)$ can be described by all points P according to

$$g: \quad \vec{r}(P) = \vec{r}(P_1) + \lambda(\vec{r}(P_2) - \vec{r}(P_1)) = \vec{r}(P_1) + \lambda \cdot \vec{a}.$$

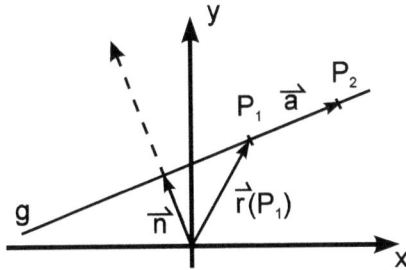

Figure 2.8. Point-direction representation of a straight line

This is the **point-direction representation** of a straight line defined by the position vector $\vec{r}(P_1)$ and the direction vector $\vec{a} := \vec{r}(P_2) - \vec{r}(P_1)$. An alternative representation of the line follows if we multiply the point-direction representation by the *normal unit vector* $\vec{n}$ (Example 2.4 ⑤).

$$\begin{pmatrix} x \\ y \end{pmatrix} \cdot \vec{n} = (\begin{pmatrix} x_1 \\ y_1 \end{pmatrix} + \lambda \cdot \vec{a}) \cdot \vec{n} = \begin{pmatrix} x_1 \\ y_1 \end{pmatrix} \vec{n} + \lambda \underbrace{\vec{a} \cdot \vec{n}}_{=0}$$

$$\Rightarrow \qquad \begin{pmatrix} x \\ y \end{pmatrix} \cdot \begin{pmatrix} n_1 \\ n_2 \end{pmatrix} = d.$$

This is the **Hesse Normal Form** of a straight line in the $\mathbb{R}^2$ and

$$d = \begin{pmatrix} x_1 \\ y_1 \end{pmatrix} \begin{pmatrix} n_1 \\ n_2 \end{pmatrix}$$

is the shortest distance of the straight line from the origin.

Example 2.5. Given are two points $P_1 = (1,1)$ and $P_2 = (4,2)$. Find the point-direction representation and the Hesse normal form of the line g defined by P_1 and P_2. What is the shortest distance from the origin?

(1) Point-direction representation:

$$g : \vec{r} = \begin{pmatrix} x \\ y \end{pmatrix} = \vec{r}(P_1) + \lambda\, (\vec{r}(P_2) - \vec{r}(P_1)) = \begin{pmatrix} 1 \\ 1 \end{pmatrix} + \lambda \begin{pmatrix} 3 \\ 1 \end{pmatrix}.$$

(2) Hesse normal form:

With the scalar product we check that the vector $\vec{N} = \begin{pmatrix} -1 \\ 3 \end{pmatrix}$ is perpendicular to $\vec{a} = \begin{pmatrix} 3 \\ 1 \end{pmatrix}$: $\vec{N} \cdot \vec{a} = -3 + 3 = 0$. Because of $|\vec{N}| = \sqrt{1+9} = \sqrt{10}$, we get $\vec{n} := \frac{1}{N}\vec{N} = \frac{1}{\sqrt{10}} \begin{pmatrix} -1 \\ 3 \end{pmatrix}$ as the normal unit vector to g.

$$\Rightarrow d = \begin{pmatrix} x \\ y \end{pmatrix} \frac{1}{\sqrt{10}} \begin{pmatrix} -1 \\ 3 \end{pmatrix} \qquad \text{is the Hesse normal form.}$$

(3) The minimum distance of the straight line from the origin is obtained by inserting the point $P_1 = (1,1)$ into the Hesse form:

$$d = \begin{pmatrix} 1 \\ 1 \end{pmatrix} \frac{1}{\sqrt{10}} \begin{pmatrix} -1 \\ 3 \end{pmatrix} = \frac{2}{\sqrt{10}} = \frac{1}{5}\sqrt{10}.$$

(4) We calculate the scalar product on the left side of the Hesse normal form

$$\frac{1}{\sqrt{10}}(-x + 3y) = \frac{2}{\sqrt{10}}.$$

When solving with respect to y, we get the usual representation of a straight line $y = m\,x + b$

$$y = \frac{1}{3}x + \frac{2}{3}. \qquad \qquad \square$$

Visualization with MAPLE: On the homepage there are MAPLE procedures which allow the representation of vectors in $\mathbb{R}^2$ as well as the visualization of the vector operations described in 2.1.1 to 2.1.4.

2.2 Vectors in the $\mathbb{R}^3$

Similar to the procedure in $\mathbb{R}^2$, we define a vector $\vec{a}$ in $\mathbb{R}^3$ from a point $P_1 = (x_1, y_1, z_1)$ to a second point $P_2 = (x_2, y_2, z_2)$ with respect to a Cartesian coordinate system:

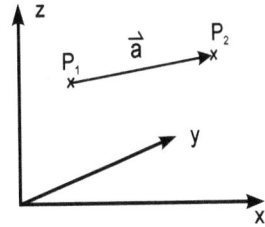

Figure 2.9.
Direction vector

$$\vec{a} := \begin{pmatrix} x_2 - x_1 \\ y_2 - y_1 \\ z_2 - z_1 \end{pmatrix}.$$

Again, $\vec{a}$ means *directional vector*. A *position vector* $\vec{r}(P)$ is a vector from the origin O to the point $P = (x, y, z)$:

$$\vec{r}(P) := \begin{pmatrix} x \\ y \\ z \end{pmatrix}.$$

2.2.1 Vector Calculation Rules

The multiplication of a vector $\vec{a}$ with a scalar λ and the addition of two vectors are done component by component:

$$\lambda \cdot \vec{a} = \lambda \begin{pmatrix} a_x \\ a_y \\ a_z \end{pmatrix} := \begin{pmatrix} \lambda\, a_x \\ \lambda\, a_y \\ \lambda\, a_z \end{pmatrix} \qquad \textbf{(Scalar Multiplication)}$$

$$\vec{a} + \vec{b} = \begin{pmatrix} a_x \\ a_y \\ a_z \end{pmatrix} + \begin{pmatrix} b_x \\ b_y \\ b_z \end{pmatrix} := \begin{pmatrix} a_x + b_x \\ a_y + b_y \\ a_z + b_z \end{pmatrix} \qquad \textbf{(Addition)}$$

The *magnitude* (or the *length*) of a vector $\vec{a}$ is defined by

$$a := |\vec{a}| = \sqrt{a_x^2 + a_y^2 + a_z^2} \qquad \textbf{(Magnitude)}$$

and corresponds to the diagonal of a block with edge lengths a_x, a_y, a_z.

Any vector $\vec{e}$ with $|\vec{e}| = 1$ is a *unit vector*. The *coordinate unit vectors* are

$$\vec{e}_1 = \begin{pmatrix} 1 \\ 0 \\ 0 \end{pmatrix}, \ \vec{e}_2 = \begin{pmatrix} 0 \\ 1 \\ 0 \end{pmatrix}, \ \vec{e}_3 = \begin{pmatrix} 0 \\ 0 \\ 1 \end{pmatrix}.$$

Any vector $\vec{a}$ can be written as a *linear combination* of these unit vectors

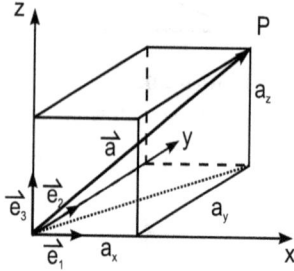

$$\vec{a} = a_x \vec{e}_1 + a_y \vec{e}_2 + a_z \vec{e}_3.$$

Figure 2.10. Linear combination of unit vectors

Examples 2.6:

① The position vector to the point $P = (5, 1, -3)$ is $\vec{r}(P) = \begin{pmatrix} 5 \\ 1 \\ -3 \end{pmatrix}$ and has the magnitude $|\vec{r}(P)| = \sqrt{5^2 + 1^2 + (-3)^2} = \sqrt{35}$.

② The direction vector from $P_1 = (3, 4, 7)$ to $P_2 = (7, 3, 1)$ is

$$\vec{a} = \overrightarrow{P_1 P_2} = \vec{r}(P_2) - \vec{r}(P_1) = \begin{pmatrix} 7 \\ 3 \\ 1 \end{pmatrix} - \begin{pmatrix} 3 \\ 4 \\ 7 \end{pmatrix} = \begin{pmatrix} 4 \\ -1 \\ -6 \end{pmatrix}.$$

③ We are looking for the coordinates of the point Q that intersects the distance from $P_1 = (3, 4, 7)$ to $P_2 = (7, 3, 1)$ at a ratio of 1:3.

$$\vec{r}(Q) = \vec{r}(P_1) + \frac{1}{3}(\vec{r}(P_2) - \vec{r}(P_1))$$

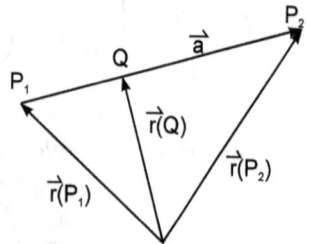

$$= \begin{pmatrix} 3 \\ 4 \\ 7 \end{pmatrix} + \frac{1}{3}\begin{pmatrix} 4 \\ -1 \\ -6 \end{pmatrix} = \frac{1}{3}\begin{pmatrix} 13 \\ 11 \\ 15 \end{pmatrix} \ \square$$

The **Scalar Product (Dot Product)** is defined in $\mathbb{R}^3$ by

$$\vec{a} \cdot \vec{b} := |\vec{a}| \cdot |\vec{b}| \cdot \cos\alpha, \qquad \alpha = \angle\left(\vec{a}, \vec{b}\right),$$

where α is the angle between the vectors $\vec{a}$ and $\vec{b}$. To evaluate the scalar product we use the same rules as in $\mathbb{R}^2$ $((S_1) - (S_3))$.

Scalar Product (Dot Product)

$$\vec{a} \cdot \vec{b} = a_x b_x + a_y b_y + a_z b_z.$$

Consequently, the angle α between two vectors is again:

Angle between two Vectors

$$\cos\alpha = \frac{\vec{a} \cdot \vec{b}}{|\vec{a}| \cdot |\vec{b}|} = \frac{a_x b_x + a_y b_y + a_z b_z}{\sqrt{a_x^2 + a_y^2 + a_z^2}\sqrt{b_x^2 + b_y^2 + b_z^2}}. \tag{3}$$

Conclusion: $\vec{a} \perp \vec{b}$

Two vectors $\vec{a}$ and $\vec{b}$ are orthogonal (**perpendicular**) if the scalar product is zero:

$$\vec{a} \perp \vec{b} \quad\Leftrightarrow\quad \vec{a} \cdot \vec{b} = 0.$$

Rule: To check whether two vectors are perpendicular to each other, it is sufficient to calculate the scalar product of the vectors.

Examples 2.7:

① **Orthonormal system:** $\vec{e}_1, \vec{e}_2, \vec{e}_3$ are an *orthonormal system* of $\mathbb{R}^3$: They are perpendicular to each other and have the magnitude 1:

$$\vec{e}_1 \cdot \vec{e}_1 = \vec{e}_2 \cdot \vec{e}_2 = \vec{e}_3 \cdot \vec{e}_3 = 1$$

$$\vec{e}_1 \cdot \vec{e}_2 = \vec{e}_2 \cdot \vec{e}_3 = \vec{e}_3 \cdot \vec{e}_1 = 0.$$

② The vectors $\vec{a}_1 = \frac{1}{\sqrt{3}}\begin{pmatrix} 1 \\ 1 \\ 1 \end{pmatrix}$, $\vec{a}_2 = \frac{1}{\sqrt{2}}\begin{pmatrix} -1 \\ 1 \\ 0 \end{pmatrix}$, $\vec{a}_3 = \frac{1}{\sqrt{6}}\begin{pmatrix} -1 \\ -1 \\ 2 \end{pmatrix}$

also form an orthonormal system.

③ **Directional cosine:** The scalar product makes it easy to calculate the

angle of a vector $\vec{a} = \begin{pmatrix} a_x \\ a_y \\ a_z \end{pmatrix}$ with the coordinate axes.

$$\vec{a} \cdot \vec{e_1} = a_x; \; |\vec{e}_1| = 1 \Rightarrow \cos\alpha = \frac{a_x}{a}$$

$$\vec{a} \cdot \vec{e}_2 = a_y; \; |\vec{e}_2| = 1 \Rightarrow \cos\beta = \frac{a_y}{a}$$

$$\vec{a} \cdot \vec{e}_3 = a_z; \; |\vec{e}_3| = 1 \Rightarrow \cos\gamma = \frac{a_z}{a}.$$

The angles α, β, γ are called *directional cosine* of $\vec{a}$. We conclude that

$$\cos^2\alpha + \cos^2\beta + \cos^2\gamma = \frac{a_x^2}{a^2} + \frac{a_y^2}{a^2} + \frac{a_z^2}{a^2} = \frac{a^2}{a^2} = 1$$

and

$$a_x = |\vec{a}|\cos\alpha, \; a_y = |\vec{a}|\cos\beta, \; a_z = |\vec{a}|\cos\gamma.$$

This equation says that for a vector $\vec{a}$ the three angles to the coordinate axes are not arbitrary. Only two out of them can be chosen independently; the third is determined by the equation.

④ **Number example:** Given are $\vec{a} = \begin{pmatrix} 1 \\ 2 \\ 3 \end{pmatrix}$ and $\vec{b} = \begin{pmatrix} -4 \\ 3 \\ -2 \end{pmatrix}$. We are

looking for the angle α between $\vec{a}$ and $\vec{b}$:

$$\left.\begin{array}{l} \vec{a} \cdot \vec{b} = \begin{pmatrix} 1 \\ 2 \\ 3 \end{pmatrix} \cdot \begin{pmatrix} -4 \\ 3 \\ -2 \end{pmatrix} = -4 + 6 - 6 = -4 \\ |\vec{a}| = \sqrt{14}, \; |\vec{b}| = \sqrt{29} \end{array}\right\} \Rightarrow \cos\alpha = \frac{-4}{\sqrt{14}\cdot\sqrt{29}}$$

$$\Rightarrow \alpha = 101, 45°.$$ □

2.2.2 Projection of a Vector

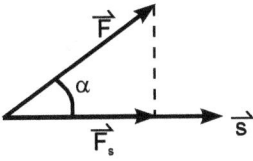

We consider the following physical problem: A point mass is clamped in a rail and can only be moved along the direction $\vec{s}$. A force $\vec{F}$ acts on this mass. $\vec{F}_s$ is the component of this force along the direction $\vec{s}$. The magnitude of $\vec{F}_s$ is given by equation (3) by

$$\left|\vec{F}_s\right| = \left|\vec{F}\right| \cdot \cos \alpha = \left|\vec{F}\right| \cdot \frac{\vec{F} \cdot \vec{s}}{\left|\vec{F}\right| \cdot \left|\vec{s}\right|}$$

and the direction by $\vec{e}_s = \frac{\vec{s}}{|\vec{s}|}$. Therefore,

$$\vec{F}_s = \left|\vec{F}_s\right| \cdot \vec{e}_s = \frac{\vec{F} \cdot \vec{s}}{|\vec{s}|} \cdot \frac{\vec{s}}{|\vec{s}|} = \frac{\vec{F} \cdot \vec{s}}{|\vec{s}|^2} \vec{s}.$$

$\vec{F}_s$ is called the *projection of $\vec{F}$ in the direction of $\vec{s}$*. We generalize this construction to two arbitrary vectors $\vec{a}$ and $\vec{b}$:

Projection of a vector along a direction

The **projection of $\vec{b}$ in the direction of $\vec{a}$** is given by the vector

$$\vec{b}_a = \frac{\vec{a} \cdot \vec{b}}{|\vec{a}|^2} \cdot \vec{a}$$

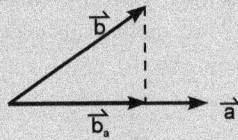

⚠ Note that $\vec{a} \cdot \vec{b}$ is the scalar product and therefore $\frac{\vec{a} \cdot \vec{b}}{a^2}$ is a real number. The second product sign is the multiplication of the vector $\vec{a}$ with this real number. The two "·" characters must not be swapped!

Remark: If we also interpret $|\vec{a}|^2$ as the usual multiplication between real numbers $|\vec{a}| \cdot |\vec{a}|$, then we have three different meanings of "·" within one expression! In programming, such an operator is called an *overloaded* operator.

Application Example 2.8 (**Force along a Direction** $\vec{s}$).

Given is the force $\vec{F} = \begin{pmatrix} 2 \\ 2 \\ -7 \end{pmatrix}$ and the direction $\vec{s} = \begin{pmatrix} -1 \\ -1 \\ -1 \end{pmatrix}$. The force affects a mass, that can only move along the direction $\vec{s}$. We are looking for the acceleration force, the magnitude of the force and the work done.

The acceleration force is the projection of the force $\vec{F}$ in the direction of motion. We first calculate the scalar product of $\vec{F}$ with $\vec{s}$:

$$\vec{F} \cdot \vec{s} = \begin{pmatrix} 2 \\ 2 \\ -7 \end{pmatrix} \cdot \begin{pmatrix} -1 \\ -1 \\ -1 \end{pmatrix} = 3, \qquad |\vec{s}|^2 = 3.$$

The acceleration force now results from the projection formula

$$\vec{F}_s = \tfrac{3}{3} \begin{pmatrix} -1 \\ -1 \\ -1 \end{pmatrix} = \begin{pmatrix} -1 \\ -1 \\ -1 \end{pmatrix} \text{ and } \left| \vec{F}_s \right| = \sqrt{3}.$$

The performed *work* W results from example 2.2 by the formula

$$W := \vec{F} \cdot \vec{s} = 3. \qquad \square$$

2.2.3 The Vector Product (Cross Product) of two Vectors

In $\mathbb{R}^3$ we define the *vector product* (also called *cross product*) of the two vectors. As the name suggests, the result is again a vector:

Definition: *The* **Vector Product (Cross Product)**

$$\vec{c} = \vec{a} \times \vec{b}$$

of two vectors $\vec{a}$ *and* $\vec{b}$ *is defined as the vector* $\vec{c}$ *with the properties:*

(1) $\vec{c}$ *is perpendicular to both to* $\vec{a}$ *and to* $\vec{b}$: $\vec{c} \cdot \vec{a} = \vec{c} \cdot \vec{b} = 0.$

(2) *The magnitude of* $\vec{c}$ *is equal to the product of the magnitudes of the vectors* $\vec{a}$ *and* $\vec{b}$ *and the sine of the enclosed angle* α:

$$|\vec{c}| = |\vec{a}| \cdot \left|\vec{b}\right| \cdot \sin\alpha.$$

(3) *The vectors* $\vec{a}, \vec{b}, \vec{c}$ *form a right-hand system.*

Remarks:

(1) Unlike the scalar product, the vector product is a vector.

(2) Instead of $\vec{a} \times \vec{b}$, the symbol $\left[\vec{a}, \vec{b}\right]$ is sometimes used.

(3) ⚠ **Caution:** The vector product is only defined in $\mathbb{R}^3$!

(4) ⚠ **Caution:** The vector product is **not** commutative!

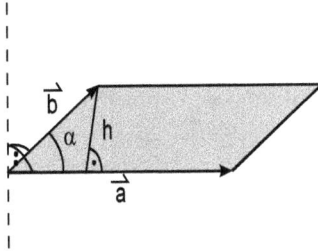

Figure 2.11. Cross product $\vec{a} \times \vec{b}$

Geometric Interpretation:

Since $\vec{c} \perp \vec{a}$ and $\vec{c} \perp \vec{b}$, the direction of the vector $\vec{c}$ is the dashed line shown in Fig. 2.11. Since $\vec{a}, \vec{b}, \vec{c}$ form a right hand system, only the upward direction remains. The area of the parallelogram spanned by $\vec{a}$ and $\vec{b}$ is defined as the product of the base line and the height:

$$A = |\vec{a}| \cdot h = |\vec{a}| \cdot |\vec{b}| \cdot \sin\alpha = \left|\vec{a} \times \vec{b}\right|.$$

Examples 2.9:

① The vector products of the unit vectors are calculated directly from their definition:

$$\vec{e_1} \times \vec{e_1} = 0, \ \vec{e_2} \times \vec{e_2} = 0, \ \vec{e_3} \times \vec{e_3} = 0;$$

$$\vec{e_1} \times \vec{e_2} = \vec{e_3}, \ \vec{e_2} \times \vec{e_3} = \vec{e_1}, \ \vec{e_3} \times \vec{e_1} = \vec{e_2}.$$

② Criterion for **collinear vectors:**
If the cross product of $\vec{a} \neq 0$ and $\vec{b} \neq 0$ is zero then either $\vec{a} \uparrow\uparrow \vec{b}$ ($\vec{a}$ parallel to $\vec{b}$) or $\vec{a} \uparrow\downarrow \vec{b}$ ($\vec{a}$ antiparallel to $\vec{b}$). □

We specify the main **calculation rules** for the vector product:

$$
\begin{array}{llll}
(V_1) & \vec{a} \times (\vec{b} + \vec{c}) &=& \vec{a} \times \vec{b} + \vec{a} \times \vec{c} \\
& (\vec{a} + \vec{b}) \times \vec{c} &=& \vec{a} \times \vec{c} + \vec{b} \times \vec{c} \quad \text{\textit{Distribution laws}} \\
(V_2) & \vec{a} \times \vec{b} &=& -\vec{b} \times \vec{a} \quad\quad\quad\ \text{\textit{Anti-symmetry}} \\
(V_3) & \lambda \cdot (\vec{a} \times \vec{b}) &=& (\lambda \vec{a}) \times \vec{b} \quad\quad\ \text{\textit{Multiplication}} \\
& &=& \vec{a} \times (\lambda \vec{b}) \quad\quad\ \text{\textit{with scalar} } \lambda
\end{array}
$$

⊘ Formula for the Vector Product

Using the cross product of the unit vectors we get a representation of the vector product over the components of the vectors:

$$
\vec{a} \times \vec{b} = (a_x \vec{e}_1 + a_y \vec{e}_2 + a_z \vec{e}_3) \times (b_x \vec{e}_1 + b_y \vec{e}_2 + b_z \vec{e}_3)
$$

$$
= a_x b_x \underbrace{(\vec{e}_1 \times \vec{e}_1)}_{=\vec{0}} + a_x b_y \underbrace{(\vec{e}_1 \times \vec{e}_2)}_{\vec{e}_3} + a_x b_z \underbrace{(\vec{e}_1 \times \vec{e}_3)}_{-\vec{e}_2} +
$$

$$
a_y b_x \underbrace{(\vec{e}_2 \times \vec{e}_1)}_{-\vec{e}_3} + a_y b_y \underbrace{(\vec{e}_2 \times \vec{e}_2)}_{=\vec{0}} + a_y b_z \underbrace{(\vec{e}_2 \times \vec{e}_3)}_{\vec{e}_1} +
$$

$$
a_z b_x \underbrace{(\vec{e}_3 \times \vec{e}_1)}_{\vec{e}_2} + a_z b_y \underbrace{(\vec{e}_3 \times \vec{e}_2)}_{-\vec{e}_1} + a_z b_z \underbrace{(\vec{e}_3 \times \vec{e}_3)}_{=\vec{0}}
$$

$$
= (a_y b_z - a_z b_y)\vec{e}_1 + (a_z b_x - a_x b_z)\vec{e}_2 + (a_x b_y - a_y b_x)\vec{e}_3.
$$

Cross Product

$$
\vec{a} \times \vec{b} = \begin{pmatrix} a_x \\ a_y \\ a_z \end{pmatrix} \times \begin{pmatrix} b_x \\ b_y \\ b_z \end{pmatrix} = \begin{pmatrix} a_y b_z - a_z b_y \\ a_z b_x - a_x b_z \\ a_x b_y - a_y b_x \end{pmatrix}.
$$

Thumb rule: Formally, the vector product can be described in terms of a *three-column determinant* ($\rightarrow$ Chapter 3.2) developed with respect to the first column:

$$
\vec{a} \times \vec{b} = \begin{vmatrix} \vec{e}_1 & a_x & b_x \\ \vec{e}_2 & a_y & b_y \\ \vec{e}_3 & a_z & b_z \end{vmatrix} := \begin{vmatrix} a_y & b_y \\ a_z & b_z \end{vmatrix} \vec{e}_1 - \begin{vmatrix} a_x & b_x \\ a_z & b_z \end{vmatrix} \vec{e}_2 + \begin{vmatrix} a_x & b_x \\ a_y & b_y \end{vmatrix} \vec{e}_3.
$$

The value of a *two-row determinant* is computed by the difference between the principal and secondary diagonal products.

$$\begin{vmatrix} a & b \\ c & d \end{vmatrix} := a \cdot d - b \cdot c.$$

$$\Rightarrow \vec{a} \times \vec{b} = (a_y\, b_z - b_y\, a_z)\, \vec{e}_1 - (a_x\, b_z - a_z\, b_x)\, \vec{e}_2 + (a_x\, b_y - a_y\, b_x)\, \vec{e}_3.$$

Example 2.10. We are looking for a vector $\vec{c}$ that is perpendicular to both vectors

$$\vec{a} = \begin{pmatrix} 2 \\ 1 \\ -2 \end{pmatrix} \text{ and } \vec{b} = \begin{pmatrix} -1 \\ 2 \\ -1 \end{pmatrix}: \qquad \vec{c} = \vec{a} \times \vec{b}.$$

$$\vec{c} = \begin{vmatrix} \vec{e}_1 & 2 & -1 \\ \vec{e}_2 & 1 & 2 \\ \vec{e}_3 & -2 & -1 \end{vmatrix} = \begin{vmatrix} 1 & 2 \\ -2 & -1 \end{vmatrix} \vec{e}_1 - \begin{vmatrix} 2 & -1 \\ -2 & -1 \end{vmatrix} \vec{e}_2 + \begin{vmatrix} 2 & -1 \\ 1 & 2 \end{vmatrix} \vec{e}_3$$

$$= \begin{pmatrix} 3 \\ 4 \\ 5 \end{pmatrix}. \qquad \qquad \square$$

Example 2.11. Given are the vectors $\vec{a} = \begin{pmatrix} 1 \\ -4 \\ 1 \end{pmatrix}$ and $\vec{b} = \begin{pmatrix} 2 \\ 0 \\ 2 \end{pmatrix}$. We want to find the cross product of these two vectors and the area of the parallelogram spanned by the two vectors.

(i) $$\vec{a} \times \vec{b} = \begin{vmatrix} \vec{e}_1 & 1 & 2 \\ \vec{e}_2 & -4 & 0 \\ \vec{e}_3 & 1 & 2 \end{vmatrix} = \begin{vmatrix} -4 & 0 \\ 1 & 2 \end{vmatrix} \vec{e}_1 - \begin{vmatrix} 1 & 2 \\ 1 & 2 \end{vmatrix} \vec{e}_2 + \begin{vmatrix} 1 & 2 \\ -4 & 0 \end{vmatrix} \vec{e}_3$$

$$= -8\vec{e}_1 - 0\vec{e}_2 + 8\vec{e}_3 = \begin{pmatrix} -8 \\ 0 \\ 8 \end{pmatrix}.$$

(ii) The area of the parallelogram is

$$A = \left| \vec{a} \times \vec{b} \right| = \left| \begin{pmatrix} -8 \\ 0 \\ 8 \end{pmatrix} \right| = \sqrt{64 + 64} = \sqrt{128} = 8\sqrt{2}. \qquad \square$$

Application Example 2.12 (Cross Product in Physics).

In physics, for example, the cross product appears in important laws:

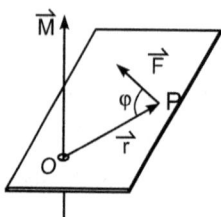

Figure 2.12. Torque

① **Torque:** A body can rotate around a fixed point O. A force $\overrightarrow{F}$ is applied to point P of this body. The magnitude M is the *torque* from $\overrightarrow{F}$ to O

$$M = \left|\overrightarrow{r}\right| \cdot \left|\overrightarrow{F}\right| \cdot \sin\varphi$$

(Force times lever arm).

The torque vector is perpendicular to $\overrightarrow{r}$ and $\overrightarrow{F}$ and can be interpreted as the direction of the axis of rotation:

$$\overrightarrow{M} = \overrightarrow{r} \times \overrightarrow{F}.$$

② **Angular Momentum:** Let O be a fixed point for reference. A mass m is at P and has the velocity $\overrightarrow{v}$. Then the *angular momentum* $\overrightarrow{L}$ of the mass point with respect to O is

$$\overrightarrow{L} = m\overrightarrow{r} \times \overrightarrow{v},$$

where $\overrightarrow{r} = \overrightarrow{OP} = \overrightarrow{r}(P)$ is the position vector of point P.

③ **Lorentz Force:** If a charged particle (charge q) moves with velocity $\overrightarrow{v}$ in a magnetic field $\overrightarrow{B}$, the *Lorentz force* acts perpendicular to $\overrightarrow{B}$ and $\overrightarrow{v}$

$$\overrightarrow{F_L} = q\overrightarrow{v} \times \overrightarrow{B}. \qquad\qquad \square$$

2.2.4 The Triple Product of Three Vectors

We look at the cross product (the vectors in brackets) and take the scalar product with $\overrightarrow{c}$. So the result is a real number.

Definition: The **Triple Product** $\left[\overrightarrow{a}, \overrightarrow{b}, \overrightarrow{c}\right]$ *of three vectors is defined as the real number*

$$\left[\overrightarrow{a}, \overrightarrow{b}, \overrightarrow{c}\right] := \left(\overrightarrow{a} \times \overrightarrow{b}\right) \cdot \overrightarrow{c}.$$

Geometric Interpretation: The volume spanned by the vectors $\overrightarrow{a}$, $\overrightarrow{b}$ and $\overrightarrow{c}$ is computed by the base area G times the height h. The base area is defined by the cross product $G = \left| \overrightarrow{a} \times \overrightarrow{b} \right|$ and the height is $h = \left| \overrightarrow{c} \right| \cos \varphi$.

$$\Rightarrow V = |a \times b| \cdot |\overrightarrow{c}| \cdot \cos \varphi = \left| (\overrightarrow{a} \times \overrightarrow{b}) \cdot \overrightarrow{c} \right|.$$

The **volume of the triple product** (cube) is equal to the absolute value of the triple product by calculating

$$\left[\overrightarrow{a}, \overrightarrow{b}, \overrightarrow{c} \right] = a_x b_y c_z + a_y b_z c_x + a_z b_x c_y - a_z b_y c_x - a_y b_x c_z - a_x b_z c_y.$$

The calculation can also be understood as the result of the determinant

$$\left[\overrightarrow{a}, \overrightarrow{b}, \overrightarrow{c} \right] = \begin{vmatrix} a_x & b_x & c_x \\ a_y & b_y & c_y \\ a_z & b_z & c_z \end{vmatrix}.$$

Example 2.13. We are looking for the triple product of the three vectors:

$$\overrightarrow{a} = \begin{pmatrix} 4 \\ 2 \\ 1 \end{pmatrix}, \quad \overrightarrow{b} = \begin{pmatrix} 3 \\ -1 \\ 2 \end{pmatrix}, \quad \overrightarrow{c} = \begin{pmatrix} 0 \\ 2 \\ -5 \end{pmatrix}.$$

In order to find the triple product of the vectors, we calculate the three-row determinant by developing it with respect to the first column:

$$\begin{vmatrix} 4 & 3 & 0 \\ 2 & -1 & 2 \\ 1 & 2 & -5 \end{vmatrix} = 4 \cdot \begin{vmatrix} -1 & 2 \\ 2 & -5 \end{vmatrix} - 2 \cdot \begin{vmatrix} 3 & 0 \\ 2 & -5 \end{vmatrix} + 1 \cdot \begin{vmatrix} 3 & 0 \\ -1 & 2 \end{vmatrix}$$

$$= 4 \left((-1)(-5) - 2 \cdot 2 \right) - 2(3 \cdot (-5) - 2 \cdot 0) + (3 \cdot 2 - (-1) \cdot 0) = 40. \ \square$$

Interpreting the triple product as the volume of $\overrightarrow{a}$, $\overrightarrow{b}$ and $\overrightarrow{c}$ leads to the following important conclusion:

Conclusion: The triple product is zero if all three vectors lie in one plane:

$$\left[\overrightarrow{a}, \overrightarrow{b}, \overrightarrow{c} \right] = 0 \quad \Leftrightarrow \quad \overrightarrow{a}, \overrightarrow{b}, \overrightarrow{c} \text{ are in the same plane.}$$

The following **rule** applies to the calculation of the triple product

$$(\overrightarrow{a} \times \overrightarrow{b}) \cdot \overrightarrow{c} = (\overrightarrow{b} \times \overrightarrow{c}) \cdot \overrightarrow{a} = (\overrightarrow{c} \times \overrightarrow{a}) \cdot \overrightarrow{b}.$$

The sign of the triple product changes if we deviate from the cyclic sequence.

2.3 Straight Lines and Planes in $\mathbb{R}^3$

This section presents some applications of vectors and vector calculus: The description of straight lines and planes in $\mathbb{R}^3$ as well as distance calculations and the relation of points, lines and planes to each other.

2.3.1 Vector Representation of Straight Lines

A *straight line* g is uniquely described by two different points $P_1 = (x_1, y_1, z_1)$ and $P_2 = (x_2, y_2, z_2)$ on this line. With the direction vector $\vec{a} := \overrightarrow{P_1 P_2}$ each point $P = (x, y, z)$ on the straight line can be described according to Fig. 2.13 by

Straight Line

$$g : \vec{r}(P) = \vec{r}(P_1) + \lambda \vec{a}, \qquad \lambda \in \mathbb{R},$$

(Point-direction form of a straight line).

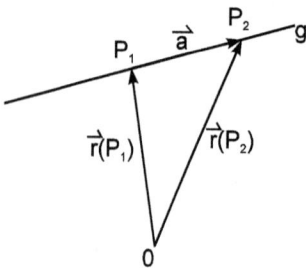

Figure 2.13. Point-direction form of a straight line

If we replace the vector $\vec{a} := \overrightarrow{P_1 P_2}$ with $\vec{r}(P_2) - \vec{r}(P_1)$, we get:

$$g : \vec{r}(P) = \vec{r}(P_1) + \lambda(\vec{r}(P_2) - \vec{r}(P_1)), \qquad \lambda \in \mathbb{R},$$

(Two-point representation of a straight line).

A point Q lies on a line g if the corresponding vector equation

$$\vec{r}(Q) = \vec{r}(P_1) + \lambda \vec{a}$$

has a solution for λ.

Example 2.14. Given are the points $P_1 = (2, 0, 4)$ and $P_2 = (2, 2, 2)$. Is the point $Q = (2, -2, 6)$ on the line g passing through points P_1 and P_2?

The equation for g with its direction vector is

$$\vec{a} = \overrightarrow{P_1 P_2} = \vec{r}(P_2) - \vec{r}(P_1) = \begin{pmatrix} 2 \\ 2 \\ 2 \end{pmatrix} - \begin{pmatrix} 2 \\ 0 \\ 4 \end{pmatrix} = \begin{pmatrix} 0 \\ 2 \\ -2 \end{pmatrix}$$

$$g : \vec{r}(P) = \vec{r}(P_1) + \lambda \vec{a} = \begin{pmatrix} 2 \\ 0 \\ 4 \end{pmatrix} + \lambda \begin{pmatrix} 0 \\ 2 \\ -2 \end{pmatrix}.$$

The point Q lies on the line g if the vector equation

$$\vec{r}(Q) = \vec{r}(P_1) + \lambda \vec{a}$$

has a solution. So we solve the equation
$\begin{pmatrix} 2 \\ -2 \\ 6 \end{pmatrix} = \begin{pmatrix} 2 \\ 0 \\ 4 \end{pmatrix} + \lambda \begin{pmatrix} 0 \\ 2 \\ -2 \end{pmatrix}$. Moving the vector $\vec{r}(P_1)$ to the left,

we get $\lambda \begin{pmatrix} 0 \\ 2 \\ -2 \end{pmatrix} = \begin{pmatrix} 0 \\ -2 \\ 2 \end{pmatrix}$. For $\lambda = -1$ this equation is satisfied and

therefore the point Q is on g.   □

2.3.2 Position of two Lines to Each Other

Two straight lines

$$g_1 : \vec{x} = \vec{r}(P_1) + \lambda \vec{a} \quad \text{and} \quad g_2 : \vec{x} = \vec{r}(P_2) + \mu \vec{b}$$

can have four different relations in $\mathbb{R}^3$:

(1) g_1 and g_2 **intersect** at exactly one point S (Fig. 2.14 (a)).

(2) g_1 and g_2 **coincide**. This is the case when $\vec{a} \parallel \vec{b}$ and $\overrightarrow{P_1 P_2} \parallel \vec{a}$.

(3) g_1 and g_2 are **parallel, but do not coincide** (Fig. 2.14 (b)). This is the case when $\vec{a} \parallel \vec{b}$ and $\overrightarrow{P_1 P_2} \nparallel \vec{a}$.

(4) g_1 and g_2 are **wind skew:** They are not parallel and do not intersect at any point (Fig. 2.14 (c)).

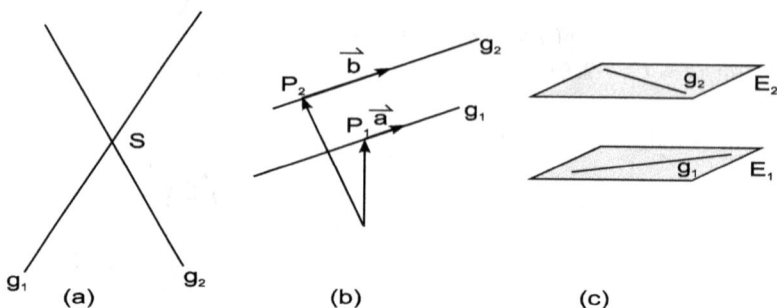

Figure 2.14. Position of two straight lines g_1 and g_2

To determine the position of two straight lines, it is sufficient to solve the vector equation $\vec{x}_{g_1} = \vec{x}_{g_2}$:

$$\Rightarrow \vec{r}(P_1) + \lambda\vec{a} = \vec{r}(P_2) + \mu\vec{b}$$
$$\Rightarrow \lambda\vec{a} - \mu\vec{b} = \vec{r}(P_2) - \vec{r}(P_1) = \overrightarrow{P_1P_2}.$$

This is a system of linear equations for the unknowns λ and μ. Because of

$$\vec{a} = \begin{pmatrix} a_x \\ a_y \\ a_z \end{pmatrix}, \vec{b} = \begin{pmatrix} b_x \\ b_y \\ b_z \end{pmatrix}, \vec{r}(P_1) = \begin{pmatrix} x_1 \\ y_1 \\ z_1 \end{pmatrix}, \vec{r}(P_2) = \begin{pmatrix} x_2 \\ y_2 \\ z_2 \end{pmatrix}.$$

The system is written in components

$$\lambda\, a_x - \mu\, b_x = x_2 - x_1$$
$$\lambda\, a_y - \mu\, b_y = y_2 - y_1$$
$$\lambda\, a_z - \mu\, b_z = z_2 - z_1$$

or in matrix notation

$$\left(\begin{array}{cc|c} a_x & -b_x & x_2 - x_1 \\ a_y & -b_y & y_2 - y_1 \\ a_z & -b_z & z_2 - z_1 \end{array} \right).$$

So the following applies

(1) If the system of linear equations for λ and μ has exactly one solution, then g_1 and g_2 intersect at exactly one point.

(2) If the system of linear equations for λ and μ has an infinite number of solutions, then g_1 and g_2 coincide.

(3) If the system of linear equations for λ and μ has no solution, then g_1 and g_2 are wind skewed, **or** they are parallel but do not coincide.

Example 2.15. g_1 is defined by the direction vector $\overrightarrow{a} = \begin{pmatrix} 1 \\ 2 \\ -1 \end{pmatrix}$ and a

point $P_1 = (3, 2, 1)$. The line g_2 is defined by two points $P_2 = (4, 0, -1)$ $P_3 = (-2, -1, -1)$. Find the relationship between the two lines.

From the point-direction representation of g_1 and the two-point representation of g_2 we set up

$$g_1 : \overrightarrow{x} = \begin{pmatrix} 3 \\ 2 \\ 1 \end{pmatrix} + \lambda \begin{pmatrix} 1 \\ 2 \\ -1 \end{pmatrix} \quad \text{and} \quad g_2 : \overrightarrow{x} = \begin{pmatrix} 4 \\ 0 \\ -1 \end{pmatrix} + \mu \begin{pmatrix} -6 \\ -1 \\ 0 \end{pmatrix}.$$

Since the direction vectors of g_1 and g_2 are not multiples of each other, they are not parallel, so the two lines can either intersect or be skewed. We set up the vector equation $\overrightarrow{x}_{g_1} = \overrightarrow{x}_{g_2}$:

$$\begin{pmatrix} 3 \\ 2 \\ 1 \end{pmatrix} + \lambda \begin{pmatrix} 1 \\ 2 \\ -1 \end{pmatrix} = \begin{pmatrix} 4 \\ 0 \\ -1 \end{pmatrix} + \mu \begin{pmatrix} -6 \\ -1 \\ 0 \end{pmatrix}$$

$$\Rightarrow \lambda \begin{pmatrix} 1 \\ 2 \\ -1 \end{pmatrix} + \mu \begin{pmatrix} 6 \\ 1 \\ 0 \end{pmatrix} = \begin{pmatrix} 4 \\ 0 \\ -1 \end{pmatrix} - \begin{pmatrix} 3 \\ 2 \\ 1 \end{pmatrix} = \begin{pmatrix} 1 \\ -2 \\ -2 \end{pmatrix}.$$

This system of linear equations is rewritten in matrix notation. We use the Gauss algorithm to solve the system of linear equations and obtain

$$\begin{pmatrix} 1 & 6 & | & 1 \\ 2 & 1 & | & -2 \\ -1 & 0 & | & -2 \end{pmatrix} \hookrightarrow \begin{pmatrix} 1 & 6 & | & 1 \\ 0 & -11 & | & -4 \\ 0 & 6 & | & -1 \end{pmatrix}.$$

From the last line follows $6\mu = -1 \Rightarrow \mu = -\frac{1}{6}$, and from the second last line $-11\mu = -4 \Rightarrow \mu = \frac{4}{11}$. This is a contradiction! So the vector equation **cannot** be solved and there is no intersection of g_1 with $g_2 \Rightarrow g_1$ and g_2 are skewed. □

2.3.3 Distances to Straight Lines

The **distance of a point Q from a straight line**

$$g : \vec{x} = \vec{r}(P_1) + \lambda \vec{a}$$

is given by the height d of the parallelogram represented by the vectors $\vec{a}$ and $\overrightarrow{P_1 Q}$ (see Fig. 2.15). The parallelogram area A is resolved according to the definition of the cross product $A = \left| \vec{a} \times \overrightarrow{P_1 Q} \right| = |\vec{a}| \cdot d$:

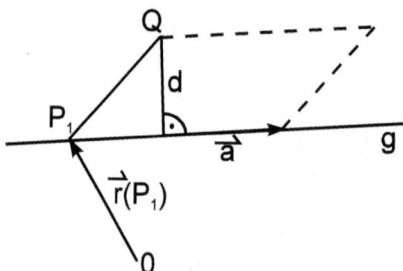

Figure 2.15. Distance of the point Q from line g

The **distance of a point Q to a straight line**

$$g : \vec{x} = \vec{r}(P_1) + \lambda \vec{a}$$

is given by

$$d = \frac{\left| \vec{a} \times \overrightarrow{P_1 Q} \right|}{|\vec{a}|}.$$

If $d = 0$, the point Q is on the straight line! From the above formula the **distance of two parallel straight lines**,

$$g_1 : \vec{x} = \vec{r}(P_1) + \lambda \vec{a} \quad \text{and} \quad g_2 : \vec{x} = \vec{r}(P_2) + \mu \vec{b},$$

can be easily calculated by selecting e.g. the point P_2 on the line g_2 and finding the distance of this point from line g_1. For $d = 0$, the straight lines coincide!

To get the **distance between two skewed lines**

$$g_1 : \vec{x} = \vec{r}(P_1) + \lambda \vec{a} \quad \text{and} \quad g_2 : \vec{x} = \vec{r}(P_2) + \mu \vec{b},$$

we define the vector $\vec{L} = \vec{a} \times \vec{b}$. $\vec{L}$ is perpendicular to $\vec{a}$ and on $\vec{b}$. If $\vec{L} = \vec{0}$ then $\vec{a}$ and $\vec{b}$ are parallel and the distance is zero.

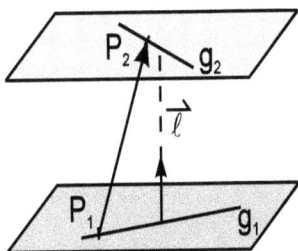

Figure 2.16. Skewed lines

For $\overrightarrow{L} \neq \overrightarrow{0}$ we go to the unit vector

$$\overrightarrow{l} = \frac{1}{|\overrightarrow{L}|} \overrightarrow{L}.$$

The distance of g_1 and g_2 is given by the projection $\overrightarrow{P_1 P_2}$ onto $\overrightarrow{l}$, so $d = |\overrightarrow{l} \cdot \overrightarrow{P_1 P_2}|$:

$$d = \left| \frac{\overrightarrow{a} \times \overrightarrow{b}}{|\overrightarrow{a} \times \overrightarrow{b}|} \cdot \overrightarrow{P_1 P_2} \right|$$

Distance of two skewed lines
$g_1 : \overrightarrow{x} = \overrightarrow{r}(P_1) + \lambda \overrightarrow{a}$ and
$g_2 : \overrightarrow{x} = \overrightarrow{r}(P_2) + \mu \overrightarrow{b}$.

If the distance is $d = 0$, then the lines intersect and the intersection angle is given by the angle between the two direction vectors $\overrightarrow{a}$ and $\overrightarrow{b}$:

$$\cos \varphi = \frac{\overrightarrow{a} \cdot \overrightarrow{b}}{|\overrightarrow{a}| |\overrightarrow{b}|}$$

Angle of intersection between two intersecting straight lines.

Example 2.16. We calculate the distance between the two parallel lines

$$g_1 : \overrightarrow{x} = \begin{pmatrix} 1 \\ 1 \\ 4 \end{pmatrix} + \lambda \begin{pmatrix} 1 \\ 1 \\ 1 \end{pmatrix} \quad \text{and} \quad g_2 : \overrightarrow{x} = \begin{pmatrix} 3 \\ 0 \\ 1 \end{pmatrix} + \mu \begin{pmatrix} 2 \\ 2 \\ 2 \end{pmatrix}.$$

Because of $\overrightarrow{a} = \begin{pmatrix} 1 \\ 1 \\ 1 \end{pmatrix}$ it is $|\overrightarrow{a}| = \sqrt{3}$ and $\overrightarrow{a} \times \overrightarrow{P_1 P_2} = \begin{pmatrix} 1 \\ 1 \\ 1 \end{pmatrix} \times$

$\begin{pmatrix} 2 \\ -1 \\ -3 \end{pmatrix} = \begin{pmatrix} -2 \\ 5 \\ -3 \end{pmatrix}$. So $d = \frac{|\overrightarrow{a} \times \overrightarrow{P_1 P_2}|}{|\overrightarrow{a}|} = \frac{\sqrt{38}}{\sqrt{3}} = 3,559.$ □

Example 2.17. Find the point of intersection S and the angle of intersection φ of the lines

$$g_1 : \vec{x} = \begin{pmatrix} 1 \\ 1 \\ 0 \end{pmatrix} + \lambda \begin{pmatrix} 2 \\ 1 \\ 1 \end{pmatrix} \quad \text{and} \quad g_2 : \vec{x} = \begin{pmatrix} 2 \\ 0 \\ 2 \end{pmatrix} + \mu \begin{pmatrix} 1 \\ -1 \\ 2 \end{pmatrix} .$$

To calculate the point of intersection S, we set $\vec{x}_{g_1} = \vec{x}_{g_2}$ in order to find the equation to solve:

$$\begin{pmatrix} 1 \\ 1 \\ 0 \end{pmatrix} + \lambda \begin{pmatrix} 2 \\ 1 \\ 1 \end{pmatrix} = \begin{pmatrix} 2 \\ 0 \\ 2 \end{pmatrix} + \mu \begin{pmatrix} 1 \\ -1 \\ 2 \end{pmatrix}$$

$$\hookrightarrow \lambda \begin{pmatrix} 2 \\ 1 \\ 1 \end{pmatrix} + \mu \begin{pmatrix} -1 \\ 1 \\ -2 \end{pmatrix} = \begin{pmatrix} 2 \\ 0 \\ 2 \end{pmatrix} - \begin{pmatrix} 1 \\ 1 \\ 0 \end{pmatrix} = \begin{pmatrix} 1 \\ -1 \\ 2 \end{pmatrix}$$

and in matrix notation

$$\left(\begin{array}{cc|c} 2 & -1 & 1 \\ 1 & 1 & -1 \\ 1 & -2 & 2 \end{array} \right) \hookrightarrow \left(\begin{array}{cc|c} 2 & -1 & 1 \\ 0 & -3 & 3 \\ 0 & 3 & -3 \end{array} \right) .$$

From both the last and the second last line we get $\mu = -1$, i.e. the LEq is solvable. From the first line we get

$$2\lambda - (-1) = 1 \quad \Rightarrow \quad \lambda = 0.$$

So $\vec{r}(S) = \begin{pmatrix} 1 \\ 1 \\ 0 \end{pmatrix} + 0 \cdot \begin{pmatrix} 2 \\ 1 \\ 1 \end{pmatrix} = \begin{pmatrix} 1 \\ 1 \\ 0 \end{pmatrix}$ and the coordinates of S are $S = (1, 1, 0)$. The angle of intersection is

$$\varphi = \arccos \left(\frac{\vec{a} \cdot \vec{b}}{|\vec{a}| \, |\vec{b}|} \right) .$$

We calculate $\vec{a} \cdot \vec{b} = \begin{pmatrix} 2 \\ 1 \\ 1 \end{pmatrix} \cdot \begin{pmatrix} 1 \\ -1 \\ 2 \end{pmatrix} = 2 - 1 + 2 = 3$ and $|\vec{a}| = \sqrt{6} = |\vec{b}|$ and obtain

$$\varphi = \arccos \left(\frac{\vec{a} \cdot \vec{b}}{|\vec{a}| \, |\vec{b}|} \right) = \arccos \left(\frac{3}{\sqrt{6}\sqrt{6}} \right) = 60°. \qquad \square$$

2.3.4 Vector Representation of Planes

A *plane* E is uniquely defined by three points $P_1 = (x_1, y_1, z_1)$, $P_2 = (x_2, y_2, z_2)$ and $P_3 = (x_3, y_3, z_3)$ which do not lie on a straight line. With these 3 points we define two direction vectors $\vec{a} = \overrightarrow{P_1 P_2}$ and $\vec{b} = \overrightarrow{P_1 P_3}$ and a position vector $\vec{r}(P_1)$. Each point $P = (x, y, z)$ of the plane corresponds to a vector $\vec{x} = \vec{r}(P)$ with

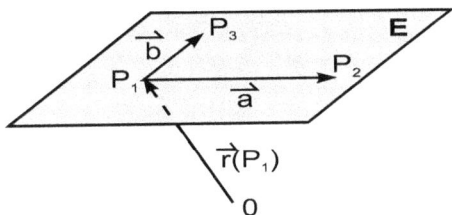

Figure 2.17. Point-direction representation of a plane E

Representation of a Plane

$$E : \vec{r}(P) = \vec{r}(P_1) + \lambda \vec{a} + \mu \vec{b} \quad (\lambda, \mu \in \mathbb{R})$$

(Point-direction representation of a plane).

Replacing the directional vectors $\vec{a} = \overrightarrow{P_1 P_2}$ and $\vec{b} = \overrightarrow{P_1 P_3}$ by the corresponding position vectors $\vec{r}(P_2) - \vec{r}(P_1)$, $\vec{r}(P_3) - \vec{r}(P_1)$, we get

$$E : \vec{r}(P) = \vec{r}(P_1) + \lambda(\vec{r}(P_2) - \vec{r}(P_1)) + \mu(\vec{r}(P_3) - \vec{r}(P_1))$$

(Three-point representation of a plane).

Example 2.18. Given are three points $P_1 = (5, 2, 1)$, $P_2 = (4, 0, -4)$ and $P_3 = (1, 1, 1)$. Then the plane defined by the 3 points is given by

$$E : \vec{r}(P) = \begin{pmatrix} 5 \\ 2 \\ 1 \end{pmatrix} + \lambda \begin{pmatrix} 4-5 \\ 0-2 \\ -4-1 \end{pmatrix} + \mu \begin{pmatrix} 1-5 \\ 1-2 \\ 1-1 \end{pmatrix}$$

$$= \begin{pmatrix} 5 \\ 2 \\ 1 \end{pmatrix} + \lambda \begin{pmatrix} -1 \\ -2 \\ -5 \end{pmatrix} + \mu \begin{pmatrix} -4 \\ -1 \\ 0 \end{pmatrix}. \qquad \square$$

⊘ Hesse Normal Form

An alternative representation of the plane is obtained by multiplying the point-direction representation of the plane with the normal vector perpendicular to E. $\overrightarrow{N} := \overrightarrow{a} \times \overrightarrow{b}$ is perpendicular to $\overrightarrow{a}$ and to $\overrightarrow{b}$ and the normal unit vector is equal to

$$\overrightarrow{n} := \frac{1}{|\overrightarrow{N}|} \overrightarrow{N} = \frac{1}{|\overrightarrow{a} \times \overrightarrow{b}|} \overrightarrow{a} \times \overrightarrow{b}.$$

This normal vector has the property $|\overrightarrow{n}| = 1$, $\overrightarrow{a} \cdot \overrightarrow{n} = 0$ and $\overrightarrow{b} \cdot \overrightarrow{n} = 0$. So for each point P on the plane E the following is true

$$\overrightarrow{r}(P) \cdot \overrightarrow{n} = \overrightarrow{r}(P_1) \cdot \overrightarrow{n} \quad \text{or} \quad (\overrightarrow{r}(P) - \overrightarrow{r}(P_1)) \cdot \overrightarrow{n} = 0.$$

If we set $\overrightarrow{r}(P) = \begin{pmatrix} x \\ y \\ z \end{pmatrix}$ and the normal vector $\overrightarrow{n} = \begin{pmatrix} n_1 \\ n_2 \\ n_3 \end{pmatrix}$ it follows:

Hesse Normal Form

$$\begin{pmatrix} x - x_1 \\ y - y_1 \\ z - z_1 \end{pmatrix} \cdot \begin{pmatrix} n_1 \\ n_2 \\ n_3 \end{pmatrix} = 0 \qquad \begin{array}{l} \textbf{Hesse normal form of a plane} \\ \text{where } P_1 \text{ is a point of the plane and} \\ \overrightarrow{n} \text{ is the normal unit vector.} \end{array}$$

We evaluate the scalar product

$$n_1(x - x_1) + n_2(y - y_1) + n_3(z - z_1) = 0.$$

So all solutions of a linear equation

$$A x + B y + C z = D$$

form a plane in $\mathbb{R}^3$.

Example 2.19. We are looking for the representation of the plane E containing the point $P = (2, -5, 3)$ and is perpendicular to the normal vector

$$\overrightarrow{N} = \begin{pmatrix} 4 \\ 2 \\ 5 \end{pmatrix}.$$

From Hesse's form we get

$$\vec{N} \cdot (\vec{r}(P) - \vec{r}(P_1)) = \begin{pmatrix} 4 \\ 2 \\ 5 \end{pmatrix} \begin{pmatrix} x - 2 \\ y + 5 \\ z - 3 \end{pmatrix} = 4(x-2) + 2(y+5) + 5(z-3) = 0$$

$$\Rightarrow \qquad 4x + 2y + 5z = 13. \qquad (*)$$

Thus, the point-direction representation of the plane is obtained by choosing $z = \mu$ (arbitrary) and $y = \lambda$ (arbitrary). The equation $(*)$ then becomes

$$4x + 2\lambda + 5\mu = 13 \quad \Rightarrow \quad x = \frac{13}{4} - \frac{1}{2}\lambda - \frac{5}{4}\mu.$$

The solution of the equation $(*)$ gives the point-directional form of the plane

$$\vec{x} = \begin{pmatrix} x \\ y \\ z \end{pmatrix} = \begin{pmatrix} \frac{13}{4} \\ 0 \\ 0 \end{pmatrix} + \lambda \begin{pmatrix} -\frac{1}{2} \\ 1 \\ 0 \end{pmatrix} + \mu \begin{pmatrix} -\frac{5}{4} \\ 0 \\ 1 \end{pmatrix}.$$

The normal vector $\vec{N} = \vec{a} \times \vec{b} = \begin{pmatrix} -\frac{1}{2} \\ 1 \\ 0 \end{pmatrix} \times \begin{pmatrix} -\frac{5}{4} \\ 0 \\ 1 \end{pmatrix} = \frac{1}{4}\begin{pmatrix} 4 \\ 2 \\ 5 \end{pmatrix}$ is

normalized by $\vec{n} = \frac{1}{|\vec{N}|}\vec{N}$. So another Hesse normal form is given by

$$\frac{1}{12\sqrt{5}}\begin{pmatrix} x - \frac{13}{4} \\ y - 0 \\ z - 0 \end{pmatrix} \cdot \begin{pmatrix} 4 \\ 2 \\ 5 \end{pmatrix} = 0. \qquad \qquad \square$$

2.3.5 Position of two Planes in Relation to Each Other

To determine the **position of two planes**

$$E_1 : \vec{x} = \vec{r}(P_1) + \lambda \vec{a}_1 + \mu \vec{b}_1 \quad \text{and} \quad E_2 : \vec{x} = \vec{r}(P_2) + \tau \vec{a}_2 + \sigma \vec{b}_2$$

the vector equation $\vec{x}_{E_1} = \vec{x}_{E_2}$ has to be solved:

$$\vec{r}(P_1) + \lambda \vec{a}_1 + \mu \vec{b}_1 = \vec{r}(P_2) + \tau \vec{a}_2 + \sigma \vec{b}_2.$$

(1) If the system cannot be solved, then the planes E_1 and E_2 are **parallel** and **do not coincide** ($E_1 \parallel E_2, E_1 \neq E_2$) (see Fig. 2.18 (a)).

(2) If the system can be solved with one parameter, then **the planes E_1 and E_2 intersect each other** in a straight line $(E_1 \nparallel E_2, E_1 \cap E_2 = g)$ (see Fig. 2.18 (b)).

(3) If the system is solvable with two parameters, then **they collapse** $(E_1 = E_2)$.

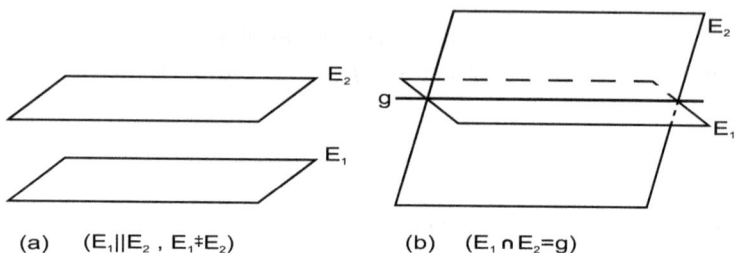

(a) $(E_1 || E_2 , E_1 \neq E_2)$ (b) $(E_1 \cap E_2 = g)$

Figure 2.18. Position of two planes E_1 and E_2 relative to each other

---

Prime Example 2.20

We determine the position of the planes E_1 and E_2 relative to each other when

$$E_1 : \vec{x} = \begin{pmatrix} 1 \\ 0 \\ 0 \end{pmatrix} + \lambda \begin{pmatrix} -1 \\ 2 \\ 0 \end{pmatrix} + \mu \begin{pmatrix} -1 \\ 0 \\ 1 \end{pmatrix} ;$$

$$E_2 : \vec{x} = \begin{pmatrix} 0 \\ 1 \\ 0 \end{pmatrix} + \tau \begin{pmatrix} 1 \\ 2 \\ 1 \end{pmatrix} + \sigma \begin{pmatrix} 0 \\ 4 \\ 0 \end{pmatrix} .$$

To determine the position relative to each other, we equate the two parameter representations of the planes, resulting in a system of linear equations. With $\vec{x}_{E_1} = \vec{x}_{E_2}$ follows

$$\begin{pmatrix} 1 \\ 0 \\ 0 \end{pmatrix} + \lambda \begin{pmatrix} -1 \\ 2 \\ 0 \end{pmatrix} + \mu \begin{pmatrix} -1 \\ 0 \\ 1 \end{pmatrix} = \begin{pmatrix} 0 \\ 1 \\ 0 \end{pmatrix} + \tau \begin{pmatrix} 1 \\ 2 \\ 1 \end{pmatrix} + \sigma \begin{pmatrix} 0 \\ 4 \\ 0 \end{pmatrix}$$

$$\hookrightarrow \lambda \begin{pmatrix} -1 \\ 2 \\ 0 \end{pmatrix} + \mu \begin{pmatrix} -1 \\ 0 \\ 1 \end{pmatrix} + \tau \begin{pmatrix} -1 \\ -2 \\ -1 \end{pmatrix} + \sigma \begin{pmatrix} 0 \\ -4 \\ 0 \end{pmatrix} = \begin{pmatrix} -1 \\ 1 \\ 0 \end{pmatrix} .$$

In matrix notation, the linear system for $\lambda, \mu, \tau, \sigma$ is

$$
\left(\begin{array}{cccc|c}
-1 & -1 & -1 & 0 & -1 \\
2 & 0 & -2 & -4 & 1 \\
0 & 1 & -1 & 0 & 0
\end{array}\right)
\hookrightarrow
\left(\begin{array}{cccc|c}
-1 & -1 & -1 & 0 & -1 \\
0 & -2 & -4 & -4 & -1 \\
0 & 1 & -1 & 0 & 0
\end{array}\right)
$$

$$
\hookrightarrow
\left(\begin{array}{cccc|c}
-1 & -1 & -1 & 0 & -1 \\
0 & -2 & -4 & -4 & -1 \\
0 & 0 & -6 & -4 & -1
\end{array}\right) .
$$

So $\sigma = t$ (arbitrary) and $-6\tau - 4t = -1 \Rightarrow \tau = \frac{1}{6} - \frac{2}{3}t$. The solution of the LEq has a free parameter t; therefore the planes E_1 and E_2 intersect in a straight line g. We obtain the representation of the equation of this straight line by substituting $\sigma = t$ and $\tau = \frac{1}{6} - \frac{2}{3}t$ into the equation E_2:

$$
g : \vec{x} = \begin{pmatrix} 0 \\ 1 \\ 0 \end{pmatrix} + \left(\frac{1}{6} - \frac{2}{3}t\right)\begin{pmatrix} 1 \\ 2 \\ 1 \end{pmatrix} + t \begin{pmatrix} 0 \\ 4 \\ 0 \end{pmatrix}
$$

$$
= \begin{pmatrix} 0 \\ 1 \\ 0 \end{pmatrix} + \frac{1}{6}\begin{pmatrix} 1 \\ 2 \\ 1 \end{pmatrix} + t \left(-\frac{2}{3}\begin{pmatrix} 1 \\ 2 \\ 1 \end{pmatrix} + \begin{pmatrix} 0 \\ 4 \\ 0 \end{pmatrix} \right)
$$

$$
g : \vec{x} = \begin{pmatrix} \frac{1}{6} \\ \frac{4}{3} \\ \frac{1}{6} \end{pmatrix} + t \begin{pmatrix} -\frac{2}{3} \\ \frac{8}{3} \\ -\frac{2}{3} \end{pmatrix} . \qquad \square
$$

⊘ **To determine the relation of a plane**

$$
E : \vec{x} = \vec{r}(P_1) + \lambda \vec{a} + \mu \vec{b}
$$

and a straight line

$$
g : \vec{x} = \vec{r}(P_2) + \tau \vec{c},
$$

we solve the vector equation $\vec{x}_E = \vec{x}_g$:

$$
\vec{r}(P_1) + \lambda \vec{a} + \mu \vec{b} = \vec{r}(P_2) + \tau \vec{c}.
$$

This is a LEq for the unknown λ, μ, τ. It applies

(1) If the system has no solution, then g is **parallel** to E but not included in E ($g \| E, g \not\subseteq E$) (see Fig. 2.19 (a)).

(2) If the system has a unique solution, then the plane E **intersects** the line g in one **point** S $(g \cap E = \{S\})$ (see Fig. 2.19 (b)).

(3) If the system is solvable with a parameter, then g is **in the plane** E $(g \subset E)$ (see Fig. 2.19 (c)).

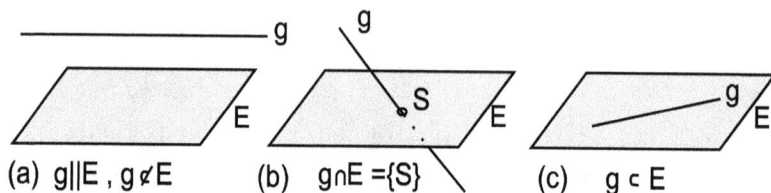

(a) g‖E , g⊄E   (b) gnE ={S}   (c) g ⊂ E

Figure 2.19. Position of a line g relative to a plane E

2.3.6 Distance Calculation for Planes

The **distance of a point** Q **from a plane**

$$E : \overrightarrow{x} = \overrightarrow{r}(P_1) + \lambda \overrightarrow{a} + \mu \overrightarrow{b}$$

is calculated by projecting the vector $\overrightarrow{P_1 Q}$ onto the normal vector $\overrightarrow{N}$: $\overrightarrow{d} = \frac{\overrightarrow{N} \cdot \overrightarrow{P_1 Q}}{|\overrightarrow{N}|^2} \cdot \overrightarrow{N}$. So the distance is $d = \left| \overrightarrow{d} \right| = \frac{|\overrightarrow{N} \cdot \overrightarrow{P_1 Q}|}{|\overrightarrow{N}|^2} \cdot \left| \overrightarrow{N} \right| = \frac{|\overrightarrow{N} \cdot \overrightarrow{P_1 Q}|}{|\overrightarrow{N}|}$:

$$d = \frac{\left| \overrightarrow{N} \cdot \overrightarrow{P_1 Q} \right|}{\left| \overrightarrow{N} \right|}$$

Distance of a point Q **from the plane** E: $\overrightarrow{x} = \overrightarrow{r}(P_1) + \lambda \overrightarrow{a} + \mu \overrightarrow{b}$ with normal vector $\overrightarrow{N} = \overrightarrow{a} \times \overrightarrow{b}$.

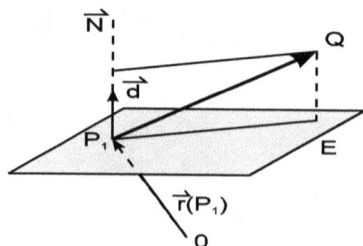

Figure 2.20. Distance point for plane

If $d = 0$, then the point Q is in the plane E.

The **distance of a parallel straight line** $g : \overrightarrow{x} = \overrightarrow{r}(P_2) + \tau \overrightarrow{c}$ to a plane E is obtained directly from the above formula by selecting any point on the line, e.g. P_2, and determining the distance of that point from the plane.

$$d = \frac{\left|\overrightarrow{N} \cdot \overrightarrow{P_1 P_2}\right|}{\left|\overrightarrow{N}\right|}$$

Distance of the line $g : \overrightarrow{x} = \overrightarrow{r}(P_2) + \tau \overrightarrow{c}$
from the plane $E : \overrightarrow{x} = \overrightarrow{r}(P_1) + \lambda \overrightarrow{a} + \mu \overrightarrow{b}$
with normal vector $\overrightarrow{N} = \overrightarrow{a} \times \overrightarrow{b}$.

For $d = 0$ the line is in the plane E.

The **distance of a plane** E **parallel to plane** $E_2 : \overrightarrow{x} = \overrightarrow{r}(P_2) + \tau \overrightarrow{c} + \sigma \overrightarrow{d}$ can also be obtained directly from the above formula by selecting any point on the plane E_2 (e.g. P_2) and inserting:

$$d = \frac{\left|\overrightarrow{N} \cdot \overrightarrow{P_1 P_2}\right|}{\left|\overrightarrow{N}\right|}$$

Distance of the plane
$E_2 : \overrightarrow{x} = \overrightarrow{r}(P_2) + \tau \overrightarrow{c} + \sigma \overrightarrow{d}$
from the plane $E : \overrightarrow{x} = \overrightarrow{r}(P_1) + \lambda \overrightarrow{a} + \mu \overrightarrow{b}$
with normal vector $\overrightarrow{N} = \overrightarrow{a} \times \overrightarrow{b}$.

If $d = 0$, both planes coincide.

Example 2.21. The distance between the point $Q = (3, 1, 5)$ and the

plane $E : \overrightarrow{x} = \begin{pmatrix} 1 \\ 2 \\ 0 \end{pmatrix} + \lambda \begin{pmatrix} 1 \\ 1 \\ 0 \end{pmatrix} + \mu \begin{pmatrix} 0 \\ 2 \\ 2 \end{pmatrix}$: Because $\overrightarrow{N} := \overrightarrow{a} \times \overrightarrow{b} =$

$\begin{pmatrix} 1 \\ 1 \\ 0 \end{pmatrix} \times \begin{pmatrix} 0 \\ 2 \\ 2 \end{pmatrix} = \begin{pmatrix} 2 \\ -2 \\ 2 \end{pmatrix}$ is $\left|\overrightarrow{N}\right| = \sqrt{12}$ and $\overrightarrow{P_1 Q} = \begin{pmatrix} 2 \\ -1 \\ 5 \end{pmatrix}$.

$$\Rightarrow d = \frac{\left|\overrightarrow{N} \cdot \overrightarrow{P_1 Q}\right|}{\left|\overrightarrow{N}\right|} = \frac{1}{\sqrt{12}} \left| \begin{pmatrix} 2 \\ -2 \\ 2 \end{pmatrix} \cdot \begin{pmatrix} 2 \\ -1 \\ 5 \end{pmatrix} \right| = \frac{16}{\sqrt{12}} = \frac{8}{3}\sqrt{3}. \qquad \square$$

2.3.7 Calculation of the Intersection of a Line with a Plane

To find the intersection of a straight line

$$g : \overrightarrow{x} = \overrightarrow{r}(P_2) + \lambda \overrightarrow{a}$$

with a plane E, we assume that the plane E is given in the Hesse form

$$E : \overrightarrow{N} \cdot (\overrightarrow{x} - \overrightarrow{r}(P_1)) = 0,$$

where P_1 is a point on the plane and $\vec{N}$ is a normal vector. We assume that the line is not parallel to the plane.

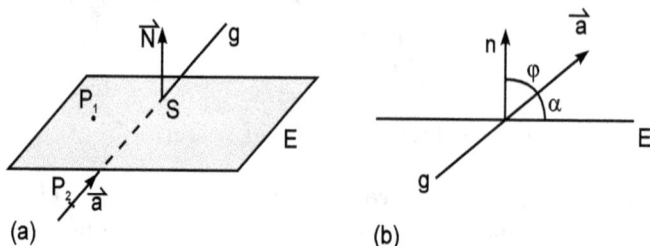

Figure 2.21. Point of intersection and angle of a line with a plane

The **intersection** S has the property that $\vec{x}_g = \vec{x}_E$, i.e. we insert $\vec{x}_g$ into the plane equation:

$$\vec{N} \cdot (\vec{r}(P_2) + \lambda \vec{a} - \vec{r}(P_1)) = \vec{N} \cdot (\vec{r}(P_2) - \vec{r}(P_1)) + \lambda \vec{N} \cdot \vec{a} = 0.$$

Since the straight line is not parallel to the plane, $\vec{N} \cdot \vec{a} \neq 0$, with $\overline{P_1 P_2} = \vec{r}(P_2) - \vec{r}(P_1)$ follows

$$\lambda = -\frac{\vec{N} \cdot \overline{P_1 P_2}}{\vec{N} \cdot \vec{a}}.$$

Substituting λ into the line equation gives us the intersection point S.

$$\vec{r}(S) = \vec{r}(P_2) - \frac{\vec{N} \cdot \overline{P_1 P_2}}{\vec{N} \cdot \vec{a}} \cdot \vec{a} \qquad \text{Position vector to the intersection point } S.$$

The angle φ between the normal of the plane and the straight line is

$$\cos \varphi = \frac{\vec{N} \cdot \vec{a}}{\left|\vec{N}\right|\left|\vec{a}\right|}.$$

φ is the complementary angle to α, $\varphi = 90° \pm \alpha$, depending on the direction of the normal vector. Therefore

$$\cos \varphi = \cos(90° \pm \alpha) = \mp \sin \alpha \quad \text{and}$$

$$\sin \alpha = \mp \frac{\vec{N} \cdot \vec{a}}{\left|\vec{N}\right|\left|\vec{a}\right|} \qquad \begin{array}{l} \text{Intersection angle between line} \\ g : \vec{x} = \vec{r}(P_2) + \lambda \vec{a} \text{ and} \\ \text{the plane } E : \vec{N}(\vec{x} - \vec{r}(P_1)) = 0. \end{array}$$

Example 2.22. We are looking for the intersection point S and the angle α of a straight line $g : \vec{x} = \begin{pmatrix} 1 \\ 2 \\ 3 \end{pmatrix} + \tau \begin{pmatrix} 2 \\ 0 \\ 2 \end{pmatrix}$ with $E : \vec{x} = \begin{pmatrix} 2 \\ -1 \\ 3 \end{pmatrix} + \lambda \begin{pmatrix} 0 \\ 1 \\ 0 \end{pmatrix} + \mu \begin{pmatrix} 4 \\ 2 \\ -1 \end{pmatrix}$.

From the direction vectors of the plane $\vec{b}_1 = \begin{pmatrix} 0 \\ 1 \\ 0 \end{pmatrix}$ and $\vec{b}_2 = \begin{pmatrix} 4 \\ 2 \\ -1 \end{pmatrix}$

the normal vector is obtained $\vec{N} = \vec{b}_1 \times \vec{b}_2 = \begin{pmatrix} -1 \\ 0 \\ -4 \end{pmatrix}$. With this

$$\vec{N} \cdot \overrightarrow{P_1 P_2} = \begin{pmatrix} -1 \\ 0 \\ -4 \end{pmatrix} \cdot \left(\begin{pmatrix} 1 \\ 2 \\ 3 \end{pmatrix} - \begin{pmatrix} 2 \\ -1 \\ 3 \end{pmatrix} \right) = 1$$

$$\vec{N} \cdot \vec{a} = \begin{pmatrix} -1 \\ 0 \\ -4 \end{pmatrix} \cdot \begin{pmatrix} 2 \\ 0 \\ 2 \end{pmatrix} = -10.$$

The intersection S is calculated from

$$\vec{r}(S) = \vec{r}(P_2) - \frac{\vec{N} \cdot \overrightarrow{P_1 P_2}}{\vec{N} \cdot \vec{a}} \cdot \vec{a} = \begin{pmatrix} 1 \\ 2 \\ 3 \end{pmatrix} + \frac{1}{10} \begin{pmatrix} 2 \\ 0 \\ 2 \end{pmatrix} = \begin{pmatrix} 1,2 \\ 2 \\ 3,2 \end{pmatrix}.$$

Then the angle of intersection α is given by

$$\sin \alpha = -\frac{\vec{N} \cdot \vec{a}}{|\vec{N}| \cdot |\vec{a}|} = \frac{10}{\sqrt{17}\sqrt{8}} \quad \Rightarrow \quad \alpha = 59,04°. \qquad \square$$

Remark: The **angle of intersection of two intersecting planes**

$$E_1 : \vec{N}_1 (\vec{r}(P) - \vec{r}(P_1)) = 0 \quad \text{and} \quad E_2 : \vec{N}_2 (\vec{r}(P) - \vec{r}(P_2)) = 0$$

is the same intersection angle as the intersection angle φ of the normal vectors. Therefore, $\quad \cos \varphi = \dfrac{\vec{N}_1 \cdot \vec{N}_2}{|\vec{N}_1| |\vec{N}_2|}.$

2.4 Vector Spaces

The definition of vectors is transferred from $\mathbb{R}^3$ to $\mathbb{R}^n$ and some elementary arithmetic operations (addition and S-multiplication) are extended to these vectors. With general considerations, the concept of a vector space is defined, which plays an important role in the description of physical systems. In particular, this formalism is needed for the representation of systems of linear equations or linear differential equations. Central to the characterization of vector spaces are the concepts of linear combination, linear independence and the basis as a minimal generating system.

2.4.1 Vector Calculus in $\mathbb{R}^n$

The vector concept is transferred from $\mathbb{R}^3$ to $\mathbb{R}^n$:

> **Definition:** *The set of all n-tuples of real numbers is called* $\mathbb{R}^n$:
> $$\mathbb{R}^n := \left\{ \begin{pmatrix} x_1 \\ x_2 \\ \vdots \\ x_n \end{pmatrix} : x_1 \in \mathbb{R},\, x_2 \in \mathbb{R}, \cdots, x_n \in \mathbb{R} \right\}.$$

Similar to the coordinate system in a plane or in space, the coordinate system in $\mathbb{R}^n$ is defined by n unit vectors perpendicular to each other.

$$\vec{e}_1 := \begin{pmatrix} 1 \\ 0 \\ \vdots \\ 0 \end{pmatrix}, \quad \vec{e}_2 := \begin{pmatrix} 0 \\ 1 \\ \vdots \\ 0 \end{pmatrix}, \quad \ldots, \quad \vec{e}_n := \begin{pmatrix} 0 \\ \vdots \\ 0 \\ 1 \end{pmatrix}.$$

Each vector $\vec{a} \in \mathbb{R}^n$ can be described by specifying its components:

$$\vec{a} = a_1 \vec{e}_1 + a_2 \vec{e}_2 + \cdots + a_n \vec{e}_n = \begin{pmatrix} a_1 \\ a_2 \\ \vdots \\ a_n \end{pmatrix}.$$

The concepts of the magnitude, equality of vectors, multiplication by a scalar, addition, the scalar product, orthogonality, etc. are transferred to $\mathbb{R}^n$.

Addition and S-Multiplication

For two vectors $\vec{a} = \begin{pmatrix} a_1 \\ \vdots \\ a_n \end{pmatrix}$, $\vec{b} = \begin{pmatrix} b_1 \\ \vdots \\ b_n \end{pmatrix}$ and a scalar $\lambda \in \mathbb{R}$

we define

$$\vec{a} + \vec{b} := \begin{pmatrix} a_1 + b_1 \\ a_2 + b_2 \\ \vdots \\ a_n + b_n \end{pmatrix}, \qquad \lambda \cdot \vec{a} := \begin{pmatrix} \lambda\, a_1 \\ \lambda\, a_2 \\ \vdots \\ \lambda\, a_n \end{pmatrix}.$$

(Addition) **(S-Multiplication)**

Both addition and S-multiplication are done component by component. For addition and S-multiplication, two *operations* are specified.

$$+ : \mathbb{R}^n \times \mathbb{R}^n \to \mathbb{R}^n \quad \text{with} \quad \left(\vec{a}, \vec{b}\right) \mapsto \vec{a} + \vec{b},$$
$$\cdot : \mathbb{R} \times \mathbb{R}^n \to \mathbb{R}^n \quad \text{with} \quad (\lambda, \vec{a}) \mapsto \lambda \cdot \vec{a}.$$

Since both addition and S-multiplication are explained component by component, the following rules apply to these vectors of the real numbers.

Rules of Addition

(A_1)	$\vec{a} + \left(\vec{b} + \vec{c}\right) = \left(\vec{a} + \vec{b}\right) + \vec{c}$	*Associative Law*
(A_2)	$\vec{a} + \vec{b} = \vec{b} + \vec{a}$	*Commutative Law*
(A_3)	The zero vector has the property $\vec{a} + \vec{0} = \vec{a}$	*Zero vector*
(A_4)	For each vector $\vec{a}$ there is a vector $(-\vec{a})$ with $\vec{a} + (-\vec{a}) = \vec{0}$	*Negative vector*

Rules of the S-Multiplication

(S_1)	$k \cdot (l \cdot \vec{a}) = (k \cdot l) \cdot \vec{a}$	*Associative Law*
(S_2)	$k \cdot \left(\vec{a} + \vec{b} \right) = k \vec{a} + k \vec{b}$	*Distribution Law 1*
(S_3)	$(k + l) \cdot \vec{a} = k \vec{a} + l \vec{a}$	*Distribution Law 2*
(S_4)	$1 \cdot \vec{a} = \vec{a}$	*Law of One*

2.4.2 Vector Spaces

The rules of addition and S-multiplication apply not only to n-tuples, but also to other objects not necessarily related to vectors (e.g. functions). To cover these objects, the concept of vector space is formally introduced for all objects that have two operations "+" and "·" that satisfy the given computational rules.

Definition: *A set* $\mathbb{V}$ *is a vector space over* $\mathbb{R}$, *if the following items are fulfilled:*

(1) *In* $\mathbb{V}$ *there is an inner operation* "+",

$$+ : \mathbb{V} \times \mathbb{V} \to \mathbb{V} \quad \text{with } \left(\vec{a}, \vec{b} \right) \mapsto \vec{a} + \vec{b} \qquad \textbf{(Addition)},$$

such that $(\mathbb{V}, +)$ *satisfies the laws of addition* $(A_1) - (A_4)$.

(2) *In* $\mathbb{V}$ *there is an outer operation* "·",

$$\cdot : \mathbb{R} \times \mathbb{V} \to \mathbb{V} \quad \text{with } (\lambda, \vec{a}) \mapsto \lambda \cdot \vec{a} \qquad \textbf{(S-Multiplication)},$$

such that $(\mathbb{V}, \cdot)$ *satisfies the laws of S-multiplication* $(S_1) - (S_4)$.

The elements of a vector space are called **vectors**, even if the vector space is not $\mathbb{R}^3$. If the number set is not $\mathbb{R}$, but some other field K, then the construction is called a *vector space over* K.

Examples 2.23:

① $\mathbb{R}^3$ is a vector space over $\mathbb{R}$ consisting of all 3-dimensional arrows (3-tuples of real numbers).

② $\mathbb{R}^n = \left\{ \begin{pmatrix} x_1 \\ \vdots \\ x_n \end{pmatrix} : x_i \in R \quad (i = 1, \ldots, n) \right\}$ is a vector space whose elements are the n-tuples. $\mathbb{R}^n$ is also called the *arithmetic vector space*.

③ The set of real-valued functions defined on the interval $[a, b]$.

$$F[a, b] := \{ f : [a, b] \to \mathbb{R} \}$$

is a vector space when defining addition and S-multiplication:

$$+ : F[a, b] \times F[a, b] \to F[a, b] \text{ with } (f, g) \mapsto f + g \text{ and}$$

$$(f + g)(x) := f(x) + g(x).$$

$$\cdot : \mathbb{R} \times F[a, b] \to F[a, b] \text{ with } (\lambda, f) \mapsto \lambda \cdot f \text{ and}$$

$$(\lambda f)(x) := \lambda \cdot f(x).$$

The laws of arithmetic are derived from the real numbers. The constant zero function 0 with $0(x) = 0$ for all $x \in [a, b]$ forms the zero vector.

④ The set of all polynomial functions with degree less than or equal to n

$$P[n] := \{ f : \mathbb{R} \to \mathbb{R} \text{ with } f(x) = \sum_{i=0}^{n} a_i x^i , a_i \in \mathbb{R} \}$$

is a vector space where "+" and "·" are declared as in ③.   □

Example 2.24. The solution set of a homogeneous system of linear equations is a vector space if "+" and "·" are defined according to 2.23 ②. As a **number example** we consider the LEq

$$-3x_1 - 5x_2 + 2x_3 = 0$$
$$4x_1 - x_2 + 3x_3 = 0.$$

We obtain the solution with the Gaussian algorithm

$$\begin{pmatrix} -3 & -5 & 2 & | & 0 \\ 4 & -1 & 3 & | & 0 \\ 0 & 0 & 0 & | & 0 \end{pmatrix} \hookrightarrow \begin{pmatrix} -3 & -5 & 2 & | & 0 \\ 0 & -23 & 17 & | & 0 \\ 0 & 0 & 0 & | & 0 \end{pmatrix}$$

$$\hookrightarrow x_3 = 23\lambda, \ x_2 = 17\lambda, \ x_1 = -13\lambda$$

$$\Rightarrow \mathbb{L} = \left\{ \overrightarrow{x} \in \mathbb{R}^3 : \overrightarrow{x} = \lambda \begin{pmatrix} -13 \\ 17 \\ 23 \end{pmatrix} ; \quad \lambda \in \mathbb{R} \quad \text{arbitrary} \right\}.$$

We only show the *closure* with respect to "+" and "·", i.e. the addition of two vectors of $\mathbb{L}$ results in another vector of $\mathbb{L}$ and $r \cdot \overrightarrow{x} \in \mathbb{L}$ if $\overrightarrow{x} \in \mathbb{L}$. The rules of addition $(A_1) - (A_4)$ and S-multiplication $(S_1) - (S_4)$ are then transferred from $\mathbb{R}^3$ to $\mathbb{L}$.

$$\overrightarrow{x}_1 + \overrightarrow{x}_2 = \lambda_1 \begin{pmatrix} -13 \\ 17 \\ 23 \end{pmatrix} + \lambda_2 \begin{pmatrix} -13 \\ 17 \\ 23 \end{pmatrix} = (\lambda_1 + \lambda_2) \begin{pmatrix} -13 \\ 17 \\ 23 \end{pmatrix} \in \mathbb{L}$$

$$r \cdot \overrightarrow{x}_1 = r \cdot \left(\lambda_1 \begin{pmatrix} -13 \\ 17 \\ 23 \end{pmatrix} \right) = (r \cdot \lambda_1) \begin{pmatrix} -13 \\ 17 \\ 23 \end{pmatrix} = \lambda \begin{pmatrix} -13 \\ 17 \\ 23 \end{pmatrix} \in \mathbb{L}.$$

We checked this with two solutions $\overrightarrow{x}_1$ and $\overrightarrow{x}_2$ of the LEq. The sum $\overrightarrow{x}_1 + \overrightarrow{x}_2$ is also a solution, and every multiple of a solution also satisfies the system of equations. Physically, this property means that the *superposition law* is true. Since the subset $\mathbb{L} \subset \mathbb{R}^3$ is itself a vector space, $\mathbb{L}$ is called a *subspace* of $\mathbb{R}^3$. This term is used whenever a subset of a vector space forms itself a vector space: □

Definition: *Let* $(\mathbb{V}, +, \cdot)$ *be a vector space. A subset* $U \subset \mathbb{V}$ *is* **vector subspace**, *if* U *referring to the linear operations "+" and "·" forms a vector space.*

Examples 2.25:

① $\mathbb{V} = \mathbb{R}^n, U = \{\text{Solutions of a homogeneous system of linear equations}\}$.

② $\mathbb{V} = \{\text{Set of all real-valued functions}\}$
$U = \{\text{Set of real-valued functions which are continuous on } [a, b]\}$.

③ $\mathbb{V} = \{$Set of all polynomial functions of degree $\leq n\}$,
  $U = \{$Set of all polynomial functions of degree $\leq n$ and $f(1) = 0\}$. ☐

For a subset $U \neq \emptyset$ of a vector space $\mathbb{V}$ it is no longer necessary to check the calculation rules to show that U itself represents a vector space. The computation rules are transferred from $\mathbb{V}$ to U when U is closed with respect to "+" and "·":

Vector Subspace Criterion
___

A non-empty subset $U \subset \mathbb{V}$ is a vector subspace if U is closed with respect to the linear operations "+" and "·".

Thus, for a subset U of a vector space $\mathbb{V}$, three properties must be examined in order to show that U itself is a vector space:

$$U \subset \mathbb{V} \text{ is vector space} \Leftrightarrow \begin{array}{l} UV1: \ \vec{0} \in U. \\ UV2: \ \vec{a}, \vec{b} \in U \Rightarrow \vec{a} + \vec{b} \in U. \\ UV3: \ \vec{a} \in U, \lambda \in \mathbb{R} \Rightarrow \lambda \cdot \vec{a} \in U. \end{array}$$

Examples 2.26:

① The set $\{0\}$ is a vector subspace of any vector space. It's the smallest possible vector space.

② $U = \{$Set of real-valued functions $f : \mathbb{R} \to \mathbb{R}$ with $f(1) = 0\}$,
   $\mathbb{V} = \{$Set of real-valued functions $f : \mathbb{R} \to \mathbb{R}\}$. $U \subset \mathbb{V}$ is a vector subspace because

   UV1: The null function $0 : \mathbb{R} \to \mathbb{R}$ with $0(x) = 0$ has the property $0(1) = 0$. Therefore, $\{0\} \in U$.

   UV2: With $f_1, f_2 \in U$ is $f_1(1) = f_2(1) = 0$ and so
   $(f_1 + f_2)(1) = f_1(1) + f_2(1) = 0 \Rightarrow f_1 + f_2 \in U$.

   UV3: With $f \in U$ is $f(1) = 0$ and so
   $(\lambda f)(1) = \lambda \cdot f(1) = \lambda \cdot 0 = 0 \Rightarrow \lambda f \in U$.

③ $U = \{$Set of real-valued functions $f : \mathbb{R} \to \mathbb{R}$ with $f(1) = 1\}$,
$V = \{$Set of real-valued functions $f : \mathbb{R} \to \mathbb{R}\}$.
$U \subset V$ is **not** a vector subspace!: For example, $f_1, f_2 \in U$, i.e. $f_1(1) = f_2(1) = 1$, results in $(f_1 + f_2)(1) = f_1(1) + f_2(1) = 2 \Rightarrow f_1 + f_2 \notin U$.
The zero vector is also not included in U.   □

2.4.3 Linear Combination and Span

In this section we assume that $(\mathbb{V}, +, \cdot)$ is a vector space and $\overrightarrow{a}_1, \overrightarrow{a}_2, \ldots, \overrightarrow{a}_n$ are vectors of this vector space.

> **Definition:** A vector $\overrightarrow{b}$ is a **linear combination** of $\overrightarrow{a}_1, \overrightarrow{a}_2, \ldots, \overrightarrow{a}_n$, if there are real numbers $\lambda_1, \lambda_2, \cdots \lambda_n$ such that $\overrightarrow{b}$ can be written as
>
> $$\overrightarrow{b} = \lambda_1 \overrightarrow{a}_1 + \lambda_2 \overrightarrow{a}_2 + \cdots + \lambda_n \overrightarrow{a}_n.$$

Examples 2.27:

① $\begin{pmatrix} 9 \\ 7 \\ 9 \end{pmatrix} = 5 \begin{pmatrix} 1 \\ 0 \\ 1 \end{pmatrix} + 3 \begin{pmatrix} 0 \\ 1 \\ 0 \end{pmatrix} + 4 \begin{pmatrix} 1 \\ 1 \\ 1 \end{pmatrix}$. The vector $\begin{pmatrix} 9 \\ 7 \\ 9 \end{pmatrix}$ is a linear

combination of the vectors $\begin{pmatrix} 1 \\ 0 \\ 1 \end{pmatrix}, \begin{pmatrix} 0 \\ 1 \\ 0 \end{pmatrix}, \begin{pmatrix} 1 \\ 1 \\ 1 \end{pmatrix}$.

② $\begin{pmatrix} 9 \\ 7 \\ 9 \end{pmatrix} = 9 \begin{pmatrix} 1 \\ 0 \\ 0 \end{pmatrix} + 7 \begin{pmatrix} 0 \\ 1 \\ 0 \end{pmatrix} + 9 \begin{pmatrix} 0 \\ 0 \\ 1 \end{pmatrix} = 9\overrightarrow{e}_1 + 7\overrightarrow{e}_2 + 9\overrightarrow{e}_3$; i.e. this

vector is also a linear combination of the vectors $\overrightarrow{e}_1, \overrightarrow{e}_2, \overrightarrow{e}_3$.   □

As we know from $\mathbb{R}^3$, every vector $\overrightarrow{b} \in \mathbb{R}^3$ can be represented as a linear combination of $\overrightarrow{e}_1, \overrightarrow{e}_2, \overrightarrow{e}_3$. So $\overrightarrow{e}_1, \overrightarrow{e}_2, \overrightarrow{e}_3$ generate the vector space. Generalizing, we define a *generating system* in $\mathbb{R}^n$.

Definition: *The set M defined by all linear combinations of vectors $\vec{a}_1, \vec{a}_2, \ldots, \vec{a}_n$ is called* **Span** *of $\vec{a}_1, \ldots, \vec{a}_n$. For this purpose we write*

$$
\begin{aligned}
M &= \{\text{Set of all linear combinations of } \vec{a}_1, \ldots, \vec{a}_n\} \\
&= [\vec{a}_1, \vec{a}_2, \ldots, \vec{a}_n] \\
&= \{\vec{b} : \vec{b} = \lambda_1 \vec{a}_1 + \lambda_2 \vec{a}_2 + \cdots + \lambda_n \vec{a}_n \qquad (\lambda_i \in \mathbb{R})\}.
\end{aligned}
$$

Example 2.28. Is the vector $\vec{b}$ in the span of $\vec{a}_1, \vec{a}_2, \vec{a}_3$ when

$$
\vec{b} = \begin{pmatrix} 1 \\ 0 \\ 1 \end{pmatrix}; \quad \vec{a}_1 = \begin{pmatrix} 0 \\ 1 \\ 1 \end{pmatrix}, \quad \vec{a}_2 = \begin{pmatrix} 1 \\ 2 \\ 3 \end{pmatrix}, \quad \vec{a}_3 = \begin{pmatrix} 1 \\ 4 \\ 5 \end{pmatrix}?
$$

We search for $\lambda_1, \lambda_2, \lambda_3 \in \mathbb{R}$, so that

$$
\vec{b} = \lambda_1 \vec{a}_1 + \lambda_2 \vec{a}_2 + \lambda_3 \vec{a}_3.
$$

Then $\vec{b}$ is a linear combination of $\vec{a}_1, \vec{a}_2, \vec{a}_3$.

Approach:
$$
\begin{pmatrix} 1 \\ 0 \\ 1 \end{pmatrix} = \lambda_1 \begin{pmatrix} 0 \\ 1 \\ 1 \end{pmatrix} + \lambda_2 \begin{pmatrix} 1 \\ 2 \\ 3 \end{pmatrix} + \lambda_3 \begin{pmatrix} 1 \\ 4 \\ 5 \end{pmatrix}.
$$

In the component representation, this corresponds to the inhomogeneous system of linear equations

$$
\begin{aligned}
0 \cdot \lambda_1 + 1 \cdot \lambda_2 + 1 \cdot \lambda_3 &= 1 \\
1 \cdot \lambda_1 + 2 \cdot \lambda_2 + 4 \cdot \lambda_3 &= 0 \\
1 \cdot \lambda_1 + 3 \cdot \lambda_2 + 5 \cdot \lambda_3 &= 1.
\end{aligned}
$$

To solve the LEq, we use the Gauss algorithm in matrix notation

$$
\left(\begin{array}{ccc|c} 0 & 1 & 1 & 1 \\ 1 & 2 & 4 & 0 \\ 1 & 3 & 5 & 1 \end{array} \right) \hookrightarrow \left(\begin{array}{ccc|c} 1 & 2 & 4 & 0 \\ 0 & 1 & 1 & 1 \\ 0 & 1 & 1 & 1 \end{array} \right) \hookrightarrow \left(\begin{array}{ccc|c} 1 & 2 & 4 & 0 \\ 0 & 1 & 1 & 1 \\ 0 & 0 & 0 & 0 \end{array} \right)
$$

$$
\Rightarrow \lambda_3 = t \text{ (arbitrary)} \quad \hookrightarrow \quad \lambda_2 = 1 - t \quad \hookrightarrow \quad \lambda_1 = -2 - 2t.
$$

For $t = 1$ follows $\lambda_3 = 1$, $\lambda_2 = 0$, $\lambda_1 = -4$ and

$$
\vec{b} = -4 \cdot \vec{a}_1 + 0 \cdot \vec{a}_2 + 1 \cdot \vec{a}_3.
$$

So $\vec{b}$ is in the span of $[\vec{a}_1, \vec{a}_2, \vec{a}_3]$. $\qquad\qquad\qquad$ □

Statement: The vector equation $\vec{b} = \lambda_1 \vec{a}_1 + \lambda_2 \vec{a}_2 + \cdots + \lambda_n \vec{a}_n$ can be solved if $\vec{b} \in [\vec{a}_1, \vec{a}_2, \ldots, \vec{a}_n]$.

Because $M = [\vec{a}_1, \vec{a}_2, \ldots, \vec{a}_n]$ is the set of all linear combinations, $M \subset V$ itself represents a vector space: This is checked immediately with the vector subspace criterion.

Statement: $M = [\vec{a}_1, \vec{a}_2, \ldots, \vec{a}_n]$ is a vector space.

If M already generates the entire vector space V, M is called a *generating system*:

Definition: *A subset of vectors* $\{\vec{a}_1, \vec{a}_2, \ldots, \vec{a}_n\} \subset V$ *is a* **generating system** *of* V, *if the span of* $\vec{a}_1, \vec{a}_2, \ldots, \vec{a}_n$ *coincides with* V, *i.e.* $[\vec{a}_1, \vec{a}_2, \ldots, \vec{a}_n] = V$.

Examples 2.29:

① $\{\vec{e}_1, \vec{e}_2, \vec{e}_3\}$ is a generating system of $\mathbb{R}^3$, since every vector $\vec{x}$ can be expressed as a linear combination of $\vec{e}_1, \vec{e}_2, \vec{e}_3$: $\vec{x} = x_1 \vec{e}_1 + x_2 \vec{e}_2 + x_3 \vec{e}_3$.

② $\left\{\vec{e}_1, \vec{e}_2, \vec{e}_3, \vec{d} := \begin{pmatrix} 1 \\ 1 \\ 1 \end{pmatrix}\right\}$ is also a generating system of $\mathbb{R}^3$, because each vector $\vec{x}$ can be represented as a linear combination of these 4 vectors:

$$\vec{x} = x_1 \vec{e}_1 + x_2 \vec{e}_2 + x_3 \vec{e}_3 + 0 \cdot \vec{d}$$

or $\vec{x} = (x_1 - 1) \vec{e}_1 + (x_2 - 1) \vec{e}_2 + (x_3 - 1) \vec{e}_3 + 1 \cdot \vec{d}$. □

Both $\{\vec{e}_1, \vec{e}_2, \vec{e}_3\}$ and $\left\{\vec{e}_1, \vec{e}_2, \vec{e}_3, \vec{d}\right\}$ form a generating system of $\mathbb{R}^3$. A criterion is needed to characterize the smallest possible generating system. For this purpose we introduce the notation *linear independence*.

2.4.4 Linear Dependence and Independence

Example 2.30. Given are the vectors

$$\vec{a} = \begin{pmatrix} 2 \\ -3 \\ 5 \end{pmatrix}, \ \vec{b} = \begin{pmatrix} 4 \\ 3 \\ -2 \end{pmatrix}, \ \vec{c} = \begin{pmatrix} 8 \\ -3 \\ 8 \end{pmatrix}. \ \text{Then}$$

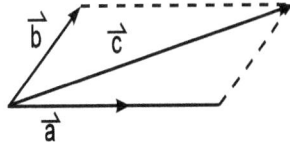

$$\vec{c} = 2 \cdot \vec{a} + 1 \cdot \vec{b}.$$

Figure 2.22. Linear dependent

So $\vec{c}$ can therefore be represented as a linear combination of the vectors $\vec{a}$ and $\vec{b}$. We call $\vec{c}$ *linearly dependent* on $\vec{a}$ and $\vec{b}$. If we rewrite $\vec{c} = 2 \cdot \vec{a} + 1 \cdot \vec{b}$ we get

$$2 \cdot \vec{a} + 1 \cdot \vec{b} - 1 \cdot \vec{c} = \vec{0}.$$

This means that the zero vector can be represented as a linear combination of $\vec{a}$, $\vec{b}$, $\vec{c}$ with non-zero coefficients. We generalize: □

Definition: *The vectors* $\vec{a}_1, \vec{a}_2, \ldots, \vec{a}_n \in \mathbb{V}$ *are called* **linearly dependent**, *if the equation*

$$k_1 \vec{a}_1 + k_2 \vec{a}_2 + \cdots + k_n \vec{a}_n = \vec{0} \quad (*)$$

can be solved with at least one $k_i \neq 0$.

Because then the equation $(*)$ can be solved with respect to $\vec{a}_i$:

$$\vec{a}_i = -\frac{1}{k_i} \left(k_1 \vec{a}_1 + \cdots + k_{i-1} \vec{a}_{i-1} + k_{i+1} \vec{a}_{i+1} + \cdots + k_n \vec{a}_n \right),$$

and $\vec{a}_i$ is represented by the remaining vectors. If the equation $(*)$ can never be solved by a vector $\vec{a}_i$ $(i \in \{1, \ldots, n\})$, then the vectors are called *linearly independent*. This happens when $k_1 = k_2 = \cdots = k_n = 0$ is the only solution.

Definition: *The vectors* $\vec{a}_1, \vec{a}_2, \ldots, \vec{a}_n \in \mathbb{V}$ *are said to be* **linearly independent**, *when we conclude:*

$$k_1 \vec{a}_1 + k_2 \vec{a}_2 + \cdots + k_n \vec{a}_n = \vec{0} \quad \Rightarrow \quad k_1 = 0, \ k_2 = 0, \ldots, \ k_n = 0.$$

Examples 2.31:

① The vectors $\vec{e}_1$, $\vec{e}_2$, $\vec{e}_3$ are linearly independent: We start with the vector equation

$$k_1\vec{e}_1 + k_2\vec{e}_2 + k_3\vec{e}_3 = \vec{0},$$

which gives the corresponding system of linear equations to be solved

$$k_1\begin{pmatrix}1\\0\\0\end{pmatrix} + k_2\begin{pmatrix}0\\1\\0\end{pmatrix} + k_3\begin{pmatrix}0\\0\\1\end{pmatrix} = \begin{pmatrix}0\\0\\0\end{pmatrix}$$

$$\Rightarrow \left(\begin{array}{ccc|c}1&0&0&0\\0&1&0&0\\0&0&1&0\end{array}\right) \quad \Rightarrow \quad k_3 = 0,\ k_2 = 0,\ k_1 = 0.$$

I.e. from $k_1\vec{e}_1 + k_2\vec{e}_2 + k_3\vec{e}_3 = \vec{0}$ follows $k_1 = k_2 = k_3 = 0$. So the vectors $\vec{e}_1$, $\vec{e}_2$, $\vec{e}_3$ are linearly **independent** according to the definition.

② The vectors $\vec{a}_1 = \begin{pmatrix}2\\-1\\3\end{pmatrix}$, $\vec{a}_2 = \begin{pmatrix}0\\1\\0\end{pmatrix}$, $\vec{a}_3 = \begin{pmatrix}3\\0\\1\end{pmatrix}$ are linearly independent: $k_1\vec{a}_1 + k_2\vec{a}_2 + k_3\vec{a}_3 = \vec{0}$ results in

$$k_1\begin{pmatrix}2\\-1\\3\end{pmatrix} + k_2\begin{pmatrix}0\\1\\0\end{pmatrix} + k_3\begin{pmatrix}3\\0\\1\end{pmatrix} = \begin{pmatrix}0\\0\\0\end{pmatrix}.$$

In matrix notation we get

$$\left(\begin{array}{ccc|c}2&0&3&0\\-1&1&0&0\\3&0&1&0\end{array}\right) \hookrightarrow \left(\begin{array}{ccc|c}-1&1&0&0\\0&2&3&0\\0&3&1&0\end{array}\right) \hookrightarrow \left(\begin{array}{ccc|c}-1&1&0&0\\0&2&3&0\\0&0&7&0\end{array}\right).$$

Solving backwards we get the following solution

$$7 \cdot k_3 = 0 \Rightarrow \boxed{k_3 = 0}; \quad 2 \cdot k_2 = 0 \Rightarrow \boxed{k_2 = 0}; \quad -1 \cdot k_1 = 0 \Rightarrow \boxed{k_1 = 0}.$$

I.e. from $k_1\vec{a}_1 + k_2\vec{a}_2 + k_3\vec{a}_3 = \vec{0}$ we conclude that $k_1 = k_2 = k_3 = 0$. So the vectors $\vec{a}_1$, $\vec{a}_2$, $\vec{a}_3$ are linearly **independent**.

③ The vectors $\vec{a}_1 = \begin{pmatrix} 2 \\ -1 \\ 3 \end{pmatrix}$, $\vec{a}_2 = \begin{pmatrix} 0 \\ 1 \\ 0 \end{pmatrix}$, $\vec{a}_3 = \begin{pmatrix} 2 \\ 0 \\ 3 \end{pmatrix}$ are linearly

dependent: $k_1 \vec{a}_1 + k_2 \vec{a}_2 + k_3 \vec{a}_3 = \vec{0}$ results in matrix notation

$$\left(\begin{array}{ccc|c} 2 & 0 & 2 & 0 \\ -1 & 1 & 0 & 0 \\ 3 & 0 & 3 & 0 \end{array}\right) \hookrightarrow \left(\begin{array}{ccc|c} -1 & 1 & 0 & 0 \\ 0 & 2 & 2 & 0 \\ 0 & 3 & 3 & 0 \end{array}\right) \hookrightarrow \left(\begin{array}{ccc|c} -1 & 1 & 0 & 0 \\ 0 & 1 & 1 & 0 \\ 0 & 0 & 0 & 0 \end{array}\right)$$

$$\Rightarrow k_3 = \lambda \text{ (arbitrary); } k_2 = -\lambda; k_1 = -\lambda.$$

For example, specifying $\lambda = 1$ will result in

$$(-1) \vec{a}_1 + (-1) \vec{a}_2 + 1\vec{a}_3 = \vec{0} \Rightarrow \vec{a}_1 = -\vec{a}_2 + \vec{a}_3.$$

So the system of equations can be solved with non-zero k_1, k_2, k_3, and thus the vector equation can be solved for the vector $\vec{a}_1$. The vectors $\vec{a}_1$, $\vec{a}_2$, $\vec{a}_3$ are linearly **dependent**.

④ $P[n] := \{$Set of all polynomial functions of degree $\leq n\}$. The power functions f_i with $f_i(x) = x^i$ $(i = 0, 1, 2, \ldots, n)$ are linearly **independent** in $P[n]$: Because

$$k_0 f_0 + k_1 f_1 + \cdots + k_n f_n = 0$$

means $k_0 f_0(x) + k_1 f_1(x) + \cdots + k_n f_n(x) = 0(x) = 0$ for all $x \in \mathbb{R}$, i.e.

$$k_0 + k_1 x + \cdots + k_n x^n = 0$$

for all $x \in \mathbb{R}$. Inserting $x = 0$ results in $k_0 = 0$. Then x can be excluded on the left side. Again, inserting $x = 0$ results in $k_1 = 0$. The result is $k_0 = k_1 = \cdots = k_n = 0$. □

Examples ① - ④ show that the vector equation

$$k_1 \vec{a}_1 + k_2 \vec{a}_2 + \cdots + k_n \vec{a}_n = \vec{0}$$

is either uniquely solvable; then $k_1 = k_2 = \cdots = k_n = 0$ and the vectors $\vec{a}_1, \vec{a}_2, \ldots, \vec{a}_n$ are linearly independent. Or, if the vector equation cannot be solved uniquely, then the vectors are linearly dependent. This behavior is generally valid, as the next characterization of linearly independent vectors states.

Linearly Independent Vectors

For the vectors $\vec{a}_1, \vec{a}_2, \ldots, \vec{a}_n \in \mathbb{V}$ it is equivalent:

(1) $\vec{a}_1, \vec{a}_2, \ldots, \vec{a}_n$ are linearly independent.

(2) Each vector $\vec{b} \in [\vec{a}_1, \ldots, \vec{a}_n]$ can be uniquely represented by the vectors $\vec{a}_1, \ldots, \vec{a}_n$ as a linear combination.

2.4.5 Basis and Dimension

One of the most important features for describing a vector space is that of a *basis*.

Definition: *A set of* **linearly independent** *vectors that generate the entire vector space is called a* **Basis** *of the vector space.*

To be a basis, two properties must be met:

Basis

$\vec{a}_1, \vec{a}_2, \ldots, \vec{a}_n \in \mathbb{V}$ is a **Basis** of $\mathbb{V}$ $\Leftrightarrow$ (B_1) $\vec{a}_1, \ldots, \vec{a}_n$ linearly independent.
(B_2) $[\vec{a}_1, \ldots, \vec{a}_n] = \mathbb{V}$.

A basis is the smallest set of vectors that generates the vector space, and it is also the largest set of linearly independent vectors of $\mathbb{V}$. For a basis, the following statements are equivalent:

Characterization of a Basis

For a subset $B := \{\vec{a}_1, \vec{a}_2, \ldots, \vec{a}_n\} \subset \mathbb{V}$ is equivalent:

(1) B is a **basis** of $\mathbb{V}$.

(2) B is a non-extensible, linearly independent subset of $\mathbb{V}$.

(3) B is a non-reducible generating system of $\mathbb{V}$.

Examples 2.32:

① $(\vec{e}_1, \vec{e}_2, \vec{e}_3)$ is a basis of $\mathbb{R}^3$: Because $\vec{e}_1, \vec{e}_2, \vec{e}_3$ are linearly independent and each $\vec{x} \in \mathbb{R}^3$ can be represented as a linear combination of $\vec{e}_1, \vec{e}_2, \vec{e}_3$: $\vec{x} = x_1 \vec{e}_1 + x_2 \vec{e}_2 + x_3 \vec{e}_3$.

② $(\vec{e}_1, \vec{e}_2, \ldots, \vec{e}_n)$ is a basis of $\mathbb{R}^n$.

③ $(1, x, x^2, x^3, x^4, \ldots, x^n)$ is a basis of the vector space of the polynomial functions of the degree $\le n$.

Prime Example 2.33

$$\vec{a}_1 = \begin{pmatrix} 0 \\ 1 \\ 1 \end{pmatrix}, \ \vec{a}_2 = \begin{pmatrix} 1 \\ 2 \\ 3 \end{pmatrix}, \ \vec{a}_3 = \begin{pmatrix} 1 \\ 1 \\ 0 \end{pmatrix} \text{ is a basis of } \mathbb{R}^3:$$

The two conditions (B_1) (the vectors are linearly independent) and (B_2) (the vectors generate $\mathbb{R}^3$) are checked.

B_1: From $k_1 \vec{a}_1 + k_2 \vec{a}_2 + k_3 \vec{a}_3 = \vec{0}$ results

$$\left(\begin{array}{ccc|c} 0 & 1 & 1 & 0 \\ 1 & 2 & 1 & 0 \\ 1 & 3 & 0 & 0 \end{array}\right) \hookrightarrow \left(\begin{array}{ccc|c} 1 & 3 & 0 & 0 \\ 0 & 1 & -1 & 0 \\ 0 & 1 & 1 & 0 \end{array}\right) \hookrightarrow \left(\begin{array}{ccc|c} 1 & 3 & 0 & 0 \\ 0 & 1 & -1 & 0 \\ 0 & 0 & -2 & 0 \end{array}\right).$$

This LEq has $k_1 = k_2 = k_3 = 0$ as its unique solution and thus $\vec{a}_1, \vec{a}_2, \vec{a}_3$ are linearly independent.

B_2: For $\vec{b} = \begin{pmatrix} b_1 \\ b_2 \\ b_3 \end{pmatrix} \in \mathbb{R}^3$ arbitrary, λ, μ, τ must be found so that

$$\lambda \vec{a}_1 + \mu \vec{a}_2 + \tau \vec{a}_3 = \vec{b}.$$

In matrix notation, the LEq for the unknowns λ, μ and τ is:

$$\left(\begin{array}{ccc|c} 0 & 1 & 1 & b_1 \\ 1 & 2 & 1 & b_2 \\ 1 & 3 & 0 & b_3 \end{array}\right) \hookrightarrow \left(\begin{array}{ccc|c} 1 & 3 & 0 & b_3 \\ 0 & 1 & -1 & b_3 - b_2 \\ 0 & 1 & 1 & b_1 \end{array}\right) \hookrightarrow \left(\begin{array}{ccc|c} 1 & 3 & 0 & b_3 \\ 0 & 1 & 1 & b_1 \\ 0 & 0 & -2 & b_3 - b_2 - b_1 \end{array}\right).$$

The quantities we are looking for are obtained by inverse resolution

$$\tau = -\frac{1}{2}(b_3 - b_2 - b_1); \quad \mu = \frac{1}{2}(b_1 - b_2 + b_3); \quad \lambda = \frac{3}{2}(-b_1 + b_2 - \frac{1}{3}b_3).$$

So for each vector $\vec{b} \in \mathbb{R}^3$ there are parameters $\lambda, \mu, \tau \in \mathbb{R}$, such that

$$\lambda \vec{a}_1 + \mu \vec{a}_2 + \tau \vec{a}_3 = \vec{b} \quad \Rightarrow \quad [\vec{a}_1, \vec{a}_2, \vec{a}_3] = V.$$

From (B_1) and (B_2) we conclude that $(\vec{a}_1, \vec{a}_2, \vec{a}_3)$ is a basis of $\mathbb{R}^3$.   □

Examples 2.34:

① The vectors $\vec{e}_1, \vec{e}_2, \vec{e}_3, \vec{a}_4$ with $\vec{a}_4 = \begin{pmatrix} 1 \\ 1 \\ 1 \end{pmatrix}$ do **not** form a basis

of $\mathbb{R}^3$, because they are linearly dependent: $\vec{a}_4 = \vec{e}_1 + \vec{e}_2 + \vec{e}_3$.

② The vectors $\vec{a}_1 = \begin{pmatrix} 2 \\ 0 \\ 1 \end{pmatrix}$, $\vec{a}_2 = \begin{pmatrix} 0 \\ 1 \\ 2 \end{pmatrix}$ do **not** form a basis of $\mathbb{R}^3$.

$\vec{a}_1, \vec{a}_2$ are linearly independent, but not every vector $\vec{b} \in \mathbb{R}^3$ can be represented as a linear combination of $\vec{a}_1, \vec{a}_2$. Because

$$\lambda \vec{a}_1 + \mu \vec{a}_2 = \vec{b} = \begin{pmatrix} b_1 \\ b_2 \\ b_3 \end{pmatrix} \text{ results in } \left(\begin{array}{cc|c} 2 & 0 & b_1 \\ 0 & 1 & b_2 \\ 1 & 2 & b_3 \end{array}\right)$$

$$\hookrightarrow \left(\begin{array}{cc|c} 2 & 0 & b_1 \\ 0 & -4 & -2b_3 + b_1 \\ 0 & 1 & b_2 \end{array}\right).$$

From the last line of the system of equation we get $1 \cdot \mu = b_2$, from the second last line $\mu = \frac{1}{4}(b_1 - 2b_3)$. The LEq is solvable if the vector $\vec{b}$ satisfies:

$$b_2 = \frac{1}{4}(b_1 - 2b_3).$$

However, this condition on the components is not true for all vectors $\vec{b} \in \mathbb{R}^3$. Therefore, $(\vec{a}_1, \vec{a}_2)$ is not a generating system of $\mathbb{R}^3$. The property (B_2) is not fulfilled!   □

There are vector spaces with an arbitrary number of basis vectors. However, if we find a finite basis $\vec{a}_1, \ldots, \vec{a}_n$, then every other basis also consists of exactly n vectors. The maximum number of linearly independent vectors is characteristic of a vector space:

Definition: *Let* $\mathbb{V}$ *be a vector space. If a basis consists of* n *vectors, then* n *is called the* **Dimension** *of the vector space* $\mathbb{V}$.

Notation: $dim(\mathbb{V}) = n$.

Note: Although every vector space has a basis, this basis does not necessarily consist of a finite number of vectors. In the following we will only consider finite-dimensional vector spaces. For these finite-dimensional vector spaces the following statement applies

Simplified Basis Check

Let $\mathbb{V}$ be an n-dimensional vector space. For n vectors $\vec{a}_1, \ldots, \vec{a}_n \in \mathbb{V}$ it is equivalent:

$\vec{a}_1, \ldots, \vec{a}_n$ are linearly independent.	$\Leftrightarrow$	$(\vec{a}_1, \ldots, \vec{a}_n)$ is a **Basis** of $\mathbb{V}$.

Examples 2.35:

① $\vec{a}_1 = \begin{pmatrix} 5 \\ 2 \end{pmatrix}$ and $\vec{a}_2 = \begin{pmatrix} 2 \\ 1 \end{pmatrix}$ is a basis of $\mathbb{R}^2$, because $\vec{a}_1, \vec{a}_2$ are linearly independent: We start with the vector equation

$$k_1 \vec{a}_1 + k_2 \vec{a}_2 = \vec{0}$$

and solve the corresponding system of linear equations

$$\hookrightarrow \begin{pmatrix} 5 & 2 & | & 0 \\ 2 & 1 & | & 0 \end{pmatrix} \hookrightarrow \begin{pmatrix} 5 & 2 & | & 0 \\ 0 & -1 & | & 0 \end{pmatrix} \Rightarrow k_2 = k_1 = 0.$$

So the only solution of the system of equations is zero and the vectors are linearly independent.

② Given are $\vec{a}_1 = \begin{pmatrix} 1 \\ 0 \\ 1 \\ 0 \end{pmatrix}, \vec{a}_2 = \begin{pmatrix} 1 \\ 1 \\ 0 \\ 0 \end{pmatrix}, \vec{a}_3 = \begin{pmatrix} 1 \\ 0 \\ 0 \\ 1 \end{pmatrix}, \vec{a}_4 = \begin{pmatrix} 0 \\ 1 \\ 0 \\ 1 \end{pmatrix}.$

They form a basis of $\mathbb{R}^4$, because $\vec{a}_1, \vec{a}_2, \vec{a}_3, \vec{a}_4$ are linearly independent: If we substitute the vectors into the vector equation

$$k_1 \vec{a}_1 + k_2 \vec{a}_2 + k_3 \vec{a}_3 + k_4 \vec{a}_4 = \vec{0}$$

we obtain the corresponding LEq, which we solve with the Gaussian algorithm:

$$\left(\begin{array}{cccc|c} 1 & 1 & 1 & 0 & 0 \\ 0 & 1 & 0 & 1 & 0 \\ 1 & 0 & 0 & 0 & 0 \\ 0 & 0 & 1 & 1 & 0 \end{array} \right) \hookrightarrow \left(\begin{array}{cccc|c} 1 & 1 & 1 & 0 & 0 \\ 0 & 1 & 0 & 1 & 0 \\ 0 & 1 & 1 & 0 & 0 \\ 0 & 0 & 1 & 1 & 0 \end{array} \right)$$

$$\hookrightarrow \left(\begin{array}{cccc|c} 1 & 1 & 1 & 0 & 0 \\ 0 & 1 & 0 & 1 & 0 \\ 0 & 0 & -1 & 1 & 0 \\ 0 & 0 & 1 & 1 & 0 \end{array} \right) \hookrightarrow \left(\begin{array}{cccc|c} 1 & 1 & 1 & 1 & 0 \\ 0 & 1 & 0 & 1 & 0 \\ 0 & 0 & -1 & 1 & 0 \\ 0 & 0 & 0 & 2 & 0 \end{array} \right)$$

$$\Rightarrow \quad k_1 = k_2 = k_3 = k_4 = 0.$$

③ $P[5] := \{f : \mathbb{R} \to \mathbb{R} : f(x) = a_0 + a_1 x + \cdots + a_5 x^5\}$ is a 6-dimensional vector space: $(x^0, x^1, x^2, \ldots, x^5)$ are linearly independent functions and each $f \in P[5]$ can be represented as a linear combination of these functions.

$\Rightarrow \quad (x^0, x^1, \ldots, x^5)$ is a basis of $P[5] \Rightarrow \dim P[5] = 6.$ □

2.5 Problems on Vector Calculus

2.1   Given are the vectors $\vec{a} = \begin{pmatrix} 2 \\ 3 \\ -1 \end{pmatrix}$, $\vec{b} = \begin{pmatrix} 0 \\ -2 \\ 4 \end{pmatrix}$, $\vec{c} = \begin{pmatrix} -5 \\ 3 \\ 1 \end{pmatrix}$.

Calculate the following vectors and their magnitudes

a) $\vec{s}_1 = 3\vec{a} - 4\vec{b} + \vec{c}$ b) $\vec{s}_2 = -3\left(5\vec{b} + \vec{c}\right) +$
$5\left(-\vec{a} + 3\vec{b}\right)$

c) $\vec{s}_3 = 3\left(\vec{a} - 2\vec{b}\right) + 5\vec{c}$ d) $\vec{s}_4 = 3\left(\vec{a} \cdot \vec{b}\right)\vec{c} - 5\left(\vec{b} \cdot \vec{c}\right)\vec{a}$

2.2   Which counter force $\vec{F}$ cancels the four single forces $\vec{F}_1$, $\vec{F}_2$, $\vec{F}_3$, $\vec{F}_4$?
(Unit of force $1N$.)
$$\vec{F}_1 = \begin{pmatrix} 200 \\ 110 \\ -50 \end{pmatrix}; \vec{F}_2 = \begin{pmatrix} -10 \\ 30 \\ -40 \end{pmatrix}; \vec{F}_3 = \begin{pmatrix} 40 \\ 85 \\ 120 \end{pmatrix}; \vec{F}_4 = -\begin{pmatrix} 30 \\ 50 \\ 40 \end{pmatrix}.$$

2.3   Normalize the following vectors:
$$\vec{a} = \begin{pmatrix} 2 \\ 3 \\ 1 \end{pmatrix}, \quad \vec{b} = 3\vec{e}_1 - 5\vec{e}_2 + 2\vec{e}_3, \quad \vec{c} = \begin{pmatrix} -1 \\ 0 \\ -1 \end{pmatrix}.$$

2.4   Determine the unit vector $\vec{e}$, which is in opposite direction to $\vec{a} = \begin{pmatrix} -4 \\ -3 \\ 0 \end{pmatrix}$.

2.5   Determine the coordinates of point Q, which is distant from the point
$P = (1, -2, 3)$ in the direction of the vector $\vec{a} = \begin{pmatrix} -2 \\ -1 \\ -1 \end{pmatrix}$ by 10 length

units.

2.6   Determine the coordinates of the center Q from $\overrightarrow{P_1 P_2}$ with $P_1 = (2, 4, 3)$
and $P_2 = (-1, 3, 2)$.

2.7   Use the vectors $\vec{a} = \begin{pmatrix} 1 \\ 0 \\ 1 \end{pmatrix}; \vec{b} = \begin{pmatrix} 2 \\ 1 \\ 2 \end{pmatrix}; \vec{c} = \begin{pmatrix} -4 \\ 2 \\ -2 \end{pmatrix}$ to compute the

scalar products:
a) $\vec{a} \cdot \vec{b}$ b) $\left(\vec{a} - 3\vec{b}\right) 4\vec{c}$ c) $\left(\vec{a} + \vec{b}\right)\left(\vec{a} - \vec{c}\right)$

2.8   What is the angle between the vectors $\vec{a}$ and $\vec{b}$?

a) $\vec{a} = \begin{pmatrix} 3 \\ 5 \\ 1 \end{pmatrix}, \vec{b} = \begin{pmatrix} 4 \\ 1 \\ 3 \end{pmatrix}$ b) $\vec{a} = \begin{pmatrix} 2 \\ -1 \\ 2 \end{pmatrix}, \vec{b} = \begin{pmatrix} -10 \\ -1 \\ -10 \end{pmatrix}$

2.9 Show that the following vectors $\vec{e}_1$, $\vec{e}_2$, $\vec{e}_3$ form an orthonornal system; i.e. the vectors are perpendicular to each other and have the length 1:

$$\vec{e}_1 = \frac{1}{\sqrt{2}} \begin{pmatrix} 1 \\ 0 \\ 1 \end{pmatrix}, \qquad \vec{e}_2 = \frac{1}{\sqrt{2}} \begin{pmatrix} 1 \\ 0 \\ -1 \end{pmatrix}, \qquad \vec{e}_3 = \begin{pmatrix} 0 \\ -1 \\ 0 \end{pmatrix}$$

2.10 Show: The three vectors

$$\vec{a} = \begin{pmatrix} 1 \\ 4 \\ -2 \end{pmatrix}, \quad \vec{b} = \begin{pmatrix} -2 \\ 2 \\ 3 \end{pmatrix}, \quad \vec{c} = \begin{pmatrix} -1 \\ 6 \\ 1 \end{pmatrix} \text{ form a right-angled triangle.}$$

2.11 Determine magnitude and the angle between the vector $\vec{a}$ and the coordinate axes:

a) $\vec{a} = \begin{pmatrix} 1 \\ 1 \\ 1 \end{pmatrix}$ b) $\vec{a} = \begin{pmatrix} 5 \\ -2 \\ 1 \end{pmatrix}$

2.12 Three points $A = (-1, 2, 4)$, $B = (5, 0, 0)$ and $C = (3, 4, -2)$ define a triangle. Calculate the lengths of the three sides, the angles in the triangle, and the area.

2.13 Calculate the component of vector $\vec{b}$ in the direction of vector $\vec{a} = \begin{pmatrix} 2 \\ -2 \\ 1 \end{pmatrix}$ for a) $\vec{b} = \begin{pmatrix} 5 \\ 1 \\ 3 \end{pmatrix}$ b) $\vec{b} = \begin{pmatrix} -2 \\ 5 \\ 0 \end{pmatrix}$

2.14 A vector $\vec{a}$ is defined by the magnitude $|\vec{a}| = 10$ and $\alpha = 30°$, $\beta = 60°$, $90° \le \gamma \le 180°$. What are the components of $\vec{a}$?

2.15 Determine the directional angles α, β, γ of the vectors

a) $\vec{a} = \begin{pmatrix} -1 \\ 1 \\ 4 \end{pmatrix}$ b) $\vec{a} = \begin{pmatrix} 4 \\ 2 \\ -3 \end{pmatrix}$

2.16 Calculate for $\vec{a} = \begin{pmatrix} 4 \\ 2 \\ 1 \end{pmatrix}$, $\vec{b} = \begin{pmatrix} 2 \\ 3 \\ 3 \end{pmatrix}$, $\vec{c} = \begin{pmatrix} 3 \\ -2 \\ 0 \end{pmatrix}$:

a) $\vec{a} \times \vec{b}$ b) $\left(\vec{a} - \vec{b}\right) \times (3\vec{c})$

c) $(-\vec{a} + 2\vec{c}) \times \left(-\vec{b}\right)$ d) $(2\vec{a}) \times \left(-\vec{b} - \vec{c}\right)$

2.17 On a power pole, 4 forces are acting on one level. Calculate the magnitude and direction of the resultant $\vec{F}_R = \vec{F}_1 + \vec{F}_2 + \vec{F}_3 + \vec{F}_4$, if $|\vec{F}_1| = 380\,N$, $|\vec{F}_2| = 400\,N$, $|\vec{F}_3| = 300\,N$, $|\vec{F}_4| = 440\,N$, and if the angle between $\vec{F}_1$ and $\vec{F}_2$ is $\alpha = 80°$, the angle between $\vec{F}_2$ and $\vec{F}_3$ is $\beta = 120°$ and the angle between $\vec{F}_3$ and $\vec{F}_4$ is $\gamma = 70°$.

2.18   Given is a body that can only move along the $\vec{a} = \begin{pmatrix} 2 \\ 1 \\ -2 \end{pmatrix}$ direction. On

this body acts a force $\vec{F} = \begin{pmatrix} 20 \\ 20 \\ 10 \end{pmatrix} N$.

a) What is the magnitude of the force $\vec{F}$?
b) Which angles include the force vector and the direction vector?
c) Which force acts on the body in the direction of $\vec{a}$?

2.19   A rigid body in shape of a circular disk is mounted so that it can rotate about its axis of symmetry. A force applied to the point P generating a

torque $\vec{M} = \vec{r} \times \vec{F}$. Let $\vec{F} = \begin{pmatrix} 1 \\ -1 \\ 2 \end{pmatrix} N$ and $\vec{r}(P) = \begin{pmatrix} 2 \\ 1 \\ 1 \end{pmatrix} m$.

a) Which angle include $\vec{r}(P)$ and $\vec{F}$?
b) Calculate the torque $\vec{M}$ and its magnitude.
c) Which force $\vec{F}_r$ acts in the direction of $\vec{r}(P)$?

2.20   Given are $A = (1, -1, 2)$, $B = (2, 1, 3)$, $C = (4, 0, 1)$. Under the influence of the constant force $\vec{F} = (1, 1, 1)$ a mass point m moves from point A to B. How much work has to be done (force unit $1\,N$, length unit $1\,m$), if
a) m moves from A to B by the shortest route?
b) m moves from A to B along the routes $\overline{AC}$ and $\overline{CB}$?

2.21   Confirm the results of tasks 2.1 - 2.20 with MAPLE.

2.22   What is the vector equation of the line g through the point P parallel to the vector $\vec{a}$? Which points belong to the parameter values $\lambda = 1$, $\lambda = 2$, $\lambda = -5$? $P = (4, 0, 3)$; $\vec{a} = \begin{pmatrix} -1 \\ 0 \\ -1 \end{pmatrix}$.

2.23   Determine the equation of the straight line g that is defined by points $P_1 = (1, 3, -2)$ and $P_2 = (6, 5, 8)$.

2.24   Are the three points $P_1 = (3, 0, 1)$, $P_2 = (1, 1, 1)$ and $P_3 = (-1, 2, -2)$ on a straight line?

2.25   Calculate the distance of the point $Q = (4, 1, 1)$ from the straight line g, which is determined by the point $P_1 = (4, 2, 3)$ and the direction vector $\vec{a} = \begin{pmatrix} 2 \\ 1 \\ 3 \end{pmatrix}$.

2.26   A straight line g runs through the point $P = (5, 3, 1)$ parallel to the vector $\vec{a}$ with the angles $\alpha = 30°$, $\beta = 90°$ and γ with $\cos \gamma < 0$. What is the equation of this straight line?

2.27 Which position do the following straight line pairs g_1, g_2 have in relation to each other? If necessary, determine the distance, intersection point and intersection angle.

a) g_1 through $P_1 = (3, 4, 6)$ and $P_2 = (-1, -2, 4)$

g_2 through $P_3 = (3, 7, -2)$ and $P_4 = (5, 15, -6)$

b) g_1 through $\vec{x} = \vec{r}_1 + \lambda \vec{a} = \begin{pmatrix} 5 \\ 1 \\ 0 \end{pmatrix} + \lambda \begin{pmatrix} -2 \\ 1 \\ 3 \end{pmatrix}$

g_2 through $\vec{x} = \vec{r}_2 + \lambda \vec{b} = \begin{pmatrix} 1 \\ 1 \\ 5 \end{pmatrix} + \lambda \begin{pmatrix} 6 \\ -3 \\ -9 \end{pmatrix}$

c) g_1 through $P_1 = (1, 2, 0)$ with direction vector $\vec{a} = \begin{pmatrix} 2 \\ 0 \\ 5 \end{pmatrix}$

g_2 through $P_2 = (6, 0, 13)$ with direction vector $\vec{b} = \begin{pmatrix} 1 \\ -2 \\ 3 \end{pmatrix}$

2.28 Show that the two straight lines g_1 and g_2 are skew and calculate their distance:

$$g_1 : \vec{x} = \vec{r}_1 + \lambda \vec{a} = \begin{pmatrix} 1 \\ -2 \\ 3 \end{pmatrix} + \lambda \begin{pmatrix} 1 \\ 1 \\ 1 \end{pmatrix}$$

$$g_2 : \vec{x} = \vec{r}_2 + \lambda \vec{b} = \begin{pmatrix} 3 \\ 3 \\ 3 \end{pmatrix} + \lambda \begin{pmatrix} 0 \\ 2 \\ 1 \end{pmatrix}$$

2.29 What is the vector equation of the plane E, which contains the point $P_1 = (3, 5, 1)$ and runs parallel to the direction vectors $\vec{a} = \begin{pmatrix} 1 \\ 1 \\ 1 \end{pmatrix}$ and $\vec{b} = \begin{pmatrix} 2 \\ 1 \\ 3 \end{pmatrix}$? Determine the normal vector $\vec{n}$ of the plane. Which point belongs to the parameter pair $\lambda = 1$, $\mu = 3$?

2.30 Determine the equation of the plane E by the points $P_1 = (3, 1, 0)$; $P_2 = (-4, 1, 1)$; $P_3 = (5, 9, 3)$.

2.31 Are the four points $P_1 = (1, 1, 1)$; $P_2 = (3, 2, 0)$; $P_3 = (4, -1, 5)$ and $P_4 = (12, -4, 12)$ in one plane?

2.32 A plane is perpendicular to the vector $\vec{n} = \begin{pmatrix} 4 \\ 3 \\ 1 \end{pmatrix}$ and contains the point $A = (5, 8, 10)$. Determine the vector equation of this plane.

2.33 What is the relationship between the straight line g and the E plane? Determine distance, intersection point and intersection angle if necessary.

a) g through $P_1 = (5,\ 1,\ 2)$ with direction vector $\vec{a} = \begin{pmatrix} 3 \\ 1 \\ 2 \end{pmatrix}$

E through $P_0 = (2,\ 1,\ 8)$ with normal vector $\vec{n} = \begin{pmatrix} -1 \\ 3 \\ 1 \end{pmatrix}$

b) $g : \vec{x}_g = \vec{r}\,(P_1) + \lambda \begin{pmatrix} 2 \\ 5 \\ 1 \end{pmatrix} = \begin{pmatrix} 5 \\ 3 \\ 6 \end{pmatrix} + \lambda \begin{pmatrix} 2 \\ 5 \\ 1 \end{pmatrix}$

$E : \vec{n}\,(\vec{r}(P) - \vec{r}\,(P_0)) = \begin{pmatrix} 3 \\ -1 \\ -1 \end{pmatrix} \begin{pmatrix} x-1 \\ y-1 \\ z-1 \end{pmatrix} = 0$

c) g through $P_1 = (2,\ 0,\ 3)$ and $P_2 = (5,\ 6,\ 18)$
E through $P_3 = (1, -2, -2)$, $P_4 = (0, -1, -1)$ and $P_5 = (-1, 0, -1)$

2.34 Show the parallelism of the two planes and calculate their distance

E_1 through $P_1 = (3,\ 5,\ 6)$ with normal vector $\vec{n}_1 = \begin{pmatrix} 1 \\ 3 \\ -2 \end{pmatrix}$

E_2 through $P_2 = (1,\ 5, -2)$ with normal vector $\vec{n}_2 = \begin{pmatrix} -3 \\ -9 \\ 6 \end{pmatrix}$.

2.35 Determine the intersection line and intersection angle of the two planes

$E_1 : \vec{n}_1\,(\vec{x}_E - \vec{r}\,(P_1)) = \begin{pmatrix} 3 \\ 1 \\ 2 \end{pmatrix} \cdot \begin{pmatrix} x-2 \\ y-5 \\ z-6 \end{pmatrix} = 0$

$E_2 : \vec{n}_2\,(\vec{x}_E - \vec{r}\,(P_2)) = \begin{pmatrix} 2 \\ 0 \\ 3 \end{pmatrix} \cdot \begin{pmatrix} x-1 \\ y-5 \\ z-1 \end{pmatrix} = 0.$

2.36 Do the vectors $\vec{a}_1 = \begin{pmatrix} 2 \\ 1 \\ 3 \end{pmatrix}$, $\vec{a}_2 = \begin{pmatrix} 1 \\ 0 \\ -2 \end{pmatrix}$, $\vec{a}_3 = \begin{pmatrix} 3 \\ 1 \\ 1 \end{pmatrix}$ generate the $\mathbb{R}^3$?

2.37 Are the following vectors of $\mathbb{R}^4$ linearly independent?

$\vec{a}_1 = \begin{pmatrix} 2 \\ -1 \\ 3 \\ 0 \end{pmatrix}$, $\vec{a}_2 = \begin{pmatrix} 0 \\ 1 \\ 0 \\ 2 \end{pmatrix}$, $\vec{a}_3 = \begin{pmatrix} 3 \\ 0 \\ 1 \\ 4 \end{pmatrix}$, $\vec{a}_4 = \begin{pmatrix} 5 \\ -2 \\ 2 \\ 3 \end{pmatrix}$.

2.38 In $\mathbb{R}^4$ the following vectors are given

$$\vec{a}_1 = \begin{pmatrix} 2 \\ 0 \\ 1 \\ 3 \end{pmatrix}, \ \vec{a}_2 = \begin{pmatrix} 0 \\ 1 \\ 2 \\ 3 \end{pmatrix}, \ \vec{a}_3 = \begin{pmatrix} 1 \\ -1 \\ 0 \\ 0 \end{pmatrix}, \ \vec{a}_4 = \begin{pmatrix} 0 \\ -2 \\ 1 \\ 0 \end{pmatrix}, \ \vec{b} = \begin{pmatrix} 0 \\ 5 \\ 2 \\ 6 \end{pmatrix}.$$

Represent $\vec{b}$ as a linear combination of $\vec{a}_1, \ \vec{a}_2, \ \vec{a}_3, \ \vec{a}_4$.

2.39 Examine the following vectors of the $\mathbb{R}^5$ for linear dependency:

$$\vec{a}_1 = \begin{pmatrix} 1 \\ 0 \\ 0 \\ 0 \\ 1 \end{pmatrix}, \ \vec{a}_2 = \begin{pmatrix} 0 \\ 0 \\ 1 \\ 1 \\ 1 \end{pmatrix}, \ \vec{a}_3 = \begin{pmatrix} 1 \\ 0 \\ 0 \\ 1 \\ 1 \end{pmatrix}, \ \vec{a}_4 = \begin{pmatrix} 0 \\ 0 \\ 0 \\ 1 \\ 0 \end{pmatrix}, \ \vec{a}_5 = \begin{pmatrix} 0 \\ 1 \\ 1 \\ 1 \\ 1 \end{pmatrix}.$$

2.40 Is the vector $\vec{b}$ in the span of the vectors $\vec{a}_1, \ \vec{a}_2, \ \vec{a}_3$?

a) $\vec{a}_1 = \begin{pmatrix} 1 \\ 1 \\ 0 \end{pmatrix}, \ \vec{a}_2 = \begin{pmatrix} 0 \\ 1 \\ 1 \end{pmatrix}, \ \vec{a}_3 = \begin{pmatrix} 1 \\ 0 \\ 0 \end{pmatrix}, \ \vec{b} = \begin{pmatrix} 2 \\ 2 \\ 1 \end{pmatrix}$

b) $\vec{a}_1 = \begin{pmatrix} 1 \\ 1 \\ 0 \end{pmatrix}, \ \vec{a}_2 = \begin{pmatrix} 0 \\ 1 \\ 1 \end{pmatrix}, \ \vec{a}_3 = \begin{pmatrix} 1 \\ 0 \\ -1 \end{pmatrix}, \ \vec{b} = \begin{pmatrix} 1 \\ 0 \\ 0 \end{pmatrix}$

2.41 Show that the vectors $\vec{a}_1, \ \vec{a}_2, \ \vec{a}_3$ form a basis of the $\mathbb{R}^3$ and represent $\vec{d}$ as a linear combination of $\vec{a}_1, \ \vec{a}_2, \ \vec{a}_3$:

$$\vec{a}_1 = \begin{pmatrix} 3 \\ 4 \\ 3 \end{pmatrix}, \ \vec{a}_2 = \begin{pmatrix} 5 \\ -1 \\ 0 \end{pmatrix}, \ \vec{a}_3 = \begin{pmatrix} -2 \\ 2 \\ -3 \end{pmatrix}, \ \vec{d} = \begin{pmatrix} 1 \\ -11 \\ -3 \end{pmatrix}$$

Chapter 3
Matrices and Determinants

3

Fundamental to this chapter is the concept of a matrix and operations on matrices. Addition and multiplication of matrices are defined, and the Gaussian-Jordan method for computing the inverse matrix is introduced.

To decide whether a matrix is invertible or not, the Gaussian-Jordan method must be applied. If the left side contains a zero line after the transformation, the matrix cannot be inverted. It would be useful to be able to decide whether the inverse matrix exists or not before applying the method. This leads to the concept of the determinant.

For systems of equations with quadratic matrices, the determinant decides whether the system is solvable for any right side. For systems with non-square matrices, the determinant is not defined. We introduce the rank of a matrix to be able to make statements about the solubility of a given right-hand side.

3

3 Matrices and Determinants

With the introduction of the coefficient matrix in Chapter 1, systems of linear equations are described in a very compact way. In this chapter, we will extend the notion of matrices and determinants associated with quadratic matrices by performing arithmetic operations on them, which we will then use to solve systems of linear equations.

3.1 Matrices

Fundamental to this chapter is the concept of a matrix and operations on matrices. Addition and multiplication of matrices are defined, and the Gauss-Jordan method for computing the inverse matrix is introduced.

3.1.1 Introduction, Special Matrices

Definition: We define an $(m \times n)$ **Matrix** $A = (a_{ij})_{mn}$ as a rectangular number scheme

$$A = \begin{pmatrix} a_{11} & a_{12} & \cdots & a_{1j} & \cdots & a_{1n} \\ \vdots & \vdots & & \vdots & & \vdots \\ a_{i1} & a_{i2} & \cdots & a_{ij} & \cdots & a_{in} \\ \vdots & \vdots & & \vdots & & \vdots \\ a_{m1} & a_{m2} & \cdots & a_{mj} & \cdots & a_{mn} \end{pmatrix} \quad \leftarrow i\text{-th row} \quad = (a_{ij})_{mn}$$

$$\uparrow$$
$$j\text{-th column}$$

where $a_{ij} \in \mathbb{R}$. a_{ij} are the matrix elements, i is the row index and j is the column index.

A matrix is composed of its *column vectors* (for short: columns) or its *row vectors* (for short: rows). An $(m \times n)$ matrix A has n columns and m rows. The index i indicates the row number and the index j the column number.

Examples 3.1:

① $A = \begin{pmatrix} 5 & 3 & \boxed{1} \\ 2 & 0 & 1 \end{pmatrix}$ is a (2×3) matrix. The element $a_{13} = 1$.

② $B = \begin{pmatrix} 0 & 1 \\ 0 & \boxed{2} \\ 4 & 2 \end{pmatrix}$ is a (3×2) matrix. The element $b_{22} = 2$.   □

⊘ Square Matrices:

The matrix A is called a **Square Matrix** if $n = m$. If $n = m$, the *matrix elements* $a_{11}, a_{22}, \ldots, a_{nn}$ are called the *main diagonal elements* (abbreviation: *diagonal*) of the matrix

$$A = \begin{pmatrix} a_{11} & \cdots & \cdots & a_{1n} \\ \vdots & \ddots & & \vdots \\ \vdots & & \ddots & \vdots \\ \vdots & & & \ddots & \vdots \\ a_{n1} & \cdots & \cdots & a_{nn} \end{pmatrix}.$$

Diagonal

⊘ Special Square Matrices:

A matrix D is called a **Diagonal Matrix** if all non-diagonal elements are zero. The **Identity Matrix** I_n is a diagonal matrix with diagonal elements 1.

$$D = \begin{pmatrix} a_{11} & 0 & \cdots & 0 \\ 0 & \ddots & \ddots & \vdots \\ \vdots & \ddots & \ddots & 0 \\ 0 & \cdots & 0 & a_{nn} \end{pmatrix}, \quad I_n = \begin{pmatrix} 1 & & 0 \\ & \ddots & \\ 0 & & 1 \end{pmatrix}.$$

A matrix R is said to be an **upper triangular matrix** if all elements below the diagonal are zero; a matrix L is called a **lower triangular matrix** if all elements above the diagonal are zero.

$$R = \begin{pmatrix} a_{11} & \cdots & a_{1n} \\ & \ddots & \vdots \\ 0 & & a_{nn} \end{pmatrix}, \quad L = \begin{pmatrix} a_{11} & & 0 \\ \vdots & \ddots & \\ a_{n1} & \cdots & a_{nn} \end{pmatrix}.$$

A square matrix is called **symmetric** if

$$a_{ij} = a_{ji} \qquad \text{for all } i, j = 1 \ldots n.$$

A symmetric matrix S consists of elements which are mirror symmetric with respect to the main diagonal. The **Zero Matrix** N has only zeros as elements.

$$S = \begin{pmatrix} 1 & 2 & 3 \\ 2 & 5 & 4 \\ 3 & 4 & 6 \end{pmatrix}, \quad N = \begin{pmatrix} 0 & \cdots & 0 \\ \vdots & \ddots & \vdots \\ 0 & \cdots & 0 \end{pmatrix}.$$

3.1.2 Matrix Operations

⊘ **(1) Multiplication of a Matrix with a Scalar**

Definition: $\alpha \in \mathbb{R}$, $A = (a_{ij})_{mn}$ is an $(m \times n)$ matrix, then

$$\alpha \cdot A = \alpha \, (a_{ij})_{mn} := (\alpha \, a_{ij})_{mn} \,.$$

A matrix is multiplied by a number by multiplying each element of the matrix by the number.

⊘ **(2) Addition of two $(m \times n)$ Matrices**

Definition: $A = (a_{ij})_{mn}$ and $B = (b_{ij})_{mn}$ are two $(m \times n)$ matrices, then the **sum** (or **difference**) of the matrices is

$$A \pm B = (a_{ij})_{mn} \pm (b_{ij})_{mn} := (a_{ij} \pm b_{ij})_{mn} \,.$$

Two $(m \times n)$ matrices are added or subtracted by adding or subtracting the respective matrix elements. Note that for the addition and the subtraction, both matrices must be $(m \times n)$ matrices.

Examples 3.2: $A = \begin{pmatrix} 5 & 0 & 1 \\ 2 & 3 & 7 \end{pmatrix}$; $B = \begin{pmatrix} 2 & -4 & 3 \\ 1 & 2 & 3 \end{pmatrix}$.

① $\quad 2A + 4B = \begin{pmatrix} 10 & 0 & 2 \\ 4 & 6 & 14 \end{pmatrix} + \begin{pmatrix} 8 & -16 & 12 \\ 4 & 8 & 12 \end{pmatrix} = \begin{pmatrix} 18 & -16 & 14 \\ 8 & 14 & 26 \end{pmatrix}$.

② $\quad 4A - 2B = \begin{pmatrix} 20 & 0 & 4 \\ 8 & 12 & 28 \end{pmatrix} - \begin{pmatrix} 4 & -8 & 6 \\ 2 & 4 & 6 \end{pmatrix} = \begin{pmatrix} 16 & 8 & -2 \\ 6 & 8 & 22 \end{pmatrix}$. $\qquad \square$

⊘ (3) Transpose a Matrix

Definition: *If the rows and columns of an $(m \times n)$ matrix $A = (a_{ij})_{mn}$ are swapped, the result is given by the **transposed matrix** A^t*

$$A^t := (a_{ji})_{nm}.$$

Example 3.3. $A = \begin{pmatrix} 5 & 0 & 1 \\ 2 & 3 & 7 \end{pmatrix} \Rightarrow A^t = \begin{pmatrix} 5 & 2 \\ 0 & 3 \\ 1 & 7 \end{pmatrix}$. $\qquad \square$

Notes:

(1) If A is an $(m \times n)$ matrix, then A^t is an $(n \times m)$ matrix.

(2) The transposed matrix is obtained by mirroring the matrix elements on the main diagonal.

(3) $(A^t)^t = A$.

(4) A square matrix is **symmetric** if $A^t = A$.

⊘ (4) Multiplication of Matrices

In order to implement the matrix multiplication, we make the following preliminary considerations. The inhomogeneous system of linear equations

$$\left. \begin{array}{l} 3x_1 + 4x_2 - 3x_3 = 4 \\ 5x_1 - x_2 + 2x_3 = 6 \\ 4x_1 - 2x_2 - 2x_3 = 1 \end{array} \right\} \qquad \left(\begin{array}{ccc|c} 3 & 4 & -3 & 4 \\ 5 & -1 & 2 & 6 \\ 4 & -2 & -2 & 1 \end{array} \right)$$

is abbreviated in matrix notation. We write it in the form

$$
\begin{pmatrix} 3 & 4 & -3 \\ 5 & -1 & 2 \\ 4 & -2 & -2 \end{pmatrix} \begin{pmatrix} x_1 \\ x_2 \\ x_3 \end{pmatrix} = \begin{pmatrix} 4 \\ 6 \\ 1 \end{pmatrix}.
$$

$A\overrightarrow{x} = \overrightarrow{b}$ is a vector equation with the matrix

$$
A = \begin{pmatrix} 3 & 4 & -3 \\ 5 & -1 & 2 \\ 4 & -2 & -2 \end{pmatrix} \quad \text{and vectors} \quad \overrightarrow{x} = \begin{pmatrix} x_1 \\ x_2 \\ x_3 \end{pmatrix}, \; \overrightarrow{b} = \begin{pmatrix} 4 \\ 6 \\ 1 \end{pmatrix}.
$$

The original equations are obtained from the matrix notation by placing the column vector $\overrightarrow{x}$ over each row of A and taking the scalar product of the row vector with the column vector $\overrightarrow{x}$. This implements the multiplication of a matrix A with a column vector $\overrightarrow{x}$. Similarly, the product of two matrices $A \cdot B$ is defined:

We have to superimpose the first *column* of the matrix B with the first *row* of the matrix A to obtain the element c_{11} of the product matrix. Then we select the first column of the matrix B and superimpose it over the second row of the matrix A to obtain the element c_{12} of the product matrix and so on. In general, this results in the following definition of the matrix product:

Definition: (Matrix Product)
Let $A = (a_{ij})_{mn}$ be an $(m \times n)$ matrix (m rows and n columns) and $B = (b_{jk})_{nl}$ be an $(n \times l)$ matrix (n rows and l columns). Then the **Matrix Product** $C = A \cdot B = (c_{ik})_{ml}$ is an $(m \times l)$ matrix defined by

$$
c_{ik} := \sum_{j-1}^{n} a_{ij}\, b_{jk} = \begin{pmatrix} a_{i1} \\ a_{i2} \\ \vdots \\ a_{in} \end{pmatrix} \begin{pmatrix} b_{1k} \\ b_{2k} \\ \vdots \\ b_{nk} \end{pmatrix} \quad \begin{matrix} i = 1 \ldots m \\ k = 1 \ldots l \end{matrix}.
$$

⚠ **Caution:** To create the product of two matrices, the column number of A must match the row number of B. The element c_{ik} of C is the scalar product of the i-th row of A with the k-th column of B, i.e. the row length of the matrix A must match the column length of the matrix B!

Examples 3.4:

① $i=1$ $i=2$
$$\begin{pmatrix} 2 & 1 \\ 0 & 4 \end{pmatrix} \cdot \begin{pmatrix} 3 & -2 & 0 \\ 0 & 4 & 2 \end{pmatrix}$$
$j=1 \quad j=2 \quad j=3$

$$= \begin{pmatrix} 2\cdot 3+1\cdot 0 & 2\cdot(-2)+1\cdot 4 & 2\cdot 0+1\cdot 2 \\ 0\cdot 3+4\cdot 0 & 0\cdot(-2)+4\cdot 4 & 0\cdot 0+4\cdot 2 \end{pmatrix} = \begin{pmatrix} 6 & 0 & 2 \\ 0 & 16 & 8 \end{pmatrix}.$$

② $i=1$ $i=2$ $i=3$
$$\begin{pmatrix} 1 & 0 & 3 \\ 4 & 2 & 1 \\ 0 & 1 & -1 \end{pmatrix} \cdot \begin{pmatrix} 2 & 2 & -1 \\ 1 & 5 & 4 \\ 3 & 1 & 2 \end{pmatrix}$$
$j=1 \quad j=2 \quad j=3$

$$= \begin{pmatrix} 2+9 & 2+3 & -1+6 \\ 8+2+3 & 8+10+1 & -4+8+2 \\ 1-3 & 5-1 & 4-2 \end{pmatrix} = \begin{pmatrix} 11 & 5 & 5 \\ 13 & 19 & 6 \\ -2 & 4 & 2 \end{pmatrix}. \quad \square$$

Note: The multiplication of the two matrices $A \cdot B$ is performed by placing the columns of B on the rows of A and then calculating the **scalar product.**

Remarks:

(1) ⚠ **Caution:** From the definition of the product, it can be seen that for the product of two matrices $A \cdot B$, the row length of A must match the column length of B. If $A \cdot B$ is defined, $B \cdot A$ does not need to be defined. If both products are defined (e.g. for square matrices), in most cases $A \cdot B \neq B \cdot A$.

(2) ⚠ **Caution:** For the two matrices $A = \begin{pmatrix} 1 & 1 \\ 1 & 1 \end{pmatrix}$ and $B = \begin{pmatrix} 1 & 1 \\ -1 & -1 \end{pmatrix}$
we can calculate $A \cdot B = \begin{pmatrix} 0 & 0 \\ 0 & 0 \end{pmatrix}$ as well as $B \cdot A = \begin{pmatrix} 2 & 2 \\ -2 & -2 \end{pmatrix} = 2\,B.$
Note that a product can lead to the zero matrix without one of the factors being the null matrix!

(3) To calculate the matrix product, it is advisable to enter the matrices into a scheme (the so-called **Falk scheme**).

The lower left segment gives the matrix A, while the upper right segment gives the second matrix B of the product.

			0	1	
			1	2	$= B$
			3	4	
2	0	2	6	10	
4	3	1	6	14	$= A \cdot B = C$

$A =$ (left: $\begin{array}{ccc} 2 & 0 & 2 \\ 4 & 3 & 1 \end{array}$)

		2	0	2	
		4	3	1	$= A$
0	1	4	3	1	
1	2	10	6	4	$= B \cdot A = D$
3	4	22	12	10	

$B =$ (left: $\begin{array}{cc} 0 & 1 \\ 1 & 2 \\ 3 & 4 \end{array}$)

The element c_{12} of the product matrix $C = A \cdot B$ is $c_{12} = 2 \cdot 1 + 0 \cdot 2 + 2 \cdot 4 = 10$; the element d_{22} of the product $D = B \cdot A$ is $d_{22} = 1 \cdot 0 + 2 \cdot 3 = 6$. From these two examples we can also see that $A \cdot B \neq B \cdot A$!

(4) The product of an $(m \times n)$ matrix A with an $(n \times k)$-matrix B can also be interpreted as follows: B consists of k column vectors $\vec{b}_1, \ldots, \vec{b}_k \Rightarrow B = \left(\vec{b}_1, \ldots, \vec{b}_k \right)$. Then the columns of the product $C = A \cdot B$ are given by

$$\vec{c}_j = A \vec{b}_j \qquad (j = 1, \ldots, k)$$

and

$$C = (\vec{c}_1, \vec{c}_2, \ldots, \vec{c}_k) = \left(A \vec{b_1}, A \vec{b_2}, \ldots, A \vec{b_k} \right).$$

Rules for calculating the product. For products of matrices A, B, C, the following rules apply (the products are assumed to be well-defined):

$$(A \cdot B) \cdot C = A \cdot (B \cdot C) \qquad \text{Associative Law}$$

$$(A \cdot B)^t = B^t \cdot A^t$$

$$A \cdot (B + C) = A \cdot B + A \cdot C \qquad \text{Distribution Law}$$

$$(A + B) \cdot C = A \cdot C + B \cdot C \qquad \text{Distribution Law}$$

3.1.3 The Inverse of a Matrix

The equation $a \cdot x = b$ has exactly one solution for $a \neq 0$. For the inverse a element a^{-1} we have

$$x = a^{-1}b = \frac{b}{a} \in \mathbb{R}, \quad \text{since} \quad a \cdot a^{-1} = a^{-1} \cdot a = 1.$$

This construction generalizes to quadratic matrices: To solve the equation

$$A\vec{x} = \vec{b}$$

with respect to $\vec{x}$, we look for a matrix A^{-1} such that

$$A \cdot A^{-1} = A^{-1} \cdot A = I_n.$$

Then the solution $\vec{x}$ is given by $\vec{x} = A^{-1}\vec{b}$. We define:

Definition: *Given is a square $(n \times n)$ matrix A. We search for an $(n \times n)$ matrix X with*

$$A \cdot X = X \cdot A = I_n = \begin{pmatrix} 1 & & 0 \\ & \ddots & \\ 0 & & 1 \end{pmatrix},$$

*X is the **Inverse Matrix** of A. It is denoted by the symbol A^{-1}.*

Remarks:

(1) If A has an inverse matrix A^{-1}, then A is called an *invertible matrix* or *regular matrix*. A^{-1} means *inverse matrix* or *inverse*.

(2) A square matrix has, if at all, exactly one *inverse*.

(3) By definition $A \cdot A^{-1} = A^{-1} \cdot A = I_n$, i.e. A and A^{-1} are commutative.

(4) ⚠ **Caution:** Not every square matrix is invertible. For example $\begin{pmatrix} 1 & 0 \\ 0 & 0 \end{pmatrix}$ is not reversible!

We will introduce a scheme to decide whether an $(n \times n)$ matrix A is invertible and how to compute the inverse A^{-1}. First, we assume that the A inverse matrix A^{-1} exists. A^{-1} is given by the column vectors $\vec{s}_1, \ldots, \vec{s}_n$, i.e.

$$A^{-1} = (\vec{s}_1, \vec{s}_2, \ldots, \vec{s}_n).$$

Because of the property of the inverse matrix

$$A \cdot A^{-1} = I_n = \begin{pmatrix} 1 & & 0 \\ & \ddots & \\ 0 & & 1 \end{pmatrix}$$

the k-th column of the product $A \cdot A^{-1}$ is the unit vector $\vec{e}_k$. The k-th column of the product is obtained by multiplying the matrix A by the k-th column vector of A^{-1}:

$$A\vec{s}_k = \vec{e}_k = \begin{pmatrix} 0 \\ \vdots \\ 0 \\ 1 \\ 0 \\ \vdots \\ 0 \end{pmatrix} \leftarrow k \qquad k = 1, 2, \ldots, n.$$

So the columns of the inverse matrix are the solutions of the system of linear equations $A\vec{s}_k = \vec{e}_k$. To find the inverse matrix we have to solve n LEq with the same matrix A. Only the right-hand sides of the LEq are different. Since the arithmetic operations of the Gaussian algorithm are determined only by the coefficients of the matrix A and not by the right-hand side, the n LEq can be solved simultaneously. This leads to the *Gauss-Jordan algorithm*:

Inverse Matrix (Gauss-Jordan Method)

(1) The matrix scheme is created by combining A and all right-hand sides:

$$(A|\, I_n) = \begin{pmatrix} a_{11} & \cdots\cdots & a_{1n} & 1 & 0 & \cdots & 0 \\ \vdots & & \vdots & 0 & \ddots & \ddots & \vdots \\ \vdots & & \vdots & \vdots & \ddots & \ddots & 0 \\ a_{n1} & \cdots\cdots & a_{nn} & 0 & \cdots & 0 & 1 \end{pmatrix}$$

(2) Using elementary row transformations, the left side of the scheme is transformed so that the unit matrix I_n takes the place of A. The inverse matrix A^{-1} then replaces I_n:

$$\begin{pmatrix} 1 & 0 & \cdots & 0 & b_{11} & \cdots & \cdots & b_{1n} \\ 0 & \ddots & \ddots & \vdots & \vdots & & & \vdots \\ \vdots & \ddots & \ddots & 0 & \vdots & & & \vdots \\ 0 & \cdots & 0 & 1 & b_{n1} & \cdots & \cdots & b_{nn} \end{pmatrix} = (I_n|\ A^{-1}).$$

Rule: The matrix A is written on the left side and the identity matrix I_n on the right side of the scheme. Then, both sides are rearranged according to the Gauss algorithm until the identity matrix appears on the left side. The inverse matrix of A is then on the right side.

⚠ If the left side of the matrix contains a zero row at the end, the matrix A **cannot** be inverted.

Examples 3.5:

① We are looking for the inverse matrix of $A = \begin{pmatrix} 1 & 0 & -2 \\ -1 & 1 & 2 \\ 0 & 4 & -1 \end{pmatrix}$.

We run the Gauss-Jordan algorithm and write the operations on the right side of the diagram:

$$(A|\ I_3) = \begin{pmatrix} 1 & 0 & -2 & 1 & 0 & 0 \\ -1 & 1 & 2 & 0 & 1 & 0 \\ 0 & 4 & -1 & 0 & 0 & 1 \end{pmatrix} \quad + Z_1$$

$$\Rightarrow \begin{pmatrix} 1 & 0 & -2 & 1 & 0 & 0 \\ 0 & 1 & 0 & 1 & 1 & 0 \\ 0 & 4 & -1 & 0 & 0 & 1 \end{pmatrix} \quad -4\,Z_2$$

$$\Rightarrow \begin{pmatrix} 1 & 0 & -2 & 1 & 0 & 0 \\ 0 & 1 & 0 & 1 & 1 & 0 \\ 0 & 0 & -1 & -4 & -4 & 1 \end{pmatrix} \quad \begin{matrix} -2\,Z_3 \\ \\ (-1) \end{matrix}$$

$$\Rightarrow \begin{pmatrix} 1 & 0 & 0 & 9 & 8 & -2 \\ 0 & 1 & 0 & 1 & 1 & 0 \\ 0 & 0 & 1 & 4 & 4 & -1 \end{pmatrix} = (I_3|\ A^{-1})$$

So the inverse matrix A^{-1} is the matrix right below the scheme.

② We are looking for the inverse of $A = \begin{pmatrix} 1 & 1 & 2 \\ 2 & 1 & -2 \\ -3 & -2 & 0 \end{pmatrix}$:

$$(A|\, I_3) = \left(\begin{array}{rrr|rrr} 1 & 1 & 2 & 1 & 0 & 0 \\ 2 & 1 & -2 & 0 & 1 & 0 \\ -3 & -2 & 0 & 0 & 0 & 1 \end{array}\right) \quad \begin{array}{l} \\ -2\,Z_1 \\ 3\,Z_1 \end{array}$$

$$\Rightarrow \left(\begin{array}{rrr|rrr} 1 & 1 & 2 & 1 & 0 & 0 \\ 0 & -1 & -6 & -2 & 1 & 0 \\ 0 & 1 & 6 & 3 & 0 & 1 \end{array}\right) \quad \begin{array}{l} \\ \\ +Z_2 \end{array}$$

$$\Rightarrow \left(\begin{array}{rrr|rrr} 1 & 1 & 2 & 1 & 0 & 0 \\ 0 & -1 & -6 & -2 & 1 & 0 \\ 0 & 0 & 0 & 1 & 1 & 1 \end{array}\right).$$

The left-hand side contains a zero line. Therefore, the system of linear equations is not solvable. A is **not invertible**. □

Rules for Inverting Matrices

Let A and B be invertible $(n \times n)$ matrices. Then it holds:

(1) The inverse of an inverse matrix is

$$\left(A^{-1}\right)^{-1} = A$$

(2) The product of two invertible matrices can be inverted with

$$(A \cdot B)^{-1} = B^{-1} \cdot A^{-1}$$

(3) The transpose of an invertible matrix can be inverted:

$$\left(A^t\right)^{-1} = \left(A^{-1}\right)^t$$

Proof: To prove the identities, we need to check that the inverse of the corresponding matrix M, which we call $(M)_{\text{inverse}}$, has the property $(M)_{\text{inverse}} \cdot M = I_n$. Multiplying the right hand side of the identities by the corresponding matrices M directly simplifies to the identity matrix.

The use of invertible matrices will be discussed in more detail in the chapter on the solvability of LEq ($\rightarrow$ Section 3.3).

3.1.4 Linear Mappings

A geometric interpretation of matrices can be given: Let $A = (a_{ij})$ be an $(m \times n)$ matrix and $\vec{x} = \begin{pmatrix} x_1 \\ \vdots \\ x_n \end{pmatrix} \in \mathbb{R}^n$ be a vector. Then the product

$$A\vec{x} = \begin{pmatrix} a_{11} & \cdots & a_{1n} \\ \vdots & & \vdots \\ a_{m1} & \cdots & a_{mn} \end{pmatrix} \begin{pmatrix} x_1 \\ \vdots \\ x_n \end{pmatrix} = \begin{pmatrix} \sum\limits_{j=1}^{n} a_{1j}\, x_j \\ \vdots \\ \sum\limits_{j=1}^{n} a_{mj}\, x_j \end{pmatrix} = \begin{pmatrix} y_1 \\ \vdots \\ y_m \end{pmatrix}$$

is a vector $\vec{y} \in \mathbb{R}^m$. So A defines a mapping $\varphi : \mathbb{R}^n \to \mathbb{R}^m$, which assigns each vector $\vec{x} \in \mathbb{R}^n$ to exactly one vector $\vec{y} \in \mathbb{R}^m$:

Theorem: The set of all $(m \times n)$ matrices corresponds exactly to all linear mappings $\varphi : \mathbb{R}^n \to \mathbb{R}^m$.

Reason:

(1) The matrix $A = (a_{ij})_{mn}$ defines a **linear mapping** from $\mathbb{R}^n$ to $\mathbb{R}^m$, since we obviously conclude that

$$A\left(\alpha\, \vec{x}_1 + \beta\, \vec{x}_2\right) = \alpha\, A\vec{x}_1 + \beta\, A\vec{x}_2$$

for arbitrary vectors $\vec{x}_1, \vec{x}_2$ and numbers α, β.

(2) If $\varphi : \mathbb{R}^n \to \mathbb{R}^m$ is a linear mapping, then this linear mapping is uniquely determined by its representations of the unit vectors $\vec{e}_j$:

$$\varphi\left(\vec{e}_j\right) = \vec{a}_j = \begin{pmatrix} a_{1j} \\ \vdots \\ a_{mj} \end{pmatrix} \qquad j = 1, \ldots, n.$$

With these vectors we build the matrix

$$A = (a_{ij})_{mn} = \left(\vec{a}_1, \vec{a}_2, \ldots, \vec{a}_n\right).$$

We calculate directly, $A\vec{e}_j = \vec{a}_j$.

The columns of A are the image vectors of the unit vectors.

Example 3.6. The mapping $f : \begin{pmatrix} x_1 \\ x_2 \\ x_3 \end{pmatrix} \mapsto \begin{pmatrix} x_1 + x_2 - x_3 \\ x_2 + x_3 - x_1 \end{pmatrix}$ is a linear mapping from $\mathbb{R}^3$ to $\mathbb{R}^2$. The images of the basis vectors are

$$f(\vec{e}_1) = f\left(\begin{pmatrix} 1 \\ 0 \\ 0 \end{pmatrix}\right) = \begin{pmatrix} 1 \\ -1 \end{pmatrix};$$

$$f(\vec{e}_2) = \begin{pmatrix} 1 \\ 1 \end{pmatrix}; \quad f(\vec{e}_3) = \begin{pmatrix} -1 \\ 1 \end{pmatrix}.$$

So the linear mapping is represented by the matrix

$$A = \begin{pmatrix} 1 & 1 & -1 \\ -1 & 1 & 1 \end{pmatrix},$$

because the columns of the matrix A are the images of the unit vectors. ☐

Application Example 3.7   (Production Chain).

In a factory, four raw materials of units r_1, r_2, r_3, r_4 are used to make three intermediate products of units z_1, z_2, z_3. The intermediate products result in three final products of units p_1, p_2, p_3. The material consumption is given by the linear equations

$$\begin{aligned} r_1 &= z_1 && + z_3 \\ r_2 &= 2z_1 + z_2 + z_3 \\ r_3 &= z_2 + z_3 \\ r_4 &= z_1 + z_2 + 2z_3 \end{aligned} \quad \text{and} \quad \begin{aligned} z_1 &= p_1 + 2p_2 + p_3 \\ z_2 &= 2p_1 + 3p_2 + p_3 \\ z_3 &= 4p_1 + 2p_2 + 2p_3. \end{aligned}$$

Introducing the matrices $A = \begin{pmatrix} 1 & 0 & 1 \\ 2 & 1 & 1 \\ 0 & 1 & 1 \\ 1 & 1 & 2 \end{pmatrix}$ and $B = \begin{pmatrix} 1 & 2 & 1 \\ 2 & 3 & 1 \\ 4 & 2 & 2 \end{pmatrix}$, we can write the two systems of equations in the form

$$\vec{r} = A\vec{z}; \quad \vec{z} = B\vec{p}.$$

How much resources are needed to produce 100 units of p_1, 80 units of p_2, and 60 units of p_3? It is

$$\vec{r} = A\vec{z} = A(B\vec{p}) = (A \cdot B)\vec{p}.$$

To calculate the amount of resources, we calculate the matrix product $A \cdot B$ and apply the product to the vector $\vec{p} = \begin{pmatrix} 100 \\ 80 \\ 60 \end{pmatrix}$:

$$\vec{r} = \begin{pmatrix} 5 & 4 & 3 \\ 8 & 9 & 5 \\ 6 & 5 & 3 \\ 11 & 9 & 6 \end{pmatrix} \begin{pmatrix} 100 \\ 80 \\ 60 \end{pmatrix} = \begin{pmatrix} 1000 \\ 1820 \\ 1180 \\ 2180 \end{pmatrix}.$$

So we need 1000 units of the first resource, 1820 units of the second resource, 1180 units of the third resource, and 2180 units of the fourth resource. □

Application Example 3.8 (Description of a Quadrapole).

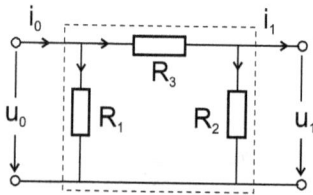

Figure 3.1. Electric quadrapole

The circuit in Fig. 3.1 is a linear *electric quadrapole*. The relationship between the input variables i_0, u_0 and the output variables i_1, u_1 is sought.

To construct the model equations we use **Kirchhoff's laws:** The *node rule* means that the sum of the incoming and outgoing currents in a node is zero. The *mesh rule* means that the sum of all voltages in a network is zero:

$$i_0 = i_{R_1} + i_{R_3}$$
$$u_0 = R_3\, i_{R_3} + u_1$$
$$i_{R_3} = i_{R_2} + i_1.$$

To obtain the desired dependence of the input variables on the output variables, we replace $i_{R_1} = \frac{1}{R_1} u_0$, $i_{R_2} = \frac{1}{R_2} u_1$ and eliminate the variable i_{R_3} from this system, leaving only i_0, u_0 and i_1, u_1:

$$u_0 = \frac{R_2 + R_3}{R_2} u_1 + R_3 i_1$$

$$i_0 = \frac{R_1 + R_2 + R_3}{R_1 R_2} u_1 + \frac{R_1 + R_3}{R_1} i_1.$$

The input variables are only on the left side of the equation; the output variables are on the right. In summary, we have two equations for the two unknowns u_1 and i_1. Introducing the link matrix

$$M = \begin{pmatrix} \frac{R_2+R_3}{R_2} & R_3 \\ \frac{R_1+R_2+R_3}{R_1\,R_2} & \frac{R_1+R_3}{R_1} \end{pmatrix}$$

the relationship between the input variables i_0, u_0 and the output variables i_1, u_1 is as follows

$$\begin{pmatrix} u_0 \\ i_0 \end{pmatrix} = M \begin{pmatrix} u_1 \\ i_1 \end{pmatrix}. \qquad \square$$

Application Example 3.9 (System of Qudrapoles).

We add a second quadrapole with the same resistances

$$\begin{pmatrix} u_0 \\ i_0 \end{pmatrix} = M \begin{pmatrix} u_1 \\ i_1 \end{pmatrix} \quad \text{and} \quad \begin{pmatrix} u_1 \\ i_1 \end{pmatrix} = M \begin{pmatrix} u_2 \\ i_2 \end{pmatrix}.$$

Combining both equations, we get a relation between the input variables u_0, i_0 and the output variables u_2, i_2

$$\begin{pmatrix} u_0 \\ i_0 \end{pmatrix} = M \cdot M \begin{pmatrix} u_2 \\ i_2 \end{pmatrix}.$$

Consequently, for a linear chain of n identical quadrapoles (see Fig. 3.2), we obtain

$$\begin{pmatrix} u_0 \\ i_0 \end{pmatrix} = M^n \begin{pmatrix} u_n \\ i_n \end{pmatrix}.$$

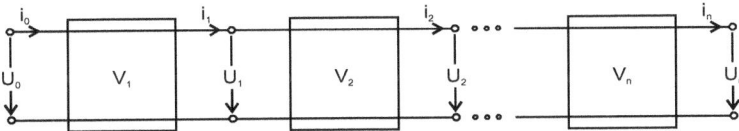

Figure 3.2. Linear chain of n quadrapoles $\qquad \square$

3.2 Determinants

To decide whether a matrix is invertible or not, so far the Gauss-Jordan method has been used. If the left side contains a zero line after the transformation, the matrix cannot be inverted. It would be useful to be able to decide whether the inverse matrix exists or not before applying the method. This leads to the concept of the determinant.

3.2.1 Introduction

In this chapter we discuss the conditions under which a quadratic system of linear equations can be solved uniquely. By using the Gaussian algorithm to solve a system of equations explicitly, we can determine at the end of the calculation whether the system of equations is solvable or not. In the case of a numerical solution of an LEq, however, it should be known in advance whether the LEq is uniquely solvable. In the following, a formalism (determinant calculation) is developed to decide whether a quadratic LEq is uniquely solvable. All matrices appearing in this section are **quadratic**.

(1) First, we consider the problem of under which condition(s) the equation

$$a_{11}\, x_1 = c_1 \qquad\qquad\qquad \text{(I)}$$

with an unknown x_1 is uniquely solvable for any c_1. The answer is: The system (I) is uniquely solvable if $a_{11} \neq 0$: $x_1 = \frac{c_1}{a_{11}}$.

(2) Now we look at the LEq

$$\begin{aligned} a_{11}\, x_1 + a_{12}\, x_2 &= c_1 \\ a_{21}\, x_1 + a_{22}\, x_2 &= c_2 \end{aligned} \qquad\qquad \text{(II)}$$

with two unknowns x_1 and x_2. What conditions must the coefficients a_{11}, a_{12}, a_{21}, a_{22} fulfill so that the system (II) is uniquely solvable for any c_1, c_2? The solution of the system is obtained by multiplying equation II.1 with $a_{22} \neq 0$ and equation II.2 with $(-a_{12}) \neq 0$ and adding both equations:

$$\left.\begin{aligned} a_{11}\, a_{22}\, x_1 + a_{12}\, a_{22}\, x_2 &= \quad c_1\, a_{22} \\ -a_{12}\, a_{21}\, x_1 - a_{12}\, a_{22}\, x_2 &= -c_2\, a_{12} \end{aligned}\right\}$$

$$\Rightarrow (a_{11}\, a_{22} - a_{12}\, a_{21})\, x_1 = c_1\, a_{22} - c_2\, a_{12}.$$

By multiplying equation II.1 with $-a_{21} \neq 0$ and equation II.2 with $a_{11} \neq 0$ and then adding the two resulting equations, we obtain the following expression

$$(a_{11} a_{22} - a_{12} a_{21}) \, x_2 = c_2 \, a_{11} - c_1 \, a_{21}.$$

Consequently, the LEq (II) is uniquely solvable for any c_1 and c_2 if

$$a_{11} a_{22} - a_{12} a_{21} \neq 0.$$

This condition also includes the special cases $a_{11} = 0$, $a_{12} = 0$, $a_{21} = 0$ or $a_{22} = 0$ which are not listed here.

Definition: (Two-Row Determinant).

Given is a (2×2) matrix $A = \begin{pmatrix} a_{11} & a_{12} \\ a_{21} & a_{22} \end{pmatrix}$. We call

$$D := \begin{vmatrix} a_{11} & a_{12} \\ a_{21} & a_{22} \end{vmatrix} = a_{11} a_{22} - a_{12} a_{21}$$

the **two-row determinant** *or* **determinant** *for short.*

Notation: The symbols D, $\det(A)$, $|A|$, $\det(a_{ij})$ are used to denote determinants.

Examples 3.10:

① $\quad \det \begin{pmatrix} 5 & 2 \\ 3 & 1 \end{pmatrix} = 5 \cdot 1 - 3 \cdot 2 = -1.$

② $\quad \det \begin{pmatrix} -2 & -2 \\ 1 & 1 \end{pmatrix} = -2 \cdot 1 - 1 \cdot (-2) = 0.$ $\hfill \square$

A two-row determinant is calculated by taking the difference between the product of the main diagonal elements $a_{11} a_{22}$ and the product of the secondary diagonal elements $a_{12} a_{21}$. So the determinant of a matrix is a real number! For two-row determinants the following rules apply. They are formulated in such a way that they can be extended to general n-row determinants.

3.2.2 Rules for Calculating Two-Row Determinants

Rule 1: The value of a determinant **doesn't** change when rows or columns are swapped: $\det A = \det A^t$.

Example: $\det \begin{pmatrix} 2 & 3 \\ -1 & 4 \end{pmatrix} = 8 + 3 = 11 = \det \begin{pmatrix} 2 & -1 \\ 3 & 4 \end{pmatrix}$.   ☐

Rule 2: The value of the determinant **changes its sign** when we swap two rows (or two columns).

Example: $\det \begin{pmatrix} 3 & 8 \\ 4 & -7 \end{pmatrix} = -21 - 32 = -53$;

$\det \begin{pmatrix} 4 & -7 \\ 3 & 8 \end{pmatrix} = 32 + 21 = 53$.   ☐

Rule 3: If the elements of **one** (!) row (or column) are multiplied by a scalar λ, the value of the determinant value is multiplied by λ.

Example: $\det \begin{pmatrix} \lambda a_{11} & \lambda a_{12} \\ a_{21} & a_{22} \end{pmatrix} = \lambda a_{11} a_{22} - a_{21} \lambda a_{12} = \lambda(a_{11}a_{22} - a_{21}a_{12})$

$$= \lambda \det \begin{pmatrix} a_{11} & a_{12} \\ a_{21} & a_{22} \end{pmatrix}.$$ ☐

Rule 4: The determinant is 0 if a column (or row) is the zero vector. More generally, when the columns (or rows) are linearly dependent.

Example: $\det \begin{pmatrix} 5 & 0 \\ 3 & 0 \end{pmatrix} = 0$; $\det \begin{pmatrix} 5 & -10 \\ 3 & -6 \end{pmatrix} = 0$, $\det \begin{pmatrix} a_{11} & a_{12} \\ \lambda a_{11} & \lambda a_{12} \end{pmatrix} = 0$. ☐

Rule 5: The determinant does **not** change if a multiple of another row (or column) is added to a row (or column) element by element.

Proof:

$$\det \begin{pmatrix} a_{11} + \lambda\, a_{21} & a_{12} + \lambda\, a_{22} \\ a_{21} & a_{22} \end{pmatrix} = (a_{11} + \lambda\, a_{21})\, a_{22} - a_{21}\, (a_{12} + \lambda\, a_{22})$$

$$= a_{11}\, a_{22} - a_{21}\, a_{12} + \underbrace{\lambda\, a_{21}\, a_{22} - \lambda\, a_{21}\, a_{22}}_{=\, 0}$$

$$= \det \begin{pmatrix} a_{11} & a_{12} \\ a_{21} & a_{22} \end{pmatrix}. \qquad \square$$

Rule 6: Multiplication Theorem for Determinants.
For two matrices A, B always holds:

$$\det (A \cdot B) = \det (A) \cdot \det (B).$$

Example: $\det \left(\begin{pmatrix} 0 & 4 \\ 3 & -1 \end{pmatrix} \begin{pmatrix} 4 & 2 \\ 3 & 2 \end{pmatrix} \right) = \det \begin{pmatrix} 12 & 8 \\ 9 & 4 \end{pmatrix} = 48 - 72 = -24,$

$\det \begin{pmatrix} 0 & 4 \\ 3 & -1 \end{pmatrix} \cdot \det \begin{pmatrix} 4 & 2 \\ 3 & 2 \end{pmatrix} = (-12) \cdot 2 = -24.$ $\qquad \square$

Rule 7: The determinant of a **triangular matrix** has the value
of the product of the main diagonal elements.

Example: $\det \begin{pmatrix} 4 & 7 \\ 0 & 2 \end{pmatrix} = 4 \cdot 2 = 8; \quad \det \begin{pmatrix} 3 & 0 \\ 5 & -1 \end{pmatrix} = 3(-1) = -3.$ $\qquad \square$

Note: The listed rules are formulated in such a way that they are also valid
for general n-row determinants.

⊚ **Transition to $(n \times n)$ Matrices:**

Let's transfer our considerations to a system of linear equations with three
unknowns x_1, x_2, x_3:

$$a_{11}x_1 + a_{12}x_2 + a_{13}x_3 = b_1$$
$$a_{21}x_1 + a_{22}x_2 + a_{23}x_3 = b_2$$
$$a_{31}x_1 + a_{32}x_2 + a_{33}x_3 = b_3.$$

This LEq can be solved formally with general coefficients according to the unknowns. E.g. using a computer algebra system, we obtain as solution

$$x_1 = \frac{a_{22}a_{33}b_1 - a_{22}b_3 a_{13} + a_{23}a_{12}b_3 - a_{23}a_{32}b_1 + b_2 a_{32}a_{13} - b_2 a_{12}a_{33}}{a_{21}a_{32}a_{13} - a_{21}a_{12}a_{33} - a_{31}a_{22}a_{13} + a_{12}a_{31}a_{23} - a_{32}a_{11}a_{23} + a_{11}a_{22}a_{33}}.$$

Similar formulas exist for x_2 and x_3. When calculating x_1, x_2 or x_3 the denominator is

$$a_{21}a_{32}a_{13} - a_{21}a_{12}a_{33} - a_{31}a_{22}a_{13} + a_{12}a_{31}a_{23} - a_{32}a_{11}a_{23} + a_{11}a_{22}a_{33}$$

in each case. The LEq is therefore uniquely solvable for each right-hand side, provided that this denominator is non-zero. This number is called the determinant of A. Similarly, for an LEq with four unknowns, there is a corresponding number. This number is called the determinant of the matrix. The general calculation of the determinant is done according to the following scheme:

3.2.3 n-Row Determinants

Definition / Theorem: (Laplace's Expansion Theorem).

Let $A = (a_{ij}) = \begin{pmatrix} a_{11} & \cdots & a_{1n} \\ \vdots & & \vdots \\ a_{n1} & \cdots & a_{nn} \end{pmatrix}$ be an $(n \times n)$ matrix. Then the **determinant of the matrix** A is defined by

$$\det A = \sum_{j=1}^{n} (-1)^{i+j}\, a_{ij}\, \det A'_{ij} \quad \text{for fixed } i \in \{1, \ldots, n\}.$$

Here, A'_{ij} is the $(n-1) \times (n-1)$ sub-matrix resulting from A by deleting the ith row and the jth column:

$$A'_{ij} = \begin{pmatrix} a_{11} & \cdots & a_{1j} & \cdots & a_{1n} \\ \vdots & & \vdots & & \vdots \\ a_{i1} & \cdots & a_{ij} & \cdots & a_{in} \\ \vdots & & \vdots & & \vdots \\ a_{n1} & \cdots & a_{nj} & \cdots & a_{nn} \end{pmatrix} \leftarrow i^{th} \text{ row}$$

This procedure is called expanding the determinant with respect to the ith row.

As A'_{ij} is an $(n-1) \times (n-1)$ matrix, we call $\det\left(A'_{ij}\right)$ a sub-determinant. By repeatedly applying the expansion formula, an n-row determinant can be reduced to 2-row determinants.

Remarks:

(1) ⚠ **Caution:** The determinant is only defined for **square** matrices.

(2) The sign $(-1)^{i+j}$ can be assumed to be a sign in the following chess-board pattern. For example, the sign of a_{43} is: $(-1)^{4+3} = -1$.

$$
\begin{array}{cccccc}
+ & - & + & - & + & - \\
- & + & - & + & - & + \\
+ & - & + & - & + & - \\
- & + & \boxed{-} & + & - & + \\
+ & - & + & - & + & - \\
- & + & - & + & - & +
\end{array}
$$

(3) The expansion row $i \in \{1, \ldots, n\}$ is arbitrary; the value of the determinant does not depend on i.

(4) The expansion of the determinant with respect to the i-th row can be viewed by deleting the i-th row and moving a crosshair over each column:

1. Column: $(-1)^{i+1} a_{i1} \det A'_{i1}$

2. Column: $(-1)^{i+2} a_{i2} \det A'_{i2}$

$$\vdots$$

n-th column: $(-1)^{i+n} a_{in} \det A'_{in}$.

Summing all the sub-determinants with the corresponding factor a_{ij} and the sign according to the chessboard pattern gives the determinant.

(5) The determinant can also be expanded with respect to the kth **column:**

$$\det A = \sum_{i=1}^{n} (-1)^{i+k} a_{ik} \det A'_{ik} \quad \text{for fixed } k \in \{1, 2, \ldots, n\}.$$

(6) For n-row determinants, the same calculation rules apply as those specified in Section 3.2.2 for 2-row determinants.

Special cases:

$n = 1:$ $\det(a_{11}) = a_{11}.$

$n = 2:$ Expansion with respect to the first column

$$\det \begin{pmatrix} a_{11} & a_{12} \\ a_{21} & a_{22} \end{pmatrix} = a_{11} \det(a_{22}) - a_{21} \det(a_{12})$$

$$= a_{11} a_{22} - a_{21} a_{12}.$$

$n = 3:$ Expansion with respect to the first column

$$\det \begin{pmatrix} a_{11} & a_{12} & a_{13} \\ a_{21} & a_{22} & a_{23} \\ a_{31} & a_{32} & a_{33} \end{pmatrix} = (+1)\, \mathbf{a_{11}} \begin{vmatrix} \cancel{a_{11}} & \cancel{a_{12}} & \cancel{a_{13}} \\ a_{21} & a_{22} & a_{23} \\ a_{31} & a_{32} & a_{33} \end{vmatrix}$$

$$+ (-1)\, \mathbf{a_{21}} \begin{vmatrix} a_{11} & a_{12} & a_{13} \\ \cancel{a_{21}} & \cancel{a_{22}} & \cancel{a_{23}} \\ a_{31} & a_{32} & a_{33} \end{vmatrix}$$

$$+ (+1)\, \mathbf{a_{31}} \begin{vmatrix} a_{11} & a_{12} & a_{13} \\ a_{21} & a_{22} & a_{23} \\ \cancel{a_{31}} & \cancel{a_{32}} & \cancel{a_{33}} \end{vmatrix}$$

$$= a_{11} \begin{vmatrix} a_{22} & a_{23} \\ a_{32} & a_{33} \end{vmatrix} - a_{21} \begin{vmatrix} a_{12} & a_{13} \\ a_{32} & a_{33} \end{vmatrix} + a_{31} \begin{vmatrix} a_{12} & a_{13} \\ a_{22} & a_{23} \end{vmatrix}$$

$$= a_{11} \left(a_{22} a_{33} - a_{32} a_{23} \right) - a_{21} \left(a_{12} a_{33} - a_{32} a_{13} \right)$$

$$+ a_{31} \left(a_{12} a_{23} - a_{22} a_{13} \right).$$

Examples 3.11:

① We calculate the determinant by expanding with respect to the second row:

$$\det \begin{pmatrix} 2 & 3 & 5 \\ 0 & 4 & 1 \\ 1 & -2 & 0 \end{pmatrix} = (-1) \cdot 0 \begin{vmatrix} 3 & 5 \\ -2 & 0 \end{vmatrix} + 4 \begin{vmatrix} 2 & 5 \\ 1 & 0 \end{vmatrix} + (-1) \cdot (-1) \begin{vmatrix} 2 & 3 \\ 1 & -2 \end{vmatrix}$$

$$= 4 \cdot (-5) + 1 \cdot (-7) = -27.$$

② We expand the following determinant according to the first column, because it has two zeros. We could also choose another row or column with two zeros:

$$\det \begin{pmatrix} 1\ 2\ 0\ 4 \\ 0\ 0\ 1\ 2 \\ 3\ 3\ 2\ 1 \\ 0\ 1\ 1\ 0 \end{pmatrix} = 1 \cdot \begin{vmatrix} 0\ 1\ 2 \\ 3\ 2\ 1 \\ 1\ 1\ 0 \end{vmatrix} + 3 \begin{vmatrix} 2\ 0\ 4 \\ 0\ 1\ 2 \\ 1\ 1\ 0 \end{vmatrix} = 3 + 3 \cdot (-8) = -21.$$

③ The identity matrix $I_n = \begin{pmatrix} 1 & & 0 \\ & \ddots & \\ 0 & & 1 \end{pmatrix}$ has the determinant:

$$\det I_n = 1.$$

④ A triangular matrix $D = \begin{pmatrix} \lambda_1 & & * \\ & \ddots & \\ 0 & & \lambda_n \end{pmatrix}$ has the determinant:

$$\det D = \lambda_1 \cdot \lambda_2 \cdots \cdots \lambda_n. \qquad \square$$

⊘ **Calculating a 3-Row Determinant using Sarrus' Rule**

⚠ For the special case of $n = 3$ (and only for this special case!), it is possible to obtain the value of the determinant using the rule of *Sarrus*:

$$\det \begin{pmatrix} a_{11}\ a_{12}\ a_{13} \\ a_{21}\ a_{22}\ a_{23} \\ a_{31}\ a_{32}\ a_{33} \end{pmatrix} \rightarrow$$

The first and second columns of the matrix A are again placed next to the determinant. The value of the determinant is then obtained by adding the three diagonal products and subtracting the antidiagonal products:

$$\det A = a_{11}\, a_{22}\, a_{33} + a_{12}\, a_{23}\, a_{31} + a_{13}\, a_{21}\, a_{32}$$
$$-a_{13}\, a_{22}\, a_{31} - a_{11}\, a_{23}\, a_{32} - a_{12}\, a_{21}\, a_{33}.$$

Example 3.12.

We calculate the determinant of $A = \begin{pmatrix} 1 & 2 & -1 \\ 1 & 0 & 1 \\ -1 & -2 & 1 \end{pmatrix}$ by Sarrus:

$$
\begin{vmatrix} 1 & 2 & -1 \\ 1 & 0 & 1 \\ -1 & -2 & 1 \end{vmatrix} \begin{matrix} 1 & 2 \\ 1 & 0 \\ -1 & -2 \end{matrix} :
$$

$$
\det A = 1 \cdot 0 \cdot 1 + 2 \cdot 1 \cdot (-1) + (-1) \cdot 1(-2)
$$
$$
-(-1) \cdot 0 \cdot (-1) - 1 \cdot 1 \cdot (-2) - 2 \cdot 1 \cdot 1 = -2 + 2 + 2 - 2 = 0. \quad \square
$$

⊘ Determinant of the Inverse Matrix

Let A be an invertible matrix. Then there exists an inverse A^{-1} of this matrix with the property

$$
A \cdot A^{-1} = I_n
$$

where I_n is the identity matrix. According to the multiplication theorem for determinants (Rule 6), we have

$$
\det(A \cdot A^{-1}) = \det(A) \cdot \det(A^{-1}) = \det I_n = 1.
$$

Then the determinant of the inverse follows:

Determinant of the Inverse

If A is an **invertible** matrix, then

$$
\det A \neq 0
$$

and the determinant of the inverse A^{-1} is given by

$$
\det A^{-1} = \frac{1}{\det A}.
$$

> **Practical Calculation of n-Row Determinants**

In the practical calculation of n-row determinants, the computational effort increases rapidly with increasing order. When we expand an n-row determinant, we create $\prod_{k=3}^{n} k = 3 \cdot 4 \cdot \ldots \cdot n$ two-row sub-determinants or $\prod_{k=4}^{n} k = 4 \cdot 5 \cdot \ldots \cdot n$ three-row sub-determinants. Therefore, the determinant should first be transformed into a simpler form according to the calculation rules listed, and then the value of the determinant is calculated.

Since the value of the determinant does not change when a multiple of one row (or column) is added to another row (or column), it is useful to first generate as many zeros as possible in a column or row using elementary manipulations to calculate $\det A$ (for $n > 3$).

Example 3.13. We subtract the fourth row from the third row and then expand with respect to the second column. For the remaining 3-row determinant, we subtract the first row from the second and third rows. Then we expand with respect to the third column:

$$
\det \begin{pmatrix} 1 & 0 & 2 & 1 \\ 2 & 0 & 1 & 1 \\ -1 & 1 & 3 & 2 \\ 2 & 1 & -1 & 1 \end{pmatrix} = \det \begin{pmatrix} 1 & 0 & 2 & 1 \\ 2 & 0 & 1 & 1 \\ -3 & 0 & 4 & 1 \\ 2 & 1 & -1 & 1 \end{pmatrix} = (+1)\det \begin{pmatrix} 1 & 2 & 1 \\ 2 & 1 & 1 \\ -3 & 4 & 1 \end{pmatrix}
$$

$$
= \det \begin{pmatrix} 1 & 2 & 1 \\ 1 & -1 & 0 \\ -4 & 2 & 0 \end{pmatrix} = (+1)(2-4) = -2. \qquad \square
$$

Example 3.14. Before calculating the determinant, we add the first row to the third and the double of the first row to the fourth, and then we develop the 5-row determinant with respect to the first column. For the 4-row determinant, we subtract the third column from the first column and then expand with respect to the last row:

$$
\det \begin{pmatrix} -1 & 1 & 0 & -2 & 0 \\ 0 & 2 & 1 & 1 & 4 \\ 1 & 2 & 4 & 3 & 2 \\ 2 & 1 & 0 & 0 & 1 \\ 0 & 4 & 0 & 4 & 0 \end{pmatrix} = \det \begin{pmatrix} -1 & 1 & 0 & -2 & 0 \\ 0 & 2 & 1 & 1 & 4 \\ 0 & 3 & 4 & 1 & 2 \\ 0 & 3 & 0 & -4 & 1 \\ 0 & 4 & 0 & 4 & 0 \end{pmatrix} = (-1)\det \begin{pmatrix} 2 & 1 & 1 & 4 \\ 3 & 4 & 1 & 2 \\ 3 & 0 & -4 & 1 \\ 4 & 0 & 4 & 0 \end{pmatrix}
$$

$$
= (-1)\det \begin{pmatrix} 1 & 1 & 1 & 4 \\ 2 & 4 & 1 & 2 \\ 7 & 0 & -4 & 1 \\ 0 & 0 & 4 & 0 \end{pmatrix} = (-1)(-4)\det \begin{pmatrix} 1 & 1 & 4 \\ 2 & 4 & 2 \\ 7 & 0 & 1 \end{pmatrix} = 4(-96) = -384. \quad \square
$$

3.2.4 Applications of Determinants

⊘ **Solving Systems of Linear Equations: Cramer's Rule.**

Cramer's Rule

Let $A = (a_{ij})$ be an $(n \times n)$ matrix with column vectors $(\vec{a}_1, \vec{a}_2, \ldots, \vec{a}_n)$ and $\det(A) \neq 0$. Then the solution of the system of linear equations

$$A \cdot \vec{x} = \vec{b}$$

with $\vec{x} = \begin{pmatrix} x_1 \\ \vdots \\ x_n \end{pmatrix}$, $\vec{b} = \begin{pmatrix} b_1 \\ \vdots \\ b_n \end{pmatrix}$ is given by

$$x_i = \frac{1}{\det(A)} \det\left(\vec{a}_1, \ldots, \vec{a}_{i-1}, \vec{b}, \vec{a}_{i+1}, \ldots, \vec{a}_n\right).$$

Replace the ith column of A with the right-hand side $\vec{b}$ of the LEq. Then the ith component of the solution vector $\vec{x}$ is the quotient of the determinant of the resulting matrix and the determinant of A.

Proof: To prove Cramer's rule, we introduce the matrix X_i. This matrix is the identity matrix I_n up to column i. Instead of the unit vector $\vec{e}_i$, this column consists of the vector of unknowns $\vec{x}$:

$$X_i = (\vec{e}_1, \vec{e}_2, \ldots, \vec{e}_{i-1}, \vec{x}, \vec{e}_{i+1}, \ldots, \vec{e}_n) = \begin{pmatrix} 1 & & & x_1 & & 0 \\ & \ddots & & \vdots & & \\ & & 1 & x_{i-1} & & \\ & & & x_i & & \\ & & & \vdots & \ddots & \\ 0 & & & x_n & & 1 \end{pmatrix}$$

On the one hand $\det(X_i) = x_i$; on the other hand

$$A X_i = (\vec{a}_1, \vec{a}_2, \ldots, A\vec{x}, \ldots, \vec{a}_n) = \left(\vec{a}_1, \vec{a}_2, \ldots, \vec{b}, \ldots, \vec{a}_n\right).$$

Using the product rule of determinants, we have

$$\det(A X_i) = \det(A) \cdot \det(X_i) = \det(A) \cdot x_i.$$

So we get $x_i = \frac{1}{\det(A)} \det\left(\vec{a}_1, \ldots, \vec{b}, \ldots, \vec{a}_n\right).$ □

Example 3.15. The solution of the LEq is searched

$$\begin{aligned} x_1 + \; x_2 \quad\;\; &= 1 \\ x_2 + x_3 &= 1. \\ 3\,x_1 + 2\,x_2 + x_3 &= 0 \end{aligned}$$

With $A = \begin{pmatrix} 1 & 1 & 0 \\ 0 & 1 & 1 \\ 3 & 2 & 1 \end{pmatrix}$ and $\overrightarrow{b} = \begin{pmatrix} 1 \\ 1 \\ 0 \end{pmatrix}$ we get $\det A = 2 \neq 0$. According to Cramer's rule, the components of the solution vector are

$$x_1 = \tfrac{1}{2}\det\begin{pmatrix} 1 & 1 & 0 \\ 1 & 1 & 1 \\ 0 & 2 & 1 \end{pmatrix} = -1; \quad x_2 = \tfrac{1}{2}\det\begin{pmatrix} 1 & 1 & 0 \\ 0 & 1 & 1 \\ 3 & 0 & 1 \end{pmatrix} = 2$$

$$x_3 = \tfrac{1}{2}\det\begin{pmatrix} 1 & 1 & 1 \\ 0 & 1 & 1 \\ 3 & 2 & 0 \end{pmatrix} = -1 \;\Rightarrow\; \overrightarrow{x} = \begin{pmatrix} x_1 \\ x_2 \\ x_3 \end{pmatrix} = \begin{pmatrix} -1 \\ 2 \\ -1 \end{pmatrix}. \qquad \square$$

⊚ **Vector Product $\overrightarrow{a} \times \overrightarrow{b}$ of two Vectors $\overrightarrow{a}$ and $\overrightarrow{b}$:**

In Chapter 2.2.3 we introduced a formula for the vector product $\overrightarrow{c} = \overrightarrow{a} \times \overrightarrow{b}$ of two vectors $\overrightarrow{a} = \begin{pmatrix} a_1 \\ a_2 \\ a_3 \end{pmatrix}$ and $\overrightarrow{b} = \begin{pmatrix} b_1 \\ b_2 \\ b_3 \end{pmatrix}$.

A **simple rule note** for this formula is obtained by formally representing it by a 3-row determinant: If $\overrightarrow{e}_1, \overrightarrow{e}_2, \overrightarrow{e}_3$ are the unit vectors, then:

$$\overrightarrow{a} \times \overrightarrow{b} = \begin{vmatrix} \overrightarrow{e}_1 & a_x & b_x \\ \overrightarrow{e}_2 & a_y & b_y \\ \overrightarrow{e}_3 & a_z & b_z \end{vmatrix} = \overrightarrow{e}_1\begin{vmatrix} a_y & b_y \\ a_z & b_z \end{vmatrix} - \overrightarrow{e}_2\begin{vmatrix} a_x & b_x \\ a_z & b_z \end{vmatrix} + \overrightarrow{e}_3\begin{vmatrix} a_x & b_x \\ a_y & b_y \end{vmatrix}.$$

Example 3.16. Calculate the torque $\overrightarrow{M}$ on a body with

$$\overrightarrow{r} = \begin{pmatrix} 1 \\ -2 \\ 5 \end{pmatrix} m \text{ and } \overrightarrow{F} = \begin{pmatrix} 10 \\ 20 \\ 50 \end{pmatrix} N: \; \overrightarrow{M} = \overrightarrow{r} \times \overrightarrow{F} = \begin{vmatrix} \overrightarrow{e}_1 & 1 & 10 \\ \overrightarrow{e}_2 & -2 & 20 \\ \overrightarrow{e}_3 & 5 & 50 \end{vmatrix}$$

$$= \overrightarrow{e}_1\begin{vmatrix} -2 & 20 \\ 5 & 50 \end{vmatrix} - \overrightarrow{e}_2\begin{vmatrix} 1 & 10 \\ 5 & 50 \end{vmatrix} + \overrightarrow{e}_3\begin{vmatrix} 1 & 10 \\ -2 & 20 \end{vmatrix} = \begin{pmatrix} -200 \\ 0 \\ 40 \end{pmatrix} [Nm]. \qquad \square$$

3.3 Solvability of Systems of Linear Equations

For systems of equations with quadratic matrices, the determinant decides whether the system is solvable for any right side. For LEq with non-square matrices, the determinant is not defined. We introduce the rank of a matrix to be able to make statements whether the LEq can be solved for a given right-hand side.

3.3.1 System of Linear Equations, Rank

We now return to the question of the conditions under which a system of linear equations

$$A \vec{x} = \vec{b}$$

can be solved when A is not a quadratic matrix, but more generally an $(m \times n)$ matrix. And if it is solvable, whether it has a unique solution or infinitely many solutions. The following terms are used to characterize LEq:

Definition: Let A be an $(m \times n)$ matrix.

(1) *The set of all solutions of the homogeneous LEq $A \vec{x} = \vec{0}$ is called* **kernel** *or* **zero space** *of A.*

(2) *The set of vectors $A \vec{x}$, $\vec{x} \in \mathbb{R}^n$, is* **the column space** *of A.*

(3) *The maximum number of linearly independent columns of A is called the* **column rank** *of A.*

(4) *The maximum number of linearly independent rows of A is called the* **row rank** *of A.*

Remark: Since each $(m \times n)$ matrix corresponds to a linear mapping $\varphi : \mathbb{R}^n \rightarrow \mathbb{R}^m$, both the kernel and the column space of A are vector spaces. The kernel is a vector subspace of $\mathbb{R}^n$ and the column space of $\mathbb{R}^m$.

The next theorem provides information on how to determine the column and row rank of a matrix.

Theorem: Elementary row and column transformations of a matrix preserve both the row rank and the column rank unchanged.

Example 3.17. We calculate the row and column rank of the following matrix by first transforming A into the upper triangular form:

$$A = \begin{pmatrix} 1 & 2 & -1 & 1 & 2 \\ 1 & 4 & -3 & -1 & -1 \\ 2 & 6 & -4 & 0 & 1 \\ 1 & 0 & 1 & 3 & 5 \\ -1 & 4 & -1 & 1 & 1 \end{pmatrix} \rightarrow \begin{pmatrix} 1 & 2 & -1 & 1 & 2 \\ 0 & 2 & -2 & -2 & -3 \\ 0 & 2 & -2 & -2 & -3 \\ 0 & -2 & 2 & 2 & 3 \\ 0 & 6 & -2 & 2 & 3 \end{pmatrix} \rightarrow \begin{pmatrix} 1 & 2 & -1 & 1 & 2 \\ 0 & 2 & -2 & -2 & -3 \\ 0 & 0 & 0 & 0 & 0 \\ 0 & 0 & 0 & 0 & 0 \\ 0 & 0 & 4 & 8 & 12 \end{pmatrix}$$

$$\rightarrow \begin{pmatrix} 1 & 0 & 1 & 3 & 5 \\ 0 & 1 & -1 & -1 & -\frac{3}{2} \\ 0 & 0 & 1 & 2 & 3 \\ 0 & 0 & 0 & 0 & 0 \\ 0 & 0 & 0 & 0 & 0 \end{pmatrix} \rightarrow \begin{pmatrix} 1 & 0 & 0 & 1 & 2 \\ 0 & 1 & 0 & 1 & \frac{3}{2} \\ 0 & 0 & 1 & 2 & 3 \\ 0 & 0 & 0 & 0 & 0 \\ 0 & 0 & 0 & 0 & 0 \end{pmatrix} = A_2.$$

We observe that A_2 has the row rank = column rank = 3. Similarly, A also has the row rank = column rank = 3. In general, we have: □

Theorem / Definition: The column rank and the row rank of an $(m \times n)$ matrix A are always the same. Therefore, the notion of the **Rank** of a matrix is introduced:

$$Rank\,(A) := Column\ rank\,(A) = Row\ rank\,(A).$$

The rank of a matrix can now be used as a criterion to decide whether a LEq $A\vec{x} = \vec{b}$ is solvable:

Theorem: Given is a LEq

$$A\vec{x} = \vec{b} \qquad\qquad (*)$$

with an $(m \times n)$ matrix A and the right side $\vec{b} \in \mathbb{R}^m$. $(A|\vec{b})$ is the matrix that results from A by appending the vector $\vec{b}$ as an extra column to A. If

$$Rank\,(A) = Rank\left(A|\vec{b}\right),$$

then the LEq $(*)$ is solvable.

Conclusions: Let $A = \begin{pmatrix} a_{11} & \cdots & a_{1n} \\ \vdots & & \vdots \\ a_{m1} & \cdots & a_{mn} \end{pmatrix}$ be an $(m \times n)$ matrix.

(1) The homogeneous LEq $A\vec{x} = \vec{0}$ has non-zero solutions if rank $(A) < n$.

(2) $A\vec{x} = \vec{0}$ always has non-zero solutions if $m < n$, i.e. the number of equations is less than the number of unknowns.

(3) The inhomogeneous LEq $A\vec{x} = \vec{b}$ is solvable if and only if

$$Rank\,(A) = Rank\,(A|\vec{b}).$$

(4) The inhomogeneous LEq $A\vec{x} = \vec{b}$ is solvable **uniquely** only if it is solvable and $A\vec{x} = \vec{0}$ has only the zero vector as solution:

$$Rank\,A = Rank(A|\vec{b}) = n.$$

(5) If A is a quadratic matrix, the theorem is equivalent to the determinant Rule 4:

$$A\vec{x} = \vec{b} \text{ is uniquely solvable} \Longleftrightarrow \det(A) \neq 0.$$

Examples 3.18 (With MAPLE-Worksheet).

① Given is the LEq $A\vec{x} = \vec{b}$ with the matrix

$$A = \begin{pmatrix} -3 & 0 & 6 & 0 \\ 1 & 1 & -2 & 5 \\ 1 & 0 & -2 & 0 \\ -2 & -2 & 4 & -10 \end{pmatrix} \text{ and the vector } \vec{b} = \begin{pmatrix} -3 \\ 2 \\ 1 \\ -4 \end{pmatrix}.$$

We compare the rank of A with the rank of the matrix appended by $\vec{b}$. By row transformations with the extended matrix we get

$$\left(\begin{array}{cccc|c} -3 & 0 & 6 & 0 & -3 \\ 1 & 1 & -2 & 5 & 2 \\ 1 & 0 & -2 & 0 & 1 \\ -2 & -2 & 4 & -10 & -4 \end{array}\right) \hookrightarrow \left(\begin{array}{cccc|c} -1 & 0 & 2 & 0 & -1 \\ 0 & 1 & 0 & 5 & 1 \\ 0 & 0 & 0 & 0 & 0 \\ 0 & 0 & 0 & 0 & 0 \end{array}\right).$$

The number of linearly independent rows of A is 2, i.e. $Rank\,(A) = 2$. The number of linearly independent rows of the extended matrix $(A|\,\overrightarrow{b}\,)$ is also 2, i.e. $Rank\,(A|\,\overrightarrow{b}\,) = 2$. Both ranks match and the LEq is therefore solvable.

Since rank $(A) = 2 < 4$, the LEq is not unique and the solution has two free parameters t_1 and t_2:

$$x_4 = t_1, \quad x_3 = t_2, \quad x_2 = 1 - 5t_1, \quad x_1 = 1 + 2t_2.$$

② For $\overrightarrow{b} = \begin{pmatrix} -3 \\ 2 \\ 1 \\ -2 \end{pmatrix}$ the LEq $A\overrightarrow{x} = \overrightarrow{b}$ is not solvable, because the rank

of the expanded matrix is $\left(A|\,\overrightarrow{b}\,\right) = 3$. $\square$

We summarize the results for quadratic systems of linear equations with the following theorem:

Fundamental Theorem for Quadratic LEq

Given is a LEq for the unknowns $(x_1, x_2, \ldots, x_n)$:

$$a_{11}\,x_1 + \cdots + a_{1n}\,x_n = b_1$$
$$a_{21}\,x_1 + \cdots + a_{2n}\,x_n = b_2$$
$$\vdots$$
$$a_{n1}\,x_1 + \cdots + a_{nn}\,x_n = b_n.$$

We define the matrix $A = \begin{pmatrix} a_{11} & \cdots & a_{1n} \\ \vdots & & \vdots \\ a_{n1} & \cdots & a_{nn} \end{pmatrix}$, the right-hand side $\overrightarrow{b} = \begin{pmatrix} b_1 \\ \vdots \\ b_n \end{pmatrix}$ and the vector $\overrightarrow{x} = \begin{pmatrix} x_1 \\ \vdots \\ x_n \end{pmatrix}$ so that $A\overrightarrow{x} = \overrightarrow{b}$ is the LEq.

Then the following statements are equivalent:

(1) The LEq $A\vec{x} = \vec{b}$ has a unique solution.

(2) A is invertible.

(3) $\det A \neq 0$.

(4) The solution of the LEq is given by $\vec{x} = A^{-1}\vec{b}$.

Example 3.19 (With MAPLE-Worksheet). Given is the inhomogeneous LEq $A\vec{x} = \vec{b}$ with the matrix $A = \begin{pmatrix} 1 & 2 & 0 \\ 1 & 7 & 4 \\ 3 & 13 & 4 \end{pmatrix}$ and the right-hand sides

$$\vec{b}_1 = \begin{pmatrix} -4 \\ 3 \\ 1 \end{pmatrix}, \vec{b}_2 = \begin{pmatrix} 1 \\ 8 \\ 8 \end{pmatrix}, \vec{b}_3 = \begin{pmatrix} 1 \\ -4 \\ 0 \end{pmatrix}.$$

(1) The LEq has a unique solution for each vector $\vec{b}$, since the determinant of the matrix $\det(A) = -8$ is non-zero.

(2) Since $\det(A) \neq 0$, the inverse matrix exists

$$A^{-1} = \begin{pmatrix} 3 & 1 & -1 \\ -1 & -\frac{1}{2} & \frac{1}{2} \\ 1 & \frac{7}{8} & -\frac{5}{8} \end{pmatrix}.$$

(3) The unique solution of the LEq is given by

$$\vec{x} = A^{-1}\vec{b}.$$

It holds for the three inhomogeneities

$$\vec{x}_1 = A^{-1}\vec{b}_1 = \begin{pmatrix} 3 & 1 & -1 \\ -1 & -\frac{1}{2} & \frac{1}{2} \\ 1 & \frac{7}{8} & -\frac{5}{8} \end{pmatrix}\begin{pmatrix} -4 \\ 3 \\ 1 \end{pmatrix} = \begin{pmatrix} -10 \\ 3 \\ -2 \end{pmatrix}$$

$$\vec{x}_2 = A^{-1}\vec{b}_2 = \begin{pmatrix} 3 & 1 & -1 \\ -1 & -\frac{1}{2} & \frac{1}{2} \\ 1 & \frac{7}{8} & -\frac{5}{8} \end{pmatrix}\begin{pmatrix} 1 \\ 8 \\ 8 \end{pmatrix} = \begin{pmatrix} 3 \\ -1 \\ 3 \end{pmatrix}$$

$$\overrightarrow{x}_3 = A^{-1}\overrightarrow{b}_3 = \begin{pmatrix} 3 & 1 & -1 \\ -1 & -\frac{1}{2} & \frac{1}{2} \\ 1 & \frac{7}{8} & -\frac{5}{8} \end{pmatrix} \begin{pmatrix} 1 \\ -4 \\ 0 \end{pmatrix} = \begin{pmatrix} -1 \\ 1 \\ -\frac{5}{2} \end{pmatrix}.$$

The advantage of solving the LEq using the inverse matrix over solving it using the Gauss algorithm is that the inverse is computed only once and then solving the LEq for different inhomogeneities is just a matrix-vector multiplication. □

3.3.2 Applications

n vectors of the n-dimensional vector space $\mathbb{R}^n$ form a basis of $\mathbb{R}^n$ according to Section 2.4.5 if they are **linearly independent**. Therefore, n vectors $\overrightarrow{a}_1, \overrightarrow{a}_2, \ldots, \overrightarrow{a}_n \in \mathbb{R}^n$ are a basis, if

$$\lambda_1 \overrightarrow{a}_1 + \lambda_2 \overrightarrow{a}_2 + \cdots + \lambda_n \overrightarrow{a}_n = \overrightarrow{0}$$

can only be solved with $\lambda_1 = \lambda_2 = \ldots = \lambda_n = 0$. However, according to the fundamental theorem of LEq, this means that

$$\det(A) = \det(\overrightarrow{a}_1, \overrightarrow{a}_2, \ldots, \overrightarrow{a}_n) \neq 0.$$

This gives us a very simple criterion to check whether n vectors of $\mathbb{R}^n$ are a basis or not:

Basis Vectors

For n vectors $\overrightarrow{a}_1, \ldots, \overrightarrow{a}_n \in \mathbb{R}^n$ the following holds:

$\det A = \det(\overrightarrow{a}_1, \ldots, \overrightarrow{a}_n) \neq 0 \Leftrightarrow (\overrightarrow{a}_1, \ldots, \overrightarrow{a}_n)$ is a basis of $\mathbb{R}^n$.

Rule Note: n vectors $\overrightarrow{a}_1, \overrightarrow{a}_2, \ldots, \overrightarrow{a}_n$ of the $\mathbb{R}^n$ are linearly independent and form a basis of $\mathbb{R}^n$, if $\det A = \det(\overrightarrow{a}_1, \overrightarrow{a}_2, \ldots, \overrightarrow{a}_n) \neq 0$.

Example 3.20 (Numerical example, with MAPLE-Worksheet).
Given are the vectors

$$\overrightarrow{a}_1 = \begin{pmatrix} 4 \\ 3 \\ 2 \\ 1 \end{pmatrix}, \ \overrightarrow{a}_2 = \begin{pmatrix} 0 \\ 1 \\ 5 \\ 4 \end{pmatrix}, \ \overrightarrow{a}_3 = \begin{pmatrix} -1 \\ -1 \\ 0 \\ 1 \end{pmatrix}, \ \overrightarrow{a}_4 = \begin{pmatrix} 3 \\ 5 \\ 0 \\ 1 \end{pmatrix} \text{ and } \overrightarrow{b} = \begin{pmatrix} 0 \\ 1 \\ 0 \\ 1 \end{pmatrix}.$$

We show that the vectors $\vec{a}_1$, $\vec{a}_2$, $\vec{a}_3$, $\vec{a}_4$ are a basis of $\mathbb{R}^4$ and represent the vector $\vec{b}$ as a linear combination of this basis.

First, we check that the 4 vectors are linearly independent by calculating the determinant of the matrix $A = (\vec{a}_1,\ \vec{a}_2,\ \vec{a}_3,\ \vec{a}_4)$:

$$\det(A) = \begin{vmatrix} 4 & 0 & -1 & 3 \\ 3 & 1 & -1 & 5 \\ 2 & 5 & 0 & 0 \\ 1 & 4 & 1 & 1 \end{vmatrix} = 62.$$

Since the determinant is non-zero, the vectors are linearly independent and form a basis of $\mathbb{R}^4$. The vector $\vec{b}$ can therefore be represented as a linear combination of the vectors $(\vec{a}_1,\ \vec{a}_2,\ \vec{a}_3,\ \vec{a}_4)$. By solving the LEq

$$\left(\begin{array}{cccc|c} 4 & 0 & -1 & 3 & 0 \\ 3 & 1 & -1 & 5 & 1 \\ 2 & 5 & 0 & 0 & 0 \\ 1 & 4 & 1 & 1 & 1 \end{array} \right)$$

using the Gauss algorithm, we obtain the following solution

$$\left(\frac{-5}{31},\ \frac{2}{31},\ \frac{16}{31},\ \frac{12}{31} \right).$$

$$\Rightarrow \vec{b} = -\frac{5}{31}\vec{a}_1 + \frac{2}{31}\vec{a}_2 + \frac{16}{31}\vec{a}_3 + \frac{12}{31}\vec{a}_4 . \qquad \square$$

Application Example 3.21   (Circle defined by 3 Points).

In CAD systems, flat surfaces are usually composed of straight lines and circular pieces. The geometry is usually captured interactively by mouse clicks by entering characteristic points for the geometry and connecting them with straight lines or circle segments. A straight line segment is defined by two points; a circular segment by the specification of three points, if the points are not on a straight line.

We will look at whether three given points (x_1, y_1), (x_2, y_2), (x_3, y_3) can be used to define a circle, and if so, what are the center coordinates and the radius of the corresponding circle. The equation describing a circle is

$$(x - x_0)^2 + (y - y_0)^2 = R^2$$

or

$$A \left(x^2 + y^2\right) + B\,x + C\,y + D = 0.$$

As the circle must pass through the given points, the following must also apply

$$A \left(x_1^2 + y_1^2\right) + B\,x_1 + C\,y_1 + D = 0$$
$$A \left(x_2^2 + y_2^2\right) + B\,x_2 + C\,y_2 + D = 0$$
$$A \left(x_3^2 + y_3^2\right) + B\,x_3 + C\,y_3 + D = 0.$$

This is a homogeneous LEq for the quantities (A, B, C, D) with the coefficient matrix

$$M = \begin{pmatrix} x^2 + y^2 & x & y & 1 \\ x_1^2 + y_1^2 & x_1 & y_1 & 1 \\ x_2^2 + y_2^2 & x_2 & y_2 & 1 \\ x_3^2 + y_3^2 & x_3 & y_3 & 1 \end{pmatrix}.$$

To obtain not only the zero solution,

$$\boxed{\det M = 0}$$

is required. We develop the determinant with respect to the first row

$$\begin{vmatrix} x_1 & y_1 & 1 \\ x_2 & y_2 & 1 \\ x_3 & y_3 & 1 \end{vmatrix} (x^2 + y^2) - \begin{vmatrix} x_1^2 + y_1^2 & y_1 & 1 \\ x_2^2 + y_2^2 & y_2 & 1 \\ x_3^2 + y_3^2 & y_3 & 1 \end{vmatrix} x +$$

$$\begin{vmatrix} x_1^2 + y_1^2 & x_1 & 1 \\ x_2^2 + y_2^2 & x_2 & 1 \\ x_3^2 + y_3^2 & x_3 & 1 \end{vmatrix} y - \begin{vmatrix} x_1^2 + y_1^2 & x_1 & y_1 \\ x_2^2 + y_2^2 & x_2 & y_2 \\ x_3^2 + y_3^2 & x_3 & y_3 \end{vmatrix} = 0.$$

So the 3-row sub-determinants are just the coefficients in the circular equation. In particular, it follows from this representation that

$$K := \begin{vmatrix} x_1 & y_1 & 1 \\ x_2 & y_2 & 1 \\ x_3 & y_3 & 1 \end{vmatrix} \neq 0$$

is necessary for the expression $(x^2 + y^2)$ to appear in the circular equation. For $K = 0$ the 3 points are on a straight line. □

Example 3.22 (With MAPLE-Worksheet). The circular equation

$$A(x^2 + y^2) + B\,x + C\,y + D = 0$$

is searched for the points $(0,0), (1,3)$ and $(2,-1)$. We insert the three points into the circular equation

$$
\begin{array}{ccccccccc}
A\cdot & 0 & + & B\cdot 0 & + & C\cdot & 0 & + & D & = & 0 \\
A\cdot & 10 & + & B\cdot 1 & + & C\cdot & 3 & + & D & = & 0 \\
A\cdot & 5 & + & B\cdot 2 & + & C\cdot & (-1) & + & D & = & 0
\end{array}
$$

and obtain the matrix M together with the circle equation

$$
M = \begin{pmatrix}
x^2 + y^2 & x & y & 1 \\
0 & 0 & 0 & 1 \\
10 & 1 & 3 & 1 \\
5 & 2 & -1 & 1
\end{pmatrix}.
$$

If we set $\det(M) = 0$, we obtain the circular equation

$$7x^2 + 7y^2 - 25\,x - 15\,y = 0.$$

The points are on the circle, whose equation is given by

$$\left(x - \frac{25}{14}\right)^2 + \left(y - \frac{15}{14}\right)^2 = \frac{425}{98}.$$

The circle has a radius of $R = \sqrt{\frac{425}{98}}$. □

Application Example 3.23 (Frequencies of a Coupled System).

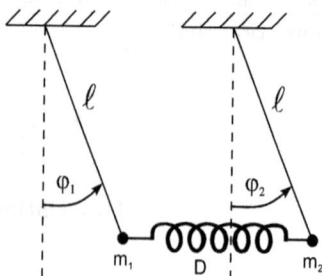

Figure 3.3. Coupled pendulums

Given are two *string pendulums* (length l) with two masses attached to their ends ($m_1 = m_2 = m$). The masses are coupled by a spring (spring constant D) (see Fig. 3.3).

$\varphi_1(t)$ is the angular displacement of the left pendulum and $\varphi_2(t)$ of the right pendulum. We apply Newton's law of motion to the masses m_1 and m_2: Mass times acceleration equals the sum of all the forces acting on the mass. If we neglect friction and assume small displacements, we obtain the equation of motion:

$$m_1 l \, \ddot{\varphi}_1 (t) = -m_1 g \, \varphi_1(t) + D l \, (\varphi_2(t) - \varphi_1(t))$$
$$m_2 l \, \ddot{\varphi}_2 (t) = -m_2 g \, \varphi_2(t) + D l \, (\varphi_1(t) - \varphi_2(t)).$$

Since $m_1 = m_2 = m$, we suppress the index. With the abbreviations $w_0^2 := \frac{g}{l}$ and $\Delta^2 := \frac{D}{m}$ we get

$$\left.\begin{array}{l} \ddot{\varphi}_1 (t) + (w_0^2 + \Delta^2) \, \varphi_1(t) = \Delta^2 \varphi_2(t) \\ \ddot{\varphi}_2 (t) + (w_0^2 + \Delta^2) \, \varphi_2(t) = \Delta^2 \varphi_1(t) \end{array}\right\} \qquad (*)$$

which is a coupled system for the displacements φ_1 and φ_2. We are looking for the modes in which both pendulums oscillate at the same frequency. Since both pendulums should oscillate at the same frequency but with different amplitudes, we choose the approach

$$\varphi_1(t) = A \, \cos(w \, t)$$
$$\varphi_2(t) = B \, \cos(w \, t).$$

So we are looking for the amplitudes A, B of the oscillations as well as the oscillation frequencies w. Substituting $\varphi_1(t)$, $\varphi_2(t)$ and $\ddot{\varphi}_1 (t)$, $\ddot{\varphi}_2 (t)$ into the equations $(*)$, we obtain

$$-A w^2 \cos(w \, t) + (w_0^2 + \Delta^2) \, A \cos(w \, t) = \Delta^2 B \cos(w \, t)$$
$$-B w^2 \cos(w \, t) + (w_0^2 + \Delta^2) \, B \cos(w \, t) = \Delta^2 A \cos(w \, t).$$

This is a system of linear equations for the amplitudes A and B. The matrix notation is

$$\begin{pmatrix} -w^2 + w_0^2 + \Delta^2 & -\Delta^2 \\ -\Delta^2 & -w^2 + w_0^2 + \Delta^2 \end{pmatrix} \begin{pmatrix} A \\ B \end{pmatrix} = \begin{pmatrix} 0 \\ 0 \end{pmatrix}.$$

According to the fundamental theorem for LEq, the determinant of the coefficient matrix must be zero. Then, the homogeneous LEq has no unique solution:

$$\det \begin{pmatrix} -w^2 + w_0^2 + \Delta^2 & -\Delta^2 \\ -\Delta^2 & -w^2 + w_0^2 + \Delta^2 \end{pmatrix} = (-w^2 + w_0^2 + \Delta^2)^2 - (\Delta^2)^2 = 0$$

$$\Rightarrow \qquad w_{1/2} = \pm\sqrt{w_0^2 + 2\Delta^2} \quad \text{or} \quad w_{3/4} = \pm w_0.$$

Since physically only positive frequencies are of interest, there are only two frequencies at which the pendulums oscillate

$$w_1 = +\sqrt{w_0^2 + 2\Delta^2} \quad \text{and} \quad w_3 = w_0. \qquad \square$$

3.4 Problems on Matrices and Determinants

3.1   Given are the matrices

$$A = \begin{pmatrix} 3 & 1 \\ 5 & 2 \\ 2 & 2 \end{pmatrix}, \qquad B = \begin{pmatrix} 5 & 2 & 1 \\ 2 & 3 & -1 \end{pmatrix}, \qquad C = \begin{pmatrix} 0 & 10 \\ 1 & 3 \\ -4 & 1 \end{pmatrix}$$

Calculate

a) $2\,A + 3\,C$ b) $A - 2\,B^t$ c) $3\,A^t + 4\,B$

d) $A + B^t + C$ e) $A^t - B + C^t$ f) $\left(A^t - B + C^t\right)^t$

3.2   Calculate $A^2 = A \cdot A$, $B^2 = B \cdot B$, $A \cdot B$, $B \cdot A$ (if they exist) and show that $A \cdot B \neq B \cdot A$.

a) $A = \begin{pmatrix} 2 & 3 & 0 \\ 0 & 4 & 1 \\ 0 & 0 & 1 \end{pmatrix}$, $B = \begin{pmatrix} 1 & 0 & 2 \\ -1 & 2 & -1 \\ 1 & -3 & 0 \end{pmatrix}$

b) $A = \begin{pmatrix} 2 & 3 & 0 & 4 \\ 0 & 1 & 0 & 1 \end{pmatrix}$, $B = \begin{pmatrix} 0 & 2 \\ -1 & 2 \\ 5 & 3 \\ 1 & 4 \end{pmatrix}$

3.3   Use the Gaussian-Jordan method to determine the inverse of the following matrices

$$A = \begin{pmatrix} 4 & 5 & -1 \\ 2 & 0 & 1 \\ 3 & 1 & 0 \end{pmatrix}, \quad B = \begin{pmatrix} 4 & 1 \\ -1 & -1 \end{pmatrix}, \quad C = \begin{pmatrix} 1 & -1 & 1 & -1 \\ -1 & 1 & 2 & 1 \\ -2 & 1 & -1 & 0 \\ 0 & 0 & 1 & 1 \end{pmatrix}$$

3.4   Check the above results with MAPLE.

3.5   A linear mapping $f : \mathbb{R}^4 \to \mathbb{R}^3$ is defined by the values of the basis vectors $f(\vec{e}_1) = (3, 5, 4)$, $f(\vec{e}_2) = (2, 1, 5)$, $f(\vec{e}_3) = (-1, 2, 1)$, $f(\vec{e}_4) = \vec{0}$. Create the underlying matrix and determine the image point of $(3, 5, 1, 2)$.

3.6   Calculate the determinants of the 2x2 matrices given below

a) $A = \begin{pmatrix} 3 & 1 \\ 1 & 3 \end{pmatrix}$ b) $B = \begin{pmatrix} 4 & 2 \\ 2 & 1 \end{pmatrix}$ c) $C = \begin{pmatrix} 2 & 3 \\ \lambda & 2\lambda \end{pmatrix}$

3.7   Check determinant rules (1) and (2) for the matrix $A = \begin{pmatrix} a & b \\ c & d \end{pmatrix}$.

3.8   What is the value of the following three row determinants?

a) $\begin{vmatrix} 5 & 4 & 0 \\ 2 & 2 & 1 \\ 1 & 4 & 2 \end{vmatrix}$ b) $\begin{vmatrix} -2 & 8 & 1 \\ 2 & 3 & 5 \\ -1 & 5 & 2 \end{vmatrix}$ c) $\begin{vmatrix} 3 & 4 & 2 \\ 5 & -1 & 0 \\ -2 & 2 & 3 \end{vmatrix}$

3.9 For which real parameters λ do the determinants disappear?

a) $\begin{vmatrix} 1-\lambda & -2 \\ 1 & -2-\lambda \end{vmatrix}$
b) $\begin{vmatrix} 1-\lambda & 2 & 1 \\ 0 & 3-\lambda & 2 \\ 0 & 0 & 2-\lambda \end{vmatrix}$

3.10 Calculate the determinants of

$$A = \begin{pmatrix} 1 & 0 & 3 & 4 \\ -2 & 1 & 0 & 3 \\ 1 & 4 & 1 & 5 \\ 0 & 2 & 2 & 0 \end{pmatrix} \quad \text{and} \quad B = \begin{pmatrix} 1 & 0 & 5 & 3 & 4 \\ 1 & 2 & 2 & 1 & 0 \\ 0 & 1 & 3 & 4 & 1 \\ -4 & 0 & -1 & 2 & 3 \\ -2 & 1 & 0 & 0 & 0 \end{pmatrix}$$

3.11 Determine the solution of LEq with Cramer's rule

$$\begin{array}{rcrcrcl}
x_1 & + & 2\,x_2 & & & = & 3 \\
x_1 & + & 7\,x_2 & + & 4\,x_3 & = & 18 \\
3\,x_1 & + & 13\,x_2 & + & 4\,x_3 & = & 30
\end{array}$$

3.12 Are the matrices

$$A = \begin{pmatrix} 4 & 5 & -1 \\ 2 & 0 & 1 \\ 3 & 1 & 0 \end{pmatrix}, \quad B = \begin{pmatrix} 3 & 4 & 2 \\ 1 & 5 & 3 \\ 0 & 1 & 0 \end{pmatrix}, \quad C = \begin{pmatrix} 1 & 1 & 2 \\ 3 & 2 & 4 \\ 2 & -5 & -1 \end{pmatrix}$$

regular matrices? If possible determine the inverse.

3.13 Determine the rank of the matrices

$$A = \begin{pmatrix} 3 & 4 & 2 \\ 1 & 5 & 3 \\ 0 & 1 & 0 \end{pmatrix}, \quad B = \begin{pmatrix} 2 & 1 & 0 & 4 \\ 3 & 1 & 0 & 2 \\ 2 & 0 & 4 & 1 \end{pmatrix},$$

$$C = \begin{pmatrix} 1 & 1 & 2 & -3 & 0 & 1 \\ -2 & 3 & 1 & 0 & 1 & 2 \\ -5 & 10 & 5 & -3 & 3 & 1 \\ -1 & 4 & 3 & -3 & 1 & 1 \end{pmatrix}, \quad D = \begin{pmatrix} 1 & 1 & 2 \\ 3 & 2 & 4 \\ 2 & -5 & -1 \end{pmatrix}.$$

3.14 Show that the LEq

$$\begin{array}{rcrcrcl}
x_1 & + & x_2 & + & x_3 & = & 0 \\
x_1 & + & 2\,x_2 & + & 4\,x_3 & = & 5 \\
x_1 & + & 3\,x_2 & + & 9\,x_3 & = & 12
\end{array}$$

is uniquely solvable and determine this solution.

3.15 Given is the matrix

$$A = \begin{pmatrix} 3 & -1 & 2 \\ 6 & 2 & 5 \\ -2 & 6 & 0 \end{pmatrix} \quad \text{and the vector } \vec{b} = \begin{pmatrix} 2 \\ -2 \\ 1 \end{pmatrix}.$$

Which system of equations $A\vec{x} = \vec{b}$, $A^T\vec{x} = \vec{b}$ has a solution? Is it unique?

3.16 a) Are the vectors

$$\vec{a} = \begin{pmatrix} 0 \\ 6 \\ -2 \end{pmatrix}, \vec{b} = \begin{pmatrix} 5 \\ 2 \\ 6 \end{pmatrix}, \vec{c} = \begin{pmatrix} 2 \\ -1 \\ 3 \end{pmatrix} \text{ linearly dependent?}$$

b) Show that the vectors $\vec{a} = \begin{pmatrix} 1 \\ 1 \\ 1 \end{pmatrix}, \vec{b} = \begin{pmatrix} 9 \\ 1 \\ 4 \end{pmatrix}, \vec{c} = \begin{pmatrix} 3 \\ 1 \\ 2 \end{pmatrix}$ form a

basis of $\mathbb{R}^3$ and represent $\vec{d} = \begin{pmatrix} 12 \\ 0 \\ 5 \end{pmatrix}$ as a linear combination of $\vec{a}, \vec{b}, \vec{c}$.

Chapter 4
Basic Functions

4

In this chapter we will introduce the basic functions we need to describe physical processes. Before doing so, we will discuss general functional properties that allow us to characterize these functions. In addition to polynomials, rational functions are discussed qualitatively. Power and root functions as well as exponential and logarithmic functions are introduced in relation to important applications. The trigonometric functions and their corresponding inverse functions are introduced to describe alternating voltages and harmonic oscillations.

4

4 Basic Functions

In this chapter we will introduce the basic functions we need to describe physical processes. Before, we will discuss general properties of functions that allow us to characterize these functions. In addition to polynomials, rational functions will be discussed qualitatively. Power and root functions as well as exponential and logarithmic functions are introduced in relation to important applications. The trigonometric functions and their corresponding inverse functions are introduced to describe alternating voltages and harmonic oscillations.

4.1 General Functional Characteristics

This section reviews basic concepts such as function and inverse function, zero and symmetric behavior, and monotonic behavior. These properties are then used to characterize basic functions.

4.1.1 Basic Concepts

The concept of a *function* is essential for the mathematical formulation of scientific laws. Almost all quantitative statements are formulated in the form of a functional relation.

Example 4.1 (Path-Time Diagram): The motion of an object is physically described in a *path-time diagram*, as shown in Fig. 4.1.

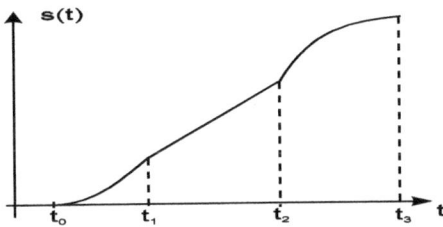

Figure 4.1. Path-Time Diagram

In the interval $t_0 \leq t \leq t_1$ we observe an accelerated motion, in the interval $t_1 \leq t \leq t_2$ a uniform motion and in the interval $t_2 \leq t \leq t_3$ a decelerated motion. It is characteristic of the motion shown that at each time $t \in [t_0, t_3]$ the position $s(t)$ is uniquely determined. This mapping defines a *function*.

Definition: (Function). *Let* $\mathbb{D} \subset \mathbb{R}$ *be a subset of* $\mathbb{R}$. **A real function** f *is a mapping*

$$
\begin{array}{rcl}
f : & \mathbb{D} & \to & \mathbb{R} \\
& x & \mapsto & f(x)
\end{array}
$$

which maps each element of $\mathbb{D}$ **uniquely** *to an element in* $\mathbb{R}$.

The set $\mathbb{D}$ is called **Domain** of f; the set $\mathbb{R}$ is called the **Target Domain (Codomain)** of f. The set

$$
\mathbb{W} := f(\mathbb{D}) := \{ f(x) : x \in \mathbb{D} \}
$$

is called the **Range** of f. We call $f(x)$ the function value (functional expression) at x.

There are special names for what can go into a function f and what can come out of it: What can go into a function is called the **Domain**. What can come out of a function is called the **Codomain**. What *actually* comes out of a function is called the **Range**. The codomain and the range are both on the output side, but they are subtly different. The codomain is the set of all possible values that can come out. For a real function, the codomain is always the set of real numbers. And the range is the set of values that actually come out.

To graph a function, we use the Cartesian coordinate system $\mathbb{R} \times \mathbb{R} = \{(x, y) : x \in \mathbb{R}, y \in \mathbb{R}\}$, where each point P of the plane is uniquely described by a pair of numbers (x, y). To illustrate the function geometrically, we draw the set of points $(x, f(x))$ in a coordinate system and obtain the graph of the function.

Definition: *The* **Graph** G_f *of a function* $f : \mathbb{D} \to \mathbb{R}$ *with* $x \mapsto f(x)$ *is the set of all pairs* (x, y):

$$
G_f := \{ (x, y) : x \in \mathbb{D} \quad and \quad y = f(x) \}.
$$

⊘ **Application Examples:**

Example 4.2 (Free Fall): The law of motion for a free fall from height s_0 with initial velocity v_0 is

$$s(t) = \frac{1}{2} g t^2 + v_0 t + s_0, \quad t \geq 0,$$

where $g = 9.81 \frac{m}{s^2}$ is the acceleration due to gravity.

Example 4.3 (Plate Capacitor): If a charge Q is applied to a plate capacitor, the induced voltage is

$$U(Q) = \frac{1}{C} Q,$$

where C is the capacitance of the plate capacitor.

Example 4.4 (DC Circuit): In a DC electric circuit, the current I depends on the DC voltage U. It holds

$$I(U) = \frac{U}{R},$$

where R is the ohmic resistance of the circuit.

Example 4.5 (Thermal Expansion of Gases (Gay-Lussac)): If V_0 is the volume of an ideal gas at temperature T_0, then the volume V at temperature T, assuming constant pressure p, is

$$V(T) = V_0 \left(1 + \gamma \left(T - T_0\right)\right), \quad T \geq T_0,$$

where γ is the coefficient of thermal expansion.

Example 4.6 (Barometric Formula): The pressure of air p at a height h above the ground is approximately

$$p(h) = p_0 e^{-\frac{h-H}{H}}, \quad h \geq H,$$

where $H = \frac{p_0}{\rho_0 \cdot g}$ and p_0, ρ_0 is the pressure and density of air at ground level.

These physical laws have in common that for an independent quantity t, Q, U, T or h the physical quantity $s(t)$, $U(Q)$, $I(U)$, $V(T)$ or $p(h)$ can be calculated **uniquely**. Therefore, variables are often referred to as the *independent variable* x and functions as the *dependent variable* $f(x)$.

⊘ **Mathematical Examples:**

Example 4.7. $f : \mathbb{R} \to \mathbb{R}$ with $x \mapsto f(x) = c$ is the *constant function*.
Example 4.8. $id_{\mathbb{R}} : \mathbb{R} \to \mathbb{R}$ with $x \mapsto id_{\mathbb{R}}(x) = x$ is the *identical function*.
Example 4.9. $abs : \mathbb{R} \to \mathbb{R}$ with $x \mapsto abs(x) = |x|$ is the *absolute function*.

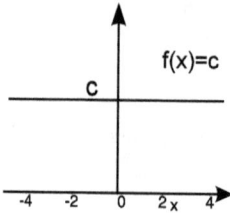

7. Constant function   8. Identical function   9. Absolute function

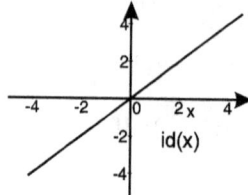

Example 4.10.
$sqr : \mathbb{R} \to \mathbb{R}$ with $x \mapsto sqr(x) = x^2$ is the *quadratic function*.
Example 4.11.
$sqrt : \mathbb{R}_{\geq 0} \to \mathbb{R}$ with $x \mapsto sqrt(x) = \sqrt{x}$ is the *root function*.
Example 4.12.
$\exp : \mathbb{R} \to \mathbb{R}$ with $x \mapsto \exp(x) = e^x$ is the *exponential function*.

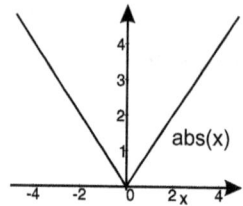

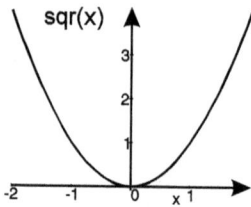

10. Quadratic function   11. Square root function   12. Exponential function

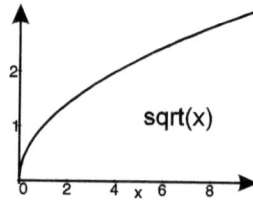

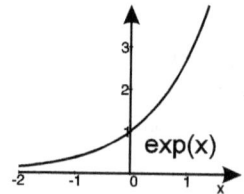

Example 4.13. $\sin : \mathbb{R} \to \mathbb{R}$ with $x \mapsto \sin(x)$ is the *sine function*.

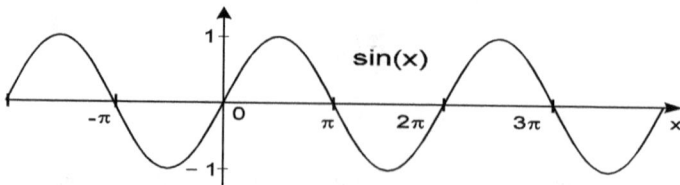

13. Sine function

⚠ **Caution:** Some functions cannot be drawn:

Example 4.14. $f : \mathbb{R} \to \mathbb{R}$ with $x \mapsto f(x) = \begin{cases} 0 & \text{for } x \text{ rational} \\ 1 & \text{for } x \text{ irrational.} \end{cases}$

Remark: In programming, a distinction must be made between a *function* and an *expression* (or functional expression): For example, *sqrt(x)* is an expression for $\sqrt{x}$, which is the square root of x for a given x. On the other hand, *sqrt* is a function that can be evaluated at a point x. A function is more than just a functional expression: Each function consists of its domain, the target domain, and a functional expression.

Application Example 4.15 (Logarithmic Representations).

In physics and engineering, logarithmic or double-logarithmic representations of the measurement results are often used to represent exponential, power or logarithmic laws.

(1) If an exponential law

$$y = c \, a^x$$

is present, a logarithmic scaling of the y-axis results in

$$\log(y) = \log(c) + x \cdot \log(a).$$

The result is a straight line with a slope of $\log(a)$ and an intersection at $\log(c)$. The function values can only be positive because the logarithm is only defined for positive real numbers. For example, the logarithmic scaling of the y-axis for a general exponential function $y = 10 \cdot 4^{-x}$ results in a straight line.

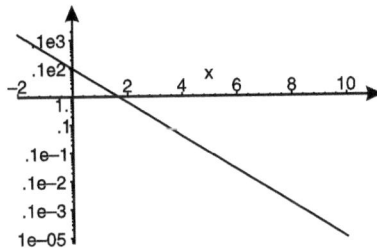

Figure 4.2. Logarithmic plot of the function $y = 10 \cdot 4^{-x}$.

(2) If a general power function

$$y = a \, x^b$$

is present, then a double logarithmic representation

$$\log(y) = \log(a) + b \cdot \log(x)$$

will give a straight line with gradient b and intersection $\log(a)$. Both, the x-range and the associated function values must be positive, since the logarithm is only defined for positive real numbers. A double-log application of the power function $y = 10\,x^{\frac{1}{3}}$ results in the following line

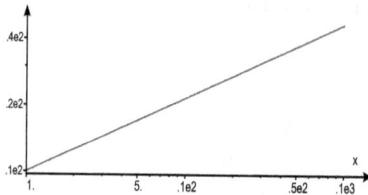

Figure 4.3. Double-logarithmic graph of the function $y = 10x^{\frac{1}{3}}$.

(3) If there is a logarithmic law of the form $y = a\,\log(x)+b$, then the logarithmic representation of the x-axis gives a straight line with gradient a and the axis segment b.   □

4.1.2 General Functional Properties

⊘ **Zero Points**

> **Definition:** $x_0 \in \mathbb{D}$ *is called* **zero (zero point, root)** *of f if*
>
> $$f(x_0) = 0.$$

Examples 4.16:
① The function $f(x) = x - 2$ intersects the x-axis at $x_0 = 2$.
② The sine function $f(x) = \sin(x)$ has an infinite number of zeros: $0, \pm\pi, \pm2\pi, \ldots$ (see Section 4.6).

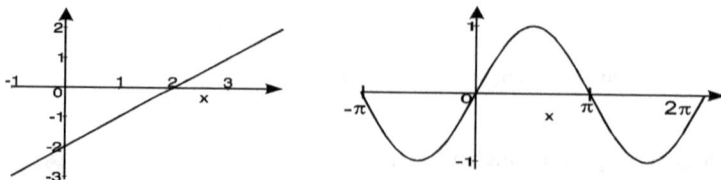

Figure 4.4. Zeros of functions

> **Symmetry Behavior**

Definition: *A function f with symmetric domain is called* **even**, *if for all $x \in \mathbb{D}$*

$$f(-x) = f(x).$$

Remark: The curve of an even function is **mirror symmetric** about the y-axis.

Examples 4.17:
① The function $f(x) = x^2$ is an even function.
② The function $f(x) = \cos(x)$ is an even function.
③ Every power function $f(x) = x^n$ with even n is even.

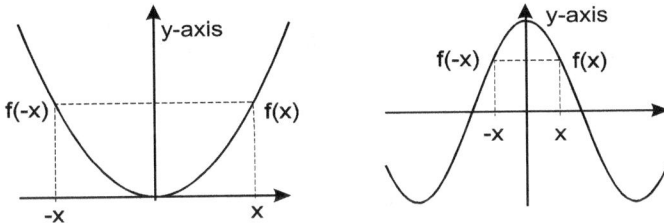

Figure 4.5. Even functions x^2 and $\cos(x)$: Mirror symmetrically about the y-axis

Definition: *A function f with symmetric domain is called* **odd**, *if for all $x \in \mathbb{D}$*

$$f(-x) = -f(x).$$

Remark: The curve of an odd function is **point symmetric** about the origin.

Examples 4.18:
① The function $f(x) = x^3$ is an odd function.
② The function $f(x) = \sin(x)$ is an odd function.
③ Any power function $f(x) = x^n$ where n is an odd number is odd.

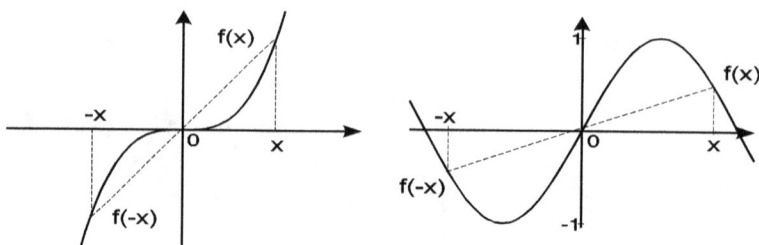

Figure 4.6. Odd functions x^3 and $\sin(x)$: Mirror symmetric to the origin

⊙ Monotone Functions

Definition: *A function* $f : \mathbb{D} \to \mathbb{R}$ *means*

increasing, if	$f(x_1) \leq f(x_2)$
strictly increasing, if	$f(x_1) < f(x_2)$
decreasing, if	$f(x_1) \geq f(x_2)$
strictly decreasing, if	$f(x_1) > f(x_2)$

applies to all $x_1, x_2 \in \mathbb{D}$ *with* $x_1 < x_2$.

Remark: A strictly increasing function f (Fig. 4.7 (a)) has the property that the smaller function value belongs to the smaller x value. For a strictly decreasing function (Fig. 4.7 (b)) it is exactly the other way round: The smaller x value belongs to the larger function value.

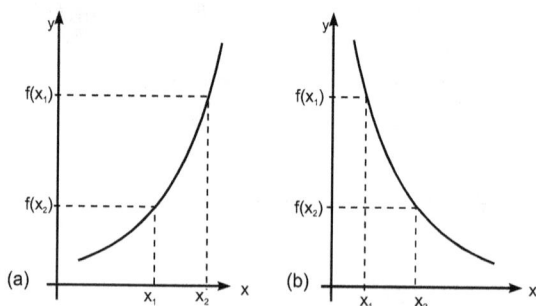

Figure 4.7. Strictly increasing (a) and strictly decreasing (b) functions

Examples 4.19:

① Strictly increasing functions:
  (i) Any straight line with a positive gradient.
  (ii) The function $f(x) = x^3$.
  (iii) The function $f : \mathbb{R}_{\geq 0} \to \mathbb{R}$ with $f(t) = y_0 \left(1 - e^{-t/t_0}\right)$.

② The charging process of a capacitor is a strictly increasing function: If a capacitor C is charged by a battery with voltage U_0, the time behavior of the voltage at the capacitor is given by

$$U(t) = U_0 \left(1 - e^{-t/RC}\right).$$

③ Strictly decreasing functions:
  (i) Any straight line with a negative gradient.
  (ii) The function $f(x) = -x^3$.
  (iii) The function $f : \mathbb{R} \to \mathbb{R}$ with $f(t) = y_0\, e^{-t/t_0}$.

③ Radioactive decay is a strictly decreasing function: In a radioactive decay, the number of atomic nuclei $n(t)$ decreases with time according to the exponential law $n(t) = n_0\, e^{-t/t_0}$.

④ The discharge of a capacitor is a strictly decreasing function: To discharge a capacitor C across an ohmic resistor R, the voltage across the capacitor decreases exponentially:

$$U(t) = U_0\, e^{-t/RC}.$$

⑤ The quadratic function $f : \mathbb{R} \to \mathbb{R}$ with $x \mapsto f(x) = x^2$ is neither decreasing nor increasing. However, if we restrict the function to one of the sub-intervals $\mathbb{R}_{\geq 0}$ or $\mathbb{R}_{\leq 0}$, it is monotonous there:
  $f_+ : \mathbb{R}_{\geq 0} \to \mathbb{R}$ with $x \mapsto f_+(x) = x^2$ is strictly increasing.
  $f_- : \mathbb{R}_{\leq 0} \to \mathbb{R}$ with $x \mapsto f_-(x) = x^2$ is strictly decreasing.

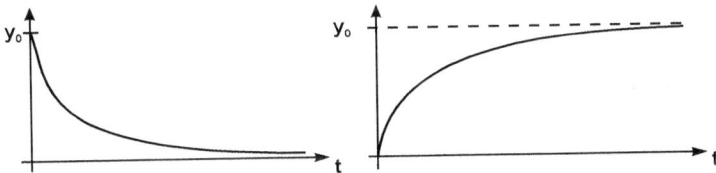

Figure 4.8. a) Strictly decreasing, b) Strictly increasing exponential functions

◎ **Periodicity**

> **Definition:** *A function* $f : \mathbb{D} \to \mathbb{R}$ *is called* **periodic** *with period* $p > 0,$ *if* $f(x+p) = f(x)$ *for all* $x \in \mathbb{D}.$

Examples 4.20:

① The sine function sin: $\mathbb{R} \to \mathbb{R}$ with $x \mapsto \sin(x)$ is periodic with the period $p = 2\pi$. The cosine function cos: $\mathbb{R} \to \mathbb{R}$ with $x \mapsto \cos(x)$ is periodic with the period $p = 2\pi$ (see §4.6).

② The tangent and cotangent functions, which we will discuss in more detail in §4.6, are periodic functions with period $p = \pi$.

4.1.3 Inverse Function

A function f uniquely assigns each number x_0 of its domain $\mathbb{D}$ to a value $y_0 = f(x_0)$. This assignment is shown in Fig. 4.9 (left) and indicated by the arrow from x_0 to y_0.

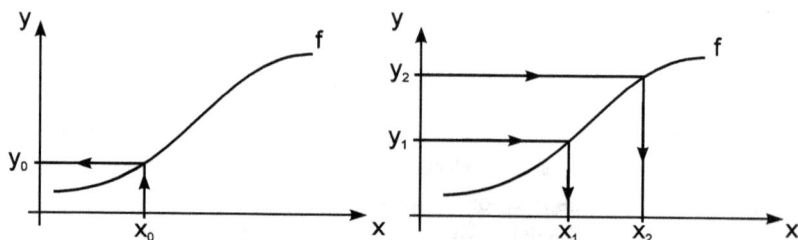

Figure 4.9. Assignment and inverse assignment

An inverse problem is often arises. For a given function value (y-value), the corresponding x-value must be determined. In Fig. 4.9 (right) this inverse problem is marked by the inverse of the arrows. Given are y_1 and y_2, to be found are the corresponding x_1 and x_2. For any given y-value from the range of f, exactly one x-value can be specified if the function has the property: If $x_1 \neq x_2$ then $f(x_1) \neq f(x_2)$.

Definition:

(1) A function $f : \mathbb{D}_f \to \mathbb{W}_f$ is called **reversible**, if $x_1 \neq x_2$ follows $f(x_1) \neq f(x_2)$.

(2) If the function f is reversible, then there is exactly one $y \in \mathbb{W}_f$ for each $x \in \mathbb{D}_f$. $y \mapsto x$ gives a function which is the **inverse function**

$$f^{-1} : \mathbb{W}_f \to \mathbb{D}_f.$$

⚠ **Caution:** Not every function is reversible, as Example 4.21 shows.

Examples 4.21:

① The function $f_1 : \mathbb{R} \to \mathbb{R}$ with $f_1(x) = x^2$ is **not** reversible, because for $x_0 \neq 0$ it returns $f(x_0) = x_0^2 = f(-x_0)$ but $x_0 \neq -x_0$ (Fig. 4.10 (a)).

② The function $f_2 : \mathbb{R}_{\geq 0} \to \mathbb{R}$ with $f_2(x) = x^2$ is **not** reversible. Although the domain has been restricted so that f_2 is a strictly increasing function, there are no corresponding x-values for the negative y-values.

③ The function $f_3 : \mathbb{R}_{\geq 0} \to \mathbb{R}_{\geq 0}$ with $f_3(x) = x^2$ is **reversible**. f_3 is a strictly increasing function on the domain and the target domain coincides with the range. Each value y from the range can be uniquely assigned to a x from the domain (Fig. 4.10 (b)).

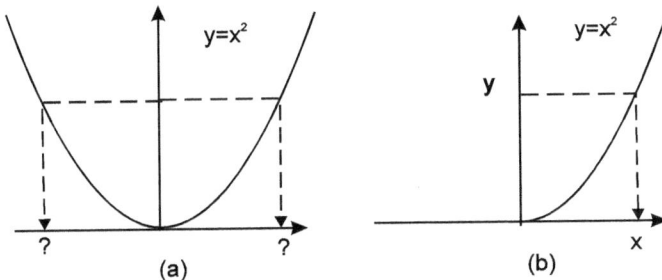

Figure 4.10. The inverse of the function $y = x^2$

Visualization: On the homepage, there is an animation where the graph of the inverse function is determined. The inverse function f^{-1} is a mapping from the range of the function f to the domain of f. However, by swapping the variables, we again get a graph with the domain on the x-axis and the range on the y-axis. Consequently, the graph of the f function is rotated so that the y-axis becomes the x-axis and vice versa to obtain the graph of the inverse function f^{-1}.

Procedure for Calculating the Inverse Function

(1) Restrict the domain of the function $\mathbb{D}_f$ so that f is strictly monotone in the restricted domain.

(2) Restrict the target domain to the range of the function $\mathbb{W}_f$. This gives a reversible function!

(3) Resolve the functional equation $y = f(x)$ with respect to the variable x and obtain the resolved form $x = g(y)$.

(4) By formally swapping the variables x and y, we obtain the expression of the inverse function

$$f^{-1}(x) = y = g(x).$$

Examples 4.22:

① $f : \mathbb{R} \to \mathbb{R}$ with $f(x) = 2x + 1$.

As we can see from Fig. 4.11, f is a straight line. f is strictly increasing on $\mathbb{R}$ with range $\mathbb{R}$; therefore f is reversible! By solving

$$y = 2x + 1$$

with respect to x, we get

$$x = \frac{1}{2}y - \frac{1}{2}.$$

By formally swapping x and y, we obtain the inverse function

$$f^{-1} : \mathbb{R} \to \mathbb{R} \text{ with } f^{-1}(x) = \frac{1}{2}x - \frac{1}{2}.$$

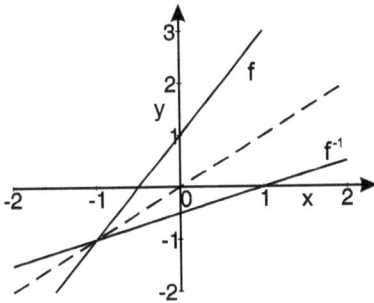

Figure 4.11. Function $f(x) = 2x + 1$ and inverse function $f^{-1}(x) = \frac{1}{2}x - \frac{1}{2}$

② $f : [0, \infty) \to [1, \infty)$ with $f(x) = 1 + x^2$.

f is strictly increasing on the domain and the target domain coincides with the range $\mathbb{W} = [1, \infty)$ (see Fig. 4.12). We solve the equation

$$y = 1 + x^2$$

with respect to x, giving

$$x = \pm\sqrt{y - 1}.$$

Since only positive x-values are allowed by the domain, the minus sign is omitted! Swapping x and y gives:

$$f^{-1} : [1, \infty) \to [0, \infty) \text{ with } f^{-1}(x) = \sqrt{x - 1}.$$

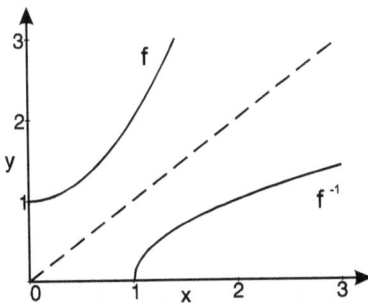

Figure 4.12. Function $f(x) = 1 + x^2$ and inverse function $f^{-1}(x) = \sqrt{x - 1}$

Summary of the Inverse Function

---

(1) Any strictly increasing or decreasing function is invertible.

(2) The domain and the range are reversed.

(3) Graphically, the graph of the inverse function is obtained by mirroring f on the line $y = x$.

To characterize the reversible functions, we introduce the following terms:

> **Definition:** Let $f : \mathbb{D}_f \to \mathbb{W}_f$ be a function. f is
>
> (1) **injective**, if $x_1 \neq x_2 \;\Rightarrow\; f(x_1) \neq f(x_2)$.
>
> (2) **surjective**, for each $y \in \mathbb{W}_f$ there is a $x \in \mathbb{D}_f$ with $y = f(x)$.
>
> (3) **bijective**, if f is both injective and surjective.

The injectivity of a function means that a parallel to the x-axis intersects the graph of the function at only one point. A function f is surjective if for every value y in the range, there is at least one x in the domain that is mapped to y.

Examples 4.23:

① The function $f : \mathbb{R} \to \mathbb{R}$ with $f(x) = x^2$ is neither injective nor surjective. Both x and $-x$ result in the same function value. The function is also not surjective, because there are no x-values mapped to negative y-values.

② The function $f : \mathbb{R}_{\geq 0} \to \mathbb{R}$ with $f(x) = x^2$ is injective, but not surjective, because e.g. for $y = (-1)$ we do not find any x from the domain with $x^2 = -1$.

③ The function $f : \mathbb{R}_{\geq 0} \to \mathbb{R}_{\geq 0}$ with $f(x) = x^2$ is bijective.

④ The function $f : \mathbb{R} \to [-1, 1]$ with $f(x) = \sin(x)$ is surjective but it is not injective: $\sin(x + 2\pi) = \sin(x)$. So as long as the domain is not restricted, it is not bijective.

⚠ **Caution:** We learn from Example 4.23 that not only the function expression is responsible for whether a function is injective or surjective, but also its domain and co-domain!

The bijective functions are exactly the reversible functions:

> **Theorem:** A function f is reversible if and only if f is bijective.

4.2 Polynomials

The section on polynomials introduces basic techniques such as zeroing, linear decomposition, Horner's scheme, and interpolation polynomials, which we will use when discussing rational functions, for example.

Polynomials play an important role in applied mathematics. Not only are they easy to represent, but they are also easy to evaluate, since only addition and multiplication need to be performed. In the chapter on Taylor Series (Volume II) we will show that complicated functions can be expressed using polynomials.

> **Definition: (Polynomial).** *A function* $f : \mathbb{R} \to \mathbb{R}$ *of the form*
>
> $$f(x) = a_n\,x^n + a_{n-1}\,x^{n-1} + \ldots + a_1\,x + a_0 \quad \text{with } a_n \neq 0$$
>
> *is called* **Polynomial** *of the degree* n. *The real numbers* $a_0, a_1, \ldots, a_n$ *are called the* **Coefficients** *of the polynomial.*

Examples 4.24:

① $p_1(x) = 4$ Polynomial of degree 0   (constant)
    $p_2(x) = 2x - 3$ Polynomial of degree 1   (linear)
    $p_3(x) = 2x^2 - 3x + 5$ Polynomial of degree 2   (quadratic)
    $p_4(x) = x^3 - x$ Polynomial of degree 3   (cubic)
    $p_5(x) = 4x^8 - x^5 - 10$ Polynomial of degree 8

② The height H of a **mercury column** depends on the temperature T:

$$H(T) = H_0\,(1 + \alpha\,(T - T_0)),$$

where H_0 is the height at the temperature T_0 and α is the coefficient of thermal expansion.

③ A **liquid or gaseous medium** (air, water) flowing around a body at an average velocity v exerts a force of

$$F_w(v) = c_w\,A\,\frac{1}{2}\,\rho\,v^2,$$

where c_w is the resistance coefficient, A is the cross-sectional area of the body, and ρ is the density of the fluid.

④   The **bending line** $y(x)$ of a beam of length l, fixed at one end at $x = 0$, is given by

$$y\left(x\right) = \frac{F}{6\,E \cdot I}\left(3\,l\,x^2 - x^3\right),$$

where $E \cdot I$ is the bending strength, F is the force at the beam's end.□

4.2.1 Determining Polynomials by Given Pairs of Values

An important property of polynomials is their determination by given pairs of values. In the case of the thermometer, 2 values are needed to define a linear scale. Weakly suspended ropes have a parabolic shape to a good approximation. To determine this relationship, 3 pairs of values must be given because 3 points define a 2nd degree polynomial. In general:

> **Interpolation Polynomial (Lagrange)**
> ___
>
> Given are $n + 1$ different pairs of values (x_1, y_1), (x_2, y_2), ..., (x_{n+1}, y_{n+1}). There is exactly one polynomial function f with $f(x_i) = y_i$ for all $i = 1, \ldots, n+1$ and a degree not greater than n.

One way of specifying the polynomial is to use the **Lagrange interpolation formula**:

$$f\left(x\right) = y_1 \cdot \frac{\left(x - x_2\right)\left(x - x_3\right) \cdot \ldots \cdot \left(x - x_{n+1}\right)}{\left(x_1 - x_2\right)\left(x_1 - x_3\right) \cdot \ldots \cdot \left(x_1 - x_{n+1}\right)} +$$
$$y_2 \cdot \frac{\left(x - x_1\right)\left(x - x_3\right) \cdot \ldots \cdot \left(x - x_{n+1}\right)}{\left(x_2 - x_1\right)\left(x_2 - x_3\right) \cdot \ldots \cdot \left(x_2 - x_{n+1}\right)} +$$
$$\vdots$$
$$y_{n+1} \cdot \frac{\left(x - x_1\right)\left(x - x_2\right) \cdot \ldots \cdot \left(x - x_n\right)}{\left(x_{n+1} - x_1\right)\left(x_{n+1} - x_2\right) \cdot \ldots \cdot \left(x_{n+1} - x_n\right)}.$$

When evaluating the function at the position x_i, all terms $k \neq i$ disappear because $(x - x_i)$ is a factor. Only the term $k = i$ remains, since the factor $(x - x_i)$ does not appear in this term:

$$f\left(x_i\right) = y_i \frac{\left(x_i - x_1\right)\left(x_i - x_2\right) \cdot \ldots \cdot \left(x_i - x_{i-1}\right)\left(x_i - x_{i+1}\right) \cdot \ldots \cdot \left(x_i - x_{n+1}\right)}{\left(x_i - x_1\right)\left(x_i - x_2\right) \cdot \ldots \cdot \left(x_i - x_{i-1}\right)\left(x_i - x_{i+1}\right) \cdot \ldots \cdot \left(x_i - x_{n+1}\right)}$$
$$= y_i.$$

Furthermore, we obtain at most a polynomial of the degree n, since all terms contain only n factors $(x - x_i)$.   □

4.2.2 Comparing Coefficients

Polynomials are identical if their coefficients are the same:

Comparison of Coefficients

Two polynomials f and g with

$$f(x) = a_n x^n + a_{n-1} x^{n-1} + \cdots + a_1 x + a_0$$

and

$$g(x) = b_m x^m + a_{m-1} x^{m-1} + \cdots + b_0$$

are exactly the same if $n = m$ and $a_i = b_i$ for all $i = 1, \ldots, n$.

This theorem states that polynomials are equal if and only if they have the same degree and the coefficients are identical.

Proof: We assume that $m \geq n$. Since $f(x) = g(x)$ for all $x \in \mathbb{R}$, it holds especially for $x = 0$: $f(0) = g(0) \Rightarrow \boxed{a_0 = b_0.}$ Hence,

$$x\left(a_n x^{n-1} + \cdots + a_1\right) = x\left(b_m x^{m-1} + \cdots + b_1\right).$$

This identity holds true for all $x \in \mathbb{R}$. We therefore conclude that

$$a_n x^{n-1} + \cdots + a_2 x + a_1 = b_m x^{m-1} + \cdots + b_2 x + b_1$$

for all $x \in \mathbb{R}$. We now evaluate this identity at $x = 0 \Rightarrow \boxed{a_1 = b_1.}$

Again, we can exclude x and repeat the process:

$$\Rightarrow a_2 = b_2, a_3 = b_3, \ldots, a_{n-1} = b_{n-1},$$

until we finally get

$$a_n = b_n + b_{n+1} x + \cdots + b_m x^{m-n+1}.$$

Inserting $x = 0$ results in $a_n = b_n$. So for all $x \in \mathbb{R}$:

$$\Rightarrow b_m x^{m-n} + \cdots + b_{n+1} = 0.$$

However, a polynomial can only be the zero polynomial if all its coefficients disappear, i.e. $b_m = b_{m-1} = \ldots = b_{n+1} = 0$. Consequently, the degree of f is equal to the degree of g and the coefficients a_i of f are identical to the coefficients b_i of g. $\qquad \square$

4.2.3 Division by a Linear Factor

One of the elementary tasks of polynomials is the **evaluation** of a polynomial at a point x_0. A simple scheme is given by the following theorem:

Theorem: For any polynomial f and any value x_0, the following decomposition is possible:

$$f(x) = a_n x^n + a_{n-1} x^{n-1} + \cdots + a_1 x + a_0$$
$$= (x - x_0)\left(b_n x^{n-1} + \cdots + b_2 x + b_1\right) + r.$$

Proof: To check this decomposition, we expand the product and compare the coefficients.

$$(x - x_0)\left(b_n x^{n-1} + \cdots + b_2 x + b_1\right) + r =$$
$$= b_n x^n + b_{n-1} x^{n-1} + b_{n-2} x^{n-2} + \cdots + b_1 x$$
$$-x_0 b_n x^{n-1} - x_0 b_{n-1} x^{n-2} - \cdots - x_0 b_2 x - x_0 b_1 + r$$
$$\overset{!}{=} a_n x^n + a_{n-1} x^{n-1} + a_{n-2} x^{n-2} + \cdots + a_1 x + a_0.$$

We compare the coefficients by descending powers of x:

$$\begin{aligned}
n: & & & & b_n = a_n \\
n-1: & \quad b_{n-1} - x_0 b_n &= a_{n-1} &\Rightarrow& \quad b_{n-1} = a_{n-1} + x_0 b_n \\
n-2: & \quad b_{n-2} - x_0 b_{n-1} &= a_{n-2} &\Rightarrow& \quad b_{n-2} = a_{n-2} + x_0 b_{n-1} \\
\vdots & & & & \\
1: & \quad b_1 - x_0 b_2 &= a_1 &\Rightarrow& \quad b_1 = a_1 + x_0 b_2 \\
0: & \quad -x_0 b_1 + r &= a_0 &\Rightarrow& \quad r = a_0 + x_0 b_1
\end{aligned}$$

With this procedure it is possible to systematically determine b_n, $b_{n-1}, \ldots, b_1$ and r. Additionally, we get

$$\boxed{f(x_0) = r.}$$

$\square$

This method is the basis for an effective evaluation of polynomials which was already known by Horner in 1819. According to his scheme a polynomial can be computed at any intermediate point x_0 using the **Horner scheme:**

a_n	a_{n-1}		a_2	a_1	a_0
+	$x_0 \cdot b_n$		$x_0 \cdot b_3$	$x_0 \cdot b_2$	$x_0 \cdot b_1$
$b_n \nearrow$	$b_{n-1} \nearrow$	$\cdots \quad \nearrow$	$b_2 \quad \nearrow$	$b_1 \quad \nearrow$	$\boxed{r = f(x_0)}$

The advantage of *Horner's scheme* is that there is no need to calculate the powers $x_0^n, x_0^{n-1}, \ldots$ but rather the intermediate values are multiplied by x_0 and summed. Therefore, this evaluation scheme is extremely effective because it avoids powers. It also shows that Horner's scheme is not susceptible to rounding errors and is therefore suitable for a numerical use.

Example 4.25. Evaluate $f(x) = 2x^4 - 6x^3 - 35x + 10$ at point $x_0 = 4$:

	2	-6	0	-35	10
$x_0 = 4$:		2·4	2·4	8·4	-3·4
	2	2	8	-3	$-2 = f(4)$

The Horner scheme also gives the decomposition of the polynomial in the form of $f(x) = q(x)(x - 4) + f(4)$:

$$f(x) = 2x^4 - 6x^3 - 35x + 10 = (2x^3 + 2x^2 + 8x - 3)(x - 4) - 2.$$

⚠ **Caution:** If a power x^k ($k < n$) - as in the example x^2 - is not present in the polynomial, 0 must be entered in the corresponding position in Horner's scheme.   □

4.2.4 Roots of Polynomials (Zeros of Polynomials)

For the special case that x_0 is a root of the polynomial $f(x)$, Horner's scheme returns the so-called *product decomposition* with $(x - x_0)$. Because in this special case the remainder is $r = f(x_0) = 0$ and we get:

Decomposition of a Linear Factor

If x_0 is a **root (zero)** of the n-th degree polynomial f, then the following decomposition applies

$$f(x) = (x - x_0)\left(b_n x^{n-1} + b_{n-1} x^{n-2} + \cdots + b_1\right).$$

Example 4.26. $x_0 = 1$ is a root of $f(x) = x^3 + 2x^2 - 13x + 10$:

	1	2	-13	10
$x_0 = 1$:		1·1	3·1	-10·1
	1	3	-10	$0 = f(1)$

Horner's scheme returns not only the functional value $f(1) = 0$, but also the coefficients b_i of the decomposition of the polynomial f into

$$x^3 + 2x^2 - 13x + 10 = (x - 1)\left(1x^2 + 3x + (-10)\right).$$

The same result is obtained with **polynomial division**:

$$
\begin{array}{l}
(x^3 \quad +2x^2 \quad -13x \quad +10) \quad : \quad (x-1) \quad = \quad 1\,x^2 + 3\,x - 10 \\
\underline{-(x^3 \quad -x^2)} \\
3x^2 \quad -13x \\
\underline{-(3x^2 \quad -3x)} \\
-10x \quad +10 \\
\underline{-(-10x \quad +10)} \\
0
\end{array}
$$

$\square$

If f is a polynomial of degree n and x_1 is a zero of f, then $f(x)$ can be divided by $(x - x_1)$ without residue, and the result is a polynomial of degree $n - 1$. However, a linear factor can be split off from a polynomial of degree n at most n times, until the remaining polynomial has only the degree zero:

> **Theorem:** Every polynomial of degree n has **at most** n different zeros.

A difficult task is the concrete computation of the roots of polynomials. Only for $n = 2$ we have the p/q-formula which is applied in the next example:

Application Example 4.27 (Ohmic Resistance).

Two resistors R_1 and R_2 are connected in series and in parallel to a voltage source $U = 220V$. In the first circuit the current is $I_1 = 0.9A$ and in the second circuit $I_2 = 6A$. What are R_1 and R_2?

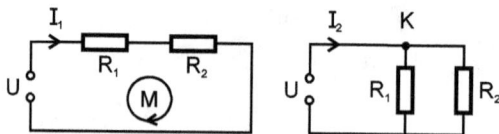

Figure 4.13. Circuit diagrams

For the first circuit, the mesh rule applies: $U = I_1 R_1 + I_1 R_2$ and for the parallel circuit the node rule applies for the node $K : I_2 = \frac{U}{R_1} + \frac{U}{R_2}$. We isolate R_2 in the first equation and insert it into the second equation,

$$
R_1^2 - \frac{U}{I_1} R_1 + \frac{U^2}{I_1 I_2} = 0.
$$

We are looking for the roots of the polynomial $f(x) = x^2 - \frac{220}{0,9}x + \frac{220^2}{0,9 \cdot 6}$.
With

$$x_{1/2} = -\frac{p}{2} \pm \sqrt{\left(\frac{p}{2}\right)^2 - q}$$

we get $x_1 = 44.9$ and $x_2 = 199.5$. If we choose $R_1 = 44.9\Omega$, then $R_2 = 199.5\Omega$. For reasons of symmetry, R_1 and R_2 can also be swapped.   □

For $n = 3$ and 4 the formulas are much more complicated and for $n \geq 5$ there are no more complete formulas. Therefore, finding the roots of higher degree polynomials depends heavily on numerical methods such as the bisection method in Section 6.4 or Newton's method in Section 7.8. However, if we can guess a root x_0, Horner's scheme provides a way to reduce the problem by splitting the linear factor $(x - x_0)$ and reducing the degree to $n-1$.

Example 4.28. To find the roots of the polynomial

$$f(x) = 3x^3 + 3x^2 - 3x - 3,$$

we assume a root to be $x_1 = -1$

	3	3	-3	-3
+		-3	0	3
	3	0	-3	$0 = f(-1)$

which gives $3x^3 + 3x^2 - 3x - 3 = (x+1)\left(3x^2 - 3\right)$.

Using the quadratic formula, the other two zeros $x_2 = 1$ and $x_3 = -1$ are calculated. This gives us the *linear factorization* of the polynomial:

$$3x^3 + 3x^2 - 3x - 3 = 3(x+1)(x+1)(x-1).$$ □

We summarize the result in the following theorem:

Complete Decomposition in Linear Factors

If $x_1, x_2, \ldots, x_n$ are the roots of a polynomial of degree n, then it can be represented as a product of n **linear factors**:

$$f(x) = a_n x^n + a_{n-1} x^{n-1} + \cdots + a_1 x + a_0$$
$$= a_n (x - x_1)(x - x_2) \cdot \ldots \cdot (x - x_n).$$

Remarks:

(1) ⚠ **Caution:** Do not forget the coefficient a_n in the product decomposition!

(2) ⚠ **Caution:** With double zeros, the corresponding linear factor appears twice, with triple zeros three times, and so on!

> **Hint:** If we are looking for integer roots of a polynomial with integer coefficients, then the roots must be a factor of a_0. If we expand the linear factorization, we get $a_0 = \pm\, a_n \cdot x_1 \cdot x_2 \cdot \ldots \cdot x_n$.

4.2.5 Newton's Interpolation Algorithm

The problem often arises in applications: $(n + 1)$ data points are given from an unknown functional context: $P_1(x_1, y_1)$, ..., $P_{n+1}(x_{n+1}, y_{n+1})$. The function values at the intermediate points are to be calculated. As already known from the first theorem, there is exactly one polynomial of degree $\leq n$ that fits all these $(n + 1)$ data values. This polynomial is called *interpolation polynomial*, because it can be used to calculate (=*interpolate*) any intermediate values of the function in the interval $[x_1, x_{n+1}]$.

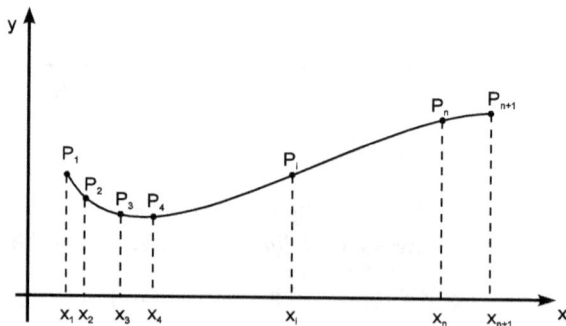

Figure 4.14. Interpolation polynomial

The Lagrange interpolation polynomial is a representation of the uniquely determined polynomial of degree at most n. However, the effort for the calculation of the coefficients is significant. A simpler scheme for determining the interpolation polynomial is based on Newton's approach:

$$f(x) = a_0 + a_1(x - x_1) + a_2(x - x_1)(x - x_2) + \cdots$$
$$+ a_n(x - x_1)(x - x_2) \cdot \ldots \cdot (x - x_n).$$

The coefficients $a_0, a_1, \ldots, a_n$ are determined *iteratively* according to:

$$y_1 = f(x_1) = a_0 \hookrightarrow \boxed{a_0 = y_1.}$$

$$y_2 = f(x_2) = a_0 + a_1(x_2 - x_1) \quad \hookrightarrow \quad \boxed{a_1 = \frac{y_2 - y_1}{x_2 - x_1}.}$$

$$y_3 = f(x_3) = a_0 + a_1(x_3 - x_1) + a_2(x_3 - x_1)(x_3 - x_2)$$

$$
\begin{aligned}
a_2 &= \frac{y_3 - a_0 - a_1(x_3 - x_1)}{(x_3 - x_1)(x_3 - x_2)} = \frac{y_3 - y_1 - \frac{y_2 - y_1}{x_2 - x_1}(x_3 - x_1)}{(x_3 - x_1)(x_3 - x_2)} \\
&= \frac{1}{x_3 - x_1}\left\{ \frac{y_3 - y_1}{x_3 - x_2} - \frac{y_2 - y_1}{x_2 - x_1}\frac{x_3 - x_1}{x_3 - x_2} \right\} \\
&= \frac{1}{x_3 - x_1}\left\{ \frac{y_3 - y_2}{x_3 - x_2} + \frac{y_2 - y_1}{x_3 - x_2} - \frac{y_2 - y_1}{x_2 - x_1}\frac{x_3 - x_1}{x_3 - x_2} \right\} \\
&= \frac{1}{x_3 - x_1}\left\{ \frac{y_3 - y_2}{x_3 - x_2} - \frac{y_2 - y_1}{x_2 - x_1} \right\}.
\end{aligned}
$$

If we take $D_{2,1} = \dfrac{y_2 - y_1}{x_2 - x_1}$, $D_{3,2} = \dfrac{y_3 - y_2}{x_3 - x_2}$ $\hookrightarrow$ $\boxed{a_2 = \dfrac{D_{3,2} - D_{2,1}}{x_3 - x_1}.}$

By mathematical induction it can be shown that with the abbreviations

$$D_{4,3,2} = \frac{D_{4,3} - D_{3,2}}{x_4 - x_2}; \ D_{3,2,1} = \frac{D_{3,2} - D_{2,1}}{x_3 - x_1} \quad \text{and so on it is valid}$$

$$\boxed{a_{k-1} := D_{k,\dots,1} \text{ for } k \geq 1.}$$

The expressions $D_{k,\dots,1}$ are called *divided differences* and the calculation is done according to the following scheme:

k	x_k	y_k				
1	x_1	$\boxed{y_1}$				
2	x_2	y_2	$\to \boxed{D_{2,1}} = \frac{y_2-y_1}{x_2-x_1}$			
3	x_3	y_3	$\to D_{3,2} = \frac{y_3-y_2}{x_3-x_2}$	$\to \boxed{D_{3,2,1}} = \frac{D_{3,2}-D_{2,1}}{x_3-x_1}$		
4	x_4	y_4	$\to D_{4,3} = \frac{y_4-y_3}{x_4-x_3}$	$\to D_{4,3,2} = \frac{D_{4,3}-D_{3,2}}{x_4-x_2}$	$\to$	
$\vdots$					$\cdots$	
$n+1$	x_{n+1}	y_{n+1}	$\to D_{n+1,n}$	$\to$	$\cdots$	$\to$

The numbers $y_1 = a_0$, $D_{2,1} = a_1$, $D_{3,2,1} = a_2$, $D_{4,3,2,1} = a_3, \dots, D_{n+1,\dots,1} = a_n$ are the coefficients of the **Newtonian interpolation polynomial**.

Remark: The main **advantage** of the Newtonian approach over the Lagrangian interpolation polynomial is that when additional data points are added, only additional rows need to be added to the scheme. The coefficients already calculated remain valid!

Example 4.29 (With MAPLE-Worksheet).

(i) The result of a measurement gives the values P_1 $(0, -12)$, P_2 $(2, 16)$ and P_3 $(5, 28)$. Find the interpolation polynomial.

Newton's Polynomial: $f(x) = a_0 + a_1 (x - x_1) + a_2 (x - x_1)(x - x_2)$.

Coefficients:

k	x_k	y_k		
1	0	$\boxed{-12}$		
2	2	16	$\frac{16+12}{2-0} = \boxed{14}$	
3	5	28	$\frac{28-16}{5-2} = 4$	$\frac{4-14}{5-0} = \boxed{-2}$

This gives the quadratic polynomial

$$f(x) = -12 + 14x - 2x(x - 2) = -2x^2 + 18x - 12.$$

(ii) Adding another point P_4 $(7, -54)$ extends the existing scheme

k	x_k	y_k			
1	0	$\boxed{-12}$			
2	2	16	$\boxed{14}$		
3	5	28	4	$\boxed{-2}$	
4	7	-54	$\frac{-54-28}{7-5} = -41$	$\frac{-41-4}{7-2} = -9$	$\frac{-9+2}{7-0} = \boxed{-1}$

and the interpolation polynomial of degree 3 can be identified

$$\begin{aligned}
f(x) &= a_0 + a_1 (x - x_1) + a_2 (x - x_1)(x - x_2) \\
&= +a_3 (x - x_1)(x - x_2)(x - x_3) \\
&= -12 + 14(x - 0) - 2(x)(x - 2) - 1(x)(x - 2)(x - 5) \\
&= -x^3 + 5x^2 + 8x - 12.
\end{aligned}$$

□

4.3 Rational Functions

The aim of this section is to characterize rational functions qualitatively by determining their zeros and poles as well as their asymptotic behavior.

Many processes in physics and applications are described by functions, which are a quotient of two polynomials. For example, for the focal length f of a lens, the image distance b is $b(x) = \frac{f\,x}{x - f}$, where x is the distance from the lens.

> **Definition:** *Any* **rational function** f *can be represented as the quotient of two polynomials* $g(x)$ *and* $h(x)$
>
> $$f : \mathbb{R} \setminus \{x \in \mathbb{R} : h(x) = 0\} \to \mathbb{R}$$
>
> *where*
>
> $$x \mapsto f(x) = \frac{g(x)}{h(x)} = \frac{a_m\, x^m + \cdots + a_1\, x + a_0}{b_n\, x^n + \cdots + b_1\, x + b_0} = \frac{\displaystyle\sum_{i=0}^{m} a_i\, x^i}{\displaystyle\sum_{j=0}^{n} b_j\, x^j}.$$

As with fractions, we distinguish between *proper rational functions* $(m < n)$ and *improper rational functions* $(m \geq n)$.

Examples 4.30:

① $f_1 : \mathbb{R} \setminus \{0\} \to \mathbb{R}$ with $x \mapsto \dfrac{1}{x}$.

② $f_2 : \mathbb{R} \setminus \{-2, 2\} \to \mathbb{R}$ with $x \mapsto \dfrac{x}{x^2 - 4}$.

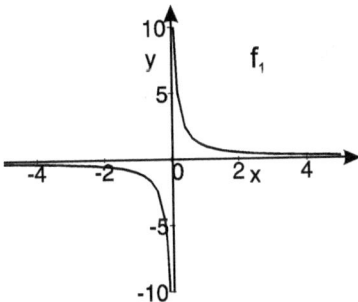

① Graph of $f_1(x) = \frac{1}{x}$

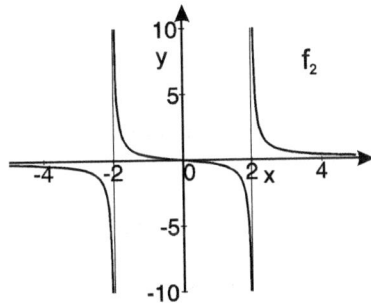

② Graph of $f_2(x) = \frac{x}{x^2-4}$

4.3.1 Gaps, Zeros, Poles

Let $f(x) = \frac{g(x)}{h(x)}$ be a rational function. x_0 is **zero** of f, if $g(x_0) = 0$ and $h(x_0) \neq 0$. If $h(x_0) = 0$ in the denominator, the function f is not defined. Therefore, we speak of a **gap in the domain**, if $h(x_0) = 0$. But not in all gaps the function goes to infinity. There is a difference between *removable gaps* and **poles**: x_0 is a *pole* if the function values in the immediate vicinity of x_0 increase or decrease beyond all limits. The graph of the function is thus adapted to a parallel of the y-axis (vertical asymptote).

If the numerator and the denominator polynomials have a common zero x_0, then both polynomials contain $(x - x_0)$ as a linear factor. Common factors are simplified. Gaps in the definition can be corrected by abbreviation and the domain can be extended. Thus, to find the roots of a rational function we only need to find the roots of the numerator, as long as the rational function is in its **Lowest Terms**. A rational expression is in its lowest terms when the upper and lower terms have no common factors.

Determining Zeros and Poles

(1) The numerator and denominator polynomials are simplified as much as possible. The rational function is then reduced to its lowest terms.

(2) The linear factors remaining in the numerator give the zeros of the function, and the linear factors remaining in the denominator give the poles.

Example 4.31. Find the gaps, zeros and poles of the rational:

$$f(x) = \frac{2x^3 + 2x^2 - 32x + 40}{x^3 + 2x^2 - 13x + 10}.$$

To find these points, we decompose both the denominator and the numerator polynomials into linear factors:

$$2x^3 + 2x^2 - 32x + 40 = 2(x - 2)^2(x + 5)$$
$$x^3 + 2x^2 - 13x + 10 = (x - 1)(x - 2)(x + 5).$$

This allows the rational function to be written in factors such as

$$f(x) = \frac{2x^3 + 2x^2 - 32x + 40}{x^3 + 2x^2 - 13x + 10} = \frac{2(x - 2)^2(x + 5)}{(x - 1)(x - 2)(x + 5)} = \frac{2(x - 2)}{x - 1}.$$

In the original formulation, all the roots of the denominator provide the gaps, whereas in the lowest terms formulation, the roots of the denominator are the poles of the function. The roots of the numerator polynomial then provide the zeros of the function. In summary, we obtain the curve of the function qualitatively:

Gaps	:	1, 2, −5
Zeros	:	(2)
Poles	:	1
Removable gaps	:	2, −5

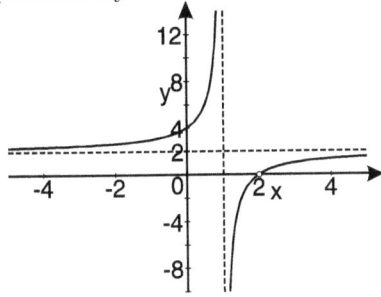

4.3.2 Asymptotic Behavior at Infinity

Especially for the discussion of the application examples, the behavior of the rationals at $x \to \pm\infty$ is of special interest. We have to consider three different cases, depending on the degree of the upper and lower polynomials, as the following examples show:

Examples 4.32.

(1.) $f_1(x) = \dfrac{5x^4 + x^3 + 4}{2x^5 + x + 2}$ $\qquad$ (2.) $f_2(x) = \dfrac{5x^4 + x^3 + 4}{2x^4 + x + 2}$

(3.) $f_3(x) = \dfrac{5x^4 + x^3 + 4}{2x^3 + x + 2}$

In all three cases, $f(x) \xrightarrow{x \to \infty} \frac{\infty}{\infty}$. The following explanation illustrates that different results are obtained. The rational expression is expanded with $\frac{1}{x^k}$, where k is the highest order:

$$\lim_{x \to \infty} f_1(x) = \lim_{x \to \infty} \frac{5x^4 + x^3 + 4}{2x^5 + x + 2} \cdot \frac{\frac{1}{x^5}}{\frac{1}{x^5}} = \lim_{x \to \infty} \frac{\frac{5}{x} + \frac{1}{x^2} + \frac{4}{x^5}}{2 + \frac{1}{x^4} + \frac{2}{x^5}} = \frac{0}{2} = 0.$$

$$\lim_{x \to \infty} f_2(x) = \lim_{x \to \infty} \frac{5x^4 + x^3 + 4}{2x^4 + x + 2} \cdot \frac{\frac{1}{x^4}}{\frac{1}{x^4}} = \lim_{x \to \infty} \frac{5 + \frac{1}{x} + \frac{4}{x^4}}{2 + \frac{1}{x^3} + \frac{2}{x^4}} = \frac{5}{2}.$$

$$\lim_{x \to \infty} f_3(x) = \lim_{x \to \infty} \frac{5x^4 + x^3 + 4}{2x^3 + x + 2} \cdot \frac{\frac{1}{x^4}}{\frac{1}{x^4}} = \lim_{x \to \infty} \frac{5 + \frac{1}{x} + \frac{4}{x^4}}{\frac{2}{x} + \frac{1}{x^3} + \frac{2}{x^4}} = \frac{5}{0} = \infty. \ \square$$

This classification can be applied to any rational function:

Behavior of Rational Functions $f(x) = \frac{g(x)}{h(x)}$ at Infinity

(1) $\mathrm{degree}(g) < \mathrm{degree}(h) \quad \Rightarrow \quad f(x) \to 0 \quad$ for $x \to \pm\infty$.

(2) $\mathrm{degree}(g) = \mathrm{degree}(h) \quad \Rightarrow \quad f(x) \to \frac{a_n}{b_n} \quad$ for $x \to \pm\infty$.

(3) $\mathrm{degree}(g) > \mathrm{degree}(h) \quad \Rightarrow \quad f(x) \to \pm\infty \quad$ for $x \to \pm\infty$
$\to$ followed by polynomial division see Example 4.33:

In the last case, the improper rational function is split by polynomial division into a polynomial function $p(x)$ and a proper rational function $r(x)$: $f(x) = p(x) + r(x)$ with $r(x) \to 0$ for $x \to \infty$. The function $f(x)$ approaches for $x \to \infty$ the function $p(x)$ because the rest of $r(x)$ converges to 0! The $p(x)$ is called the **Asymptote** of f.

Example 4.33. $\qquad f(x) = \dfrac{\frac{1}{2}x^3 - \frac{3}{2}x + 1}{x^2 + 3x + 2}.$

We decompose the numerator and denominator into linear factors:
The roots of the numerator are 1 (double) and -2.
The roots of the denominator are -1 and -2.

$$\Rightarrow f(x) = \frac{\frac{1}{2}(x-1)^2(x+2)}{(x+1)(x+2)} = \frac{\frac{1}{2}(x-1)^2}{x+1}.$$

The roots and poles of the function can now be identified in the lowest term representation:

Root : $x = 1$ (double)
Pole : $x = -1$ (single).

To determine the behavior at infinity, we split the function f into a polynomial $p(x)$ and a proper rational function $r(x)$ by polynomial division:

$$
\begin{array}{l}
(\frac{1}{2}x^2 \quad -1\,x \quad +\frac{1}{2}) \quad : \quad (x+1) = \quad \frac{1}{2}x - \frac{3}{2} + \frac{2}{x+1} \\
\underline{-(\frac{1}{2}x^2 \quad +\frac{1}{2}x)} \\
\qquad\quad -\frac{3}{2}x \quad +\frac{1}{2} \\
\qquad\quad \underline{-(-\frac{3}{2}x \quad -\frac{3}{2})} \\
\qquad\qquad\qquad\quad 2
\end{array}
$$

This gives $f(x) = p(x) + r(x)$ with the asymptote $\boxed{p(x) = \frac{1}{2}x - \frac{3}{2}}$

and the residual $r(x) = \dfrac{2}{x+1} \xrightarrow{x\to\infty} 0.$

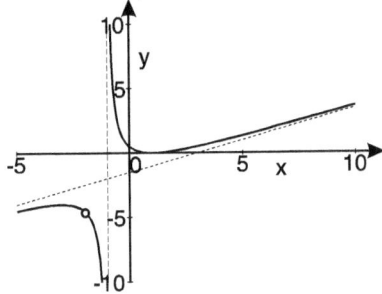

Graph of the rational function $\dfrac{1/2\,x^3 - 3/2x + 1}{x^2 + 3x + 2}$

Application Example 4.34   (Transfer Behavior of Circuits).

Like any RCL circuit, the LC circuit on the right has the following property: If the input voltage $U_E(t)$ is an AC voltage of frequency ω, then the output voltage is also an AC voltage with frequency ω, but phase-shifted and with a different amplitude. The amplitude depends on the frequency of the input voltage. The amplitude ratio $H(\omega)$ is given by $\left|\dfrac{U_A(t)}{U_E(t)}\right| = |H(\omega)|$ with

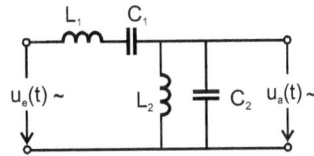

Figure 4.15. LC parallel circuit

$$H(\omega) = \frac{-\omega^2 LC}{\omega^4 L^2 C^2 - 3\omega^2 LC + 1}.$$

$H(\omega)$ is a proper rational function in ω. The zero of the $H(\omega)$ is at $\omega = 0$. To find the poles of the function we set its denominator equal to zero:

$$\omega^4 L^2 C^2 - 3\omega^2 LC + 1 = 0.$$

With the substitution $z = \omega^2$ we get $z^2 (LC)^2 - 3z\,LC + 1 = 0$:

$$z_{1/2} = \frac{3}{2}\frac{1}{LC} \pm \sqrt{\frac{9}{4}\left(\frac{1}{LC}\right)^2 - \left(\frac{1}{LC}\right)^2} = \left(\frac{3}{2} \pm \frac{\sqrt{5}}{2}\right)\frac{1}{LC}$$

$$z_1 = 2,62\,\frac{1}{LC} \quad\Rightarrow\quad \omega_{1/2} = \pm 1,61\sqrt{\frac{1}{LC}}$$

$$z_2 = 0,76\,\frac{1}{LC} \quad\Rightarrow\quad \omega_{3/4} = \pm 0,87\sqrt{\frac{1}{LC}}.$$

$H(\omega)$ is symmetric about the y-axis, since

$$H(-\omega) = \frac{-(-\omega)^2 LC}{(-\omega)^4 L^2 C^2 - 3(-\omega)^2 LC + 1} = \frac{-\omega^2 LC}{\omega^4 L^2 C^2 - 3\omega^2 LC + 1} = H(\omega).$$

The degree of the numerator polynomial is less than the degree of the denominator polynomial. Therefore: $H(\omega) \to 0$ for $\omega \to \infty$. For the graphical representation we choose $L = C = 1$ and draw $|H(\omega)|$ for positive frequencies as shown in Fig. 4.16.

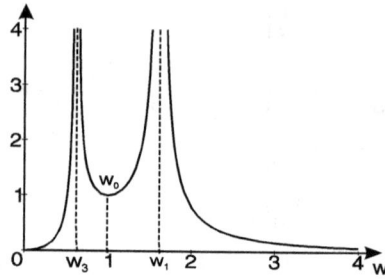

Figure 4.16. Transfer function $|H(\omega)|$

Discussion: This is the typical behavior of a bandpass filter blocking low and high frequencies and passing frequencies in a band between ω_1 and ω_2. □

4.4 Power and Root Functions

Definition: *A polynomial of the form*

$$p : \mathbb{R} \to \mathbb{R} \qquad with \qquad x \mapsto x^n \quad (n \in \mathbb{N})$$

is called **Power Function.**

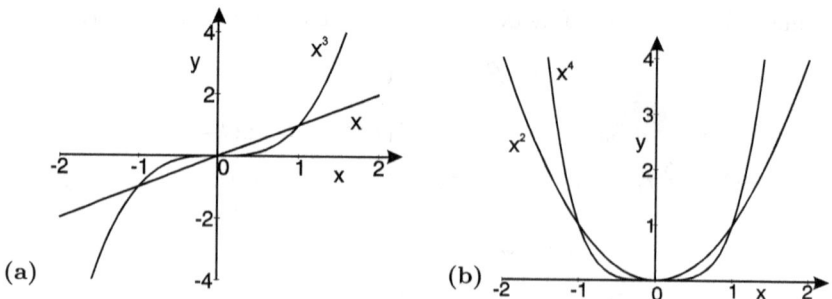

Figure 4.17. Odd (a) and even (b) power functions

With the power functions, we have to distinguish between n even and n odd. For odd n the power function is strictly increasing and point symmetric about the origin (shown in Fig. 4.17 left). For even n there is no monotonic behavior, the power function is axis symmetric with respect to the y-axis (shown in Fig. 4.17 right).

If we restrict the domain of the power functions to the non-negative real numbers, then

$$p : \quad \mathbb{R}_{\geq 0} \quad \to \quad \mathbb{R}_{\geq 0}$$
$$x \quad \mapsto \quad x^n$$

is a strictly increasing function with range $\mathbb{R}_{\geq 0}$. So this function is reversible. With the help of

$$y = x^n \to x = \sqrt[n]{y} \to y = \sqrt[n]{x}$$

we get the inverse function

$$p^{-1} : \mathbb{R}_{\geq 0} \to \mathbb{R}_{\geq 0} \qquad \text{with} \qquad x \mapsto \sqrt[n]{x}.$$

Definition: *The function*

$$\omega : \mathbb{R}_{\geq 0} \to \mathbb{R}_{\geq 0} \ \text{with} \ x \mapsto \sqrt[n]{x}$$

means **n-th root function** $(n \in \mathbb{N})$.

Special case: Power functions with an odd exponent $p(x) = x^{2m+1}$, $m \in \mathbb{N}$, are strictly increasing on $\mathbb{R}$. Therefore, the inverse function exists on $\mathbb{R}$

$$\omega : \mathbb{R} \to \mathbb{R} \quad \text{with} \quad x \mapsto \sqrt[2m+1]{x}.$$

Visualization: The animation shows the transition of a power function with an odd exponent to the corresponding root function as the inverse function. It can be seen that the inverse function is now defined for all real numbers.

Examples 4.35:
① $p_2 : \mathbb{R}_{\geq 0} \to \mathbb{R}_{\geq 0}$ with $x \mapsto x^2$ has the inverse function

$$\omega_2 : \mathbb{R}_{\geq 0} \to \mathbb{R}_{\geq 0} \quad \text{with} \quad x \mapsto \sqrt[2]{x} = x^{\frac{1}{2}}.$$

② $p_3 : \mathbb{R} \to \mathbb{R}$ with $x \mapsto x^3$ has the inverse function

$$\omega_3 : \mathbb{R} \to \mathbb{R} \quad \text{with} \quad x \mapsto \sqrt[3]{x} = x^{\frac{1}{3}}.$$

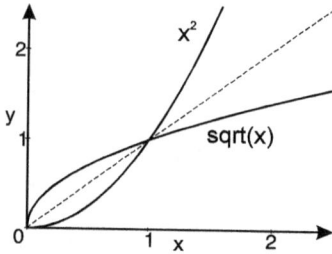

① Graph of x^2 and $\sqrt{x}$

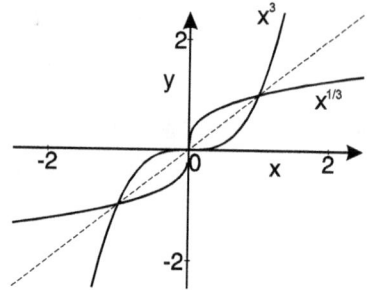

② Graph of x^3 and $x^{\frac{1}{3}}$

Application Example 4.36 (Velocity of a Free Fall)

A mass m falls freely from a height h_0 with an initial velocity $v_0 = 0$. At each time t of the motion, the total energy (sum of kinetic and potential energy) remains constant:

$$\left. \begin{array}{l} E\,(t = 0) = m \cdot g \cdot h_0 + \frac{1}{2} m\, v_0^2 = m \cdot g \cdot h_0 \\ E\,(t > 0) = m \cdot g \cdot h + \frac{1}{2} m\, v^2 \end{array} \right\} \quad E\,(t = 0) = E\,(t > 0).$$

$$\Rightarrow m \cdot g \cdot h_0 = m \cdot g \cdot h + \frac{1}{2} m\, v^2.$$

The velocity v at the height h is given by the square root function

$$\boxed{v = \sqrt{2g\,(h_0 - h)}.}$$

$\square$

⊘ Power Function with Rational Exponent

Definition: *A function*

$$f : \mathbb{R}_{>0} \to \mathbb{R} \quad \text{with} \quad f\,(x) = \sqrt[n]{x^m} = x^{\frac{m}{n}} \qquad m \in \mathbb{Z},\, n \in \mathbb{N}$$

means power function with rational exponent.

In Section §4.5 we will extend this concept of the power function to any real exponent using the exponential and logarithmic functions.

4.5 Exponential and Logarithmic Functions

One of the most important functions in physics is the exponential function. We will introduce the exponential function and its inverse function, the logarithmic function, and discuss their most important properties.

4.5.1 Exponential Function

Definition: *The function*

$$\exp : \mathbb{R} \to \mathbb{R} \quad with \quad x \mapsto e^x$$

is the **Exponential Function.** *$e \approx 2.718281828$ is Euler's number.*

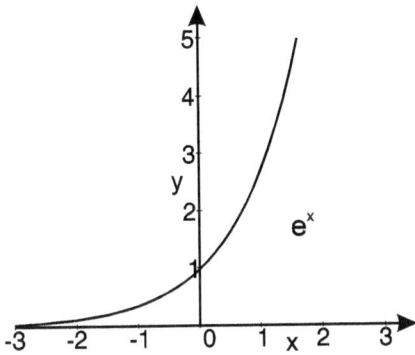

Figure 4.18.
Graph of the exponential function e^x

Properties of the Exponential Function:

	e^x
Domain	$\mathbb{R}$
Range	$\mathbb{R}_{>0}$
Monotonicity	strictly increasing
Asymptote	$y = 0$ for $x \to -\infty$

The following **rules** apply to the exponential function:

Properties of the Exponential Function

(1) $e^0 = 1$

(2) $e^{x+y} = e^x \cdot e^y$

(3) $e^{-x} = (e^x)^{-1}, \quad e^{nx} = (e^x)^n$

Examples 4.37:

① The functions $f_a(x) = e^{ax}$ with $a > 0$ behave qualitatively like the exponential function e^x: For $x \to \infty$ these functions go to infinity and for $x \to -\infty$ these functions go to zero.

② The functions $f_a(x) = e^{-ax}$ with $a > 0$ behave qualitatively like the exponential function e^{-x}: For $x \to \infty$ these functions go to zero and for $x \to -\infty$ these functions go to infinity.

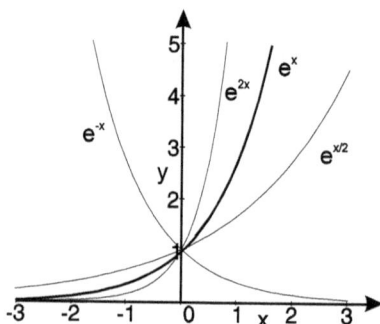

Figure 4.19.
Graphs of exponential functions

Application Example 4.38 (Exp. Functions in Applications).

① **Radioactive Decay:** When radioactive nuclei decay, the number $n(t)$ of nuclei that have not yet decayed at time t is described by the decay law

$$n(t) = n_0 \, e^{-\lambda t}.$$

Where n_0 is the number of atomic nuclei initially present at $t = 0$. $\lambda > 0$ is the decay constant which characterizes the time behavior of the decay.

② **Discharge of a Capacitor Plate:** When a plate capacitor is discharged, the voltage across the capacitor $U(t)$ at the time t is given by

$$U(t) = U_0 \, e^{-\frac{1}{RC}t}.$$

Where U_0 is the capacitor voltage at time $t = 0$, C is the capacitance, and R is the ohmic resistance.

4.5.2 Logarithmic Functions

The exponential function $\exp : \mathbb{R} \to \mathbb{R}_{>0}$ with $x \mapsto e^x$ is strictly increasing on the whole domain. Therefore, the inverse function exists on its range $\mathbb{R}_{>0}$.

Definition: *The inverse function of the exponential function is called* **Natural Logarithm**:

$$\ln : \mathbb{R}_{>0} \to \mathbb{R} \quad with \quad x \mapsto \ln(x).$$

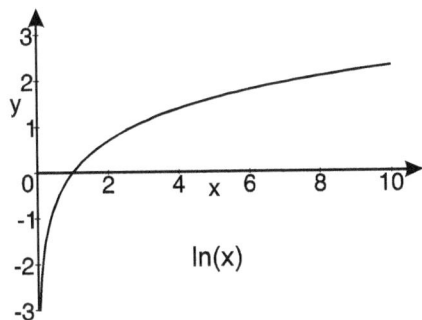

ln(x)

Figure 4.20. Graph of the logarithmic function $\ln(x)$

Properties of the Logarithmic Function are:

	$\ln(x)$
Domain	$\mathbb{R}_{>0}$
Range	$\mathbb{R}$
Zeros	$x_0 = 1$
Monotony	strictly increasing
Asymptotes	$x = 0$

Properties of the Logarithmic Function

(1) $\ln(1) = 0$

(2) $\ln(x \cdot y) = \ln x + \ln y$

(3) $\ln(x^n) = n \ln x$

(4) $\ln(e^x) = x$ or $e^{\ln x} = x$

Calculation Rules. The calculation rules follow directly from those for the exponential function for $x, y \in \mathbb{R}_{>0}$.

Application Example 4.39 (Logarithmic Function).

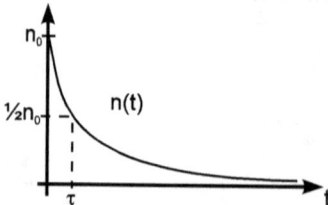

① **Half-life time τ of a radioactive substance:** The half-life time τ of a radioactive substance is the time it takes for half of the radioactive nuclei to decay: $n(\tau) = \frac{1}{2}n_0$. According to the Example 4.38 ① it is $n(t) = n_0 e^{-\lambda t}$ for $t = \tau$:

$$\frac{1}{2}n_0 = n_0 e^{-\lambda \tau}.$$

We divide by n_0, take the logarithm and solve for τ

$$\ln \frac{1}{2} = \ln\left(e^{-\lambda \tau}\right) = -\lambda \tau$$

$$\hookrightarrow \tau = -\frac{1}{\lambda}\ln\frac{1}{2} = -\frac{1}{\lambda}\left(\ln 1 - \ln 2\right).$$

The half-life time τ is then calculated as follows

$$\tau = \frac{1}{\lambda}\ln 2.$$

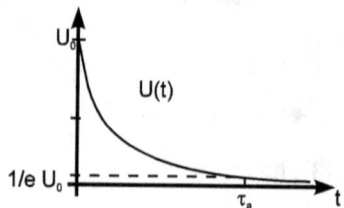

② **Discharge Time of a Capacitor:** The discharge time τ_a of a capacitor is the time it takes for the voltage across the capacitor to drop to, say, $\frac{1}{e}$-th of the maximum voltage value: $U(\tau_a) = \frac{1}{e}U_0$. According to Example 4.38 ② it is $U(t) = U_0 e^{-\frac{1}{RC}t}$ for $t = \tau_a$:

$$\frac{1}{e}U_0 = U_0 e^{-\frac{1}{RC}\tau_a} \hookrightarrow \ln\frac{1}{e} = \ln(e^{-\frac{1}{RC}\tau_a}) = -\frac{1}{RC}\tau_a.$$

The discharge time of the capacitor τ_a is therefore

$$\tau_a = -RC\ln(\frac{1}{e}) = RC.$$

① Half-life time τ ② Discharge time of a capacitor

Example 4.40. We look for the inverse function of

$$f : \mathbb{R} \to \mathbb{R}_{>0} \quad \text{with} \quad x \mapsto f(x) = 3e^{2x-1}.$$

The function f is strictly increasing on its entire domain, so the inverse function exists on its range $\mathbb{R}_{>0}$. We define

$$y = 3e^{2x-1}$$

and solve the expression in terms of x:

$$\frac{1}{3}y = e^{2x-1} \hookrightarrow \ln\frac{1}{3}y = 2x - 1 \hookrightarrow x = \frac{1}{2}(\ln(\tfrac{1}{3}y) + 1).$$

After swapping the variables, we obtain the inverse function

$$g : \mathbb{R}_{>0} \to \mathbb{R} \quad \text{with} \quad x \mapsto g(x) = \frac{1}{2}(\ln(\tfrac{1}{3}x) + 1). \qquad \square$$

4.5.3 General Power and Exponential Function

The exponential and logarithmic functions allow us to define the general power and exponential functions. Consider the general power $y = x^{\alpha}$. Applying the logarithmic function and its power rule, we get

$$\ln(y) = \ln(x^{\alpha}) = \alpha \ln(x).$$

Applying the exponential function to both sides, and remembering that the exponential and logarithm are inverse functions of each other, we get

$$e^{\ln(y)} = e^{\alpha \ln(x)} \quad \Rightarrow \quad y = e^{\alpha \ln(x)}.$$

We proceed in the same way with the general exponential function $y = a^{x}$:

$$\ln(y) = \ln(a^{x}) = x \ln(a) \quad \Rightarrow \quad e^{\ln(y)} = e^{x \ln(a)} \quad \Rightarrow \quad y = e^{x \ln(a)}.$$

General Power and Exponential Functions

(1) *The General Power Function is*

$$f : \mathbb{R}_{>0} \to \mathbb{R} \quad \text{with} \quad x \mapsto f(x) = x^{\alpha} := e^{\alpha \ln x}.$$

(2) *The General Exponential Function is*

$$f : \mathbb{R} \to \mathbb{R}_{>0} \quad \text{with} \quad x \mapsto f(x) = a^{x} := e^{x \ln a} \quad (a > 0).$$

4.6 Trigonometric Functions

The sine and cosine functions are used to describe periodic processes. To represent an alternating voltage $u(t) = u_0 \sin(\omega t + \varphi)$, the general sine function must be used. In this section we will introduce the trigonometric functions sin, cos, tan, cot together with their inverse functions, discuss important properties and illustrate the use of the trigonometric functions with application examples.

4.6.1 Fundamental Concepts

In science and technology *periodic processes* play an important role. They are characterized by the fact that a certain state repeats itself regularly, e.g. in acoustic and electromagnetic oscillations; oscillations of a string or spring; orbits of a satellite. Periodic functions of particular importance are the **Trigonometric Functions** or *angle functions*.

The measurement of angles is based on different units:

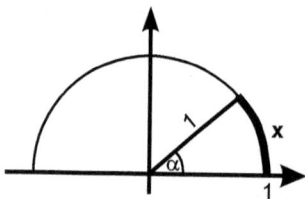

Figure 4.21. Angle in degree and radian measure

- *Degree measure* α ($360°$ for the full circle)

- *Radian measure* x (2π for the full circle)
  = Length of the segment on the unit circle, that cuts the angle α.

The angle value, measured in radians, is called x. x is positive if the angle is measured counter-clockwise and negative if it is measured clockwise. The unit of angle measurement is called *Radian (rad)*.

The relationship between the angle unit α in degrees and x in radians is as follows

$$\frac{\alpha}{360°} = \frac{x}{2\pi}.$$

4.6.2 Sine and Cosine Functions

For a given angle x (in radians), two important trigonometric functions can be associated according to the following definition:

Definition: *The **Sine** or **Cosine** of an angle x ($\sin x$ or $\cos x$) means the ordinate or abscissa of the intersection of the free side of the angle x with the unit circle.*

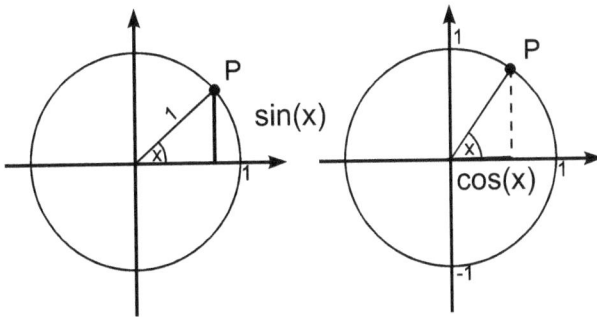

Figure 4.22. Sine (ordinate of P) and cosine (abscissa of P) on the unit circle

With this definition in the unit circle we get the trigonometric functions sine and cosine as functions on $\mathbb{R}$:

$$\sin : \quad \mathbb{R} \to \mathbb{R} \quad \text{with} \quad x \mapsto \sin(x)$$
$$\cos : \quad \mathbb{R} \to \mathbb{R} \quad \text{with} \quad x \mapsto \cos(x).$$

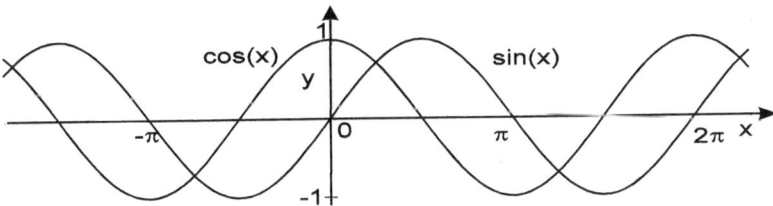

Figure 4.23. Sine and cosine functions

Visualization: The animation shows how the point of rotation P is projected onto the x-axis and y-axis, respectively. The sine function is then the projection on the x-axis and the cosine on the y-axis.

Properties of the Sine and Cosine:

	$f(x) = \sin x$	$f(x) = \cos x$	
Domain	$\mathbb{R}$	$\mathbb{R}$	
Range	$[-1,\,1]$	$[-1,\,1]$	
Period	2π	2π	
Symmetry	odd	even	
Zeros	$x_k = k \cdot \pi$	$x_k = \frac{\pi}{2} + k\,\pi$	
Relative maximum	$x_k = \frac{\pi}{2} + k \cdot 2\pi$	$x_k = k \cdot 2\pi$	
Relative minimum	$x_k = \frac{3}{2}\pi + k \cdot 2\pi$	$x_k = \pi + k\,2\pi$	$k \in \mathbb{Z}$

The General Sine Function. In the applications, sine and cosine occur not only with the argument x, but also in a more general form

$$y(x) = a \sin(bx + c) + d \quad \text{or} \quad y(x) = a \cos(bx + c) + d.$$

We will discuss the meaning of each of these parameters below:

(1) The Meaning of a: The factor a in

$$y(x) = a \cdot \sin(x)$$

gives the **maximum amplitude** of the function. So the range of this function is $\mathbb{W} = [-a,\,a]$.

Example 4.41. $y(x) = 2 \cdot \sin(x) \quad \Rightarrow \quad$ Amplitude 2.

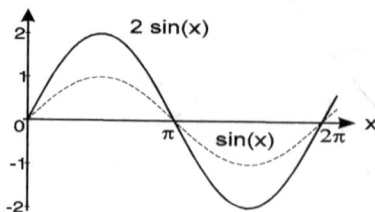

Figure 4.24. Amplitude a

(2) The Meaning of b**:** The factor b in

$$y\,(x) = \sin\,(b\,x)$$

results in a change of the period compared to the pure sine function. The period p of $\sin\,(b\,x)$ is obtained when the argument of the sine returns the third zero, i.e. for

$$b\,p = 2\pi \quad \Rightarrow \quad \boxed{p = \tfrac{2\pi}{b}} \text{ Period.}$$

⚠ **Caution:** For $b > 1$ the period decreases and for $b < 1$ it increases.

Example 4.42. $y\,(x) = \sin\,(2x) \quad \Rightarrow \quad$ Period: $p = \tfrac{2\pi}{2} = \pi$.

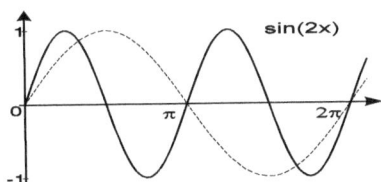

Figure 4.25. Period $p = \tfrac{2\pi}{b}$

(3) The Meaning of c**:** The constant c in

$$y\,(x) = \sin\,(x + c)$$

results in a **shift** of the sine function **along the** x**-axis.** c is also called the *phase* or *zero phase*. The first zero of $\sin(x + c)$ is found when the argument $x + c$ becomes zero:

$$\sin\,(x + c) = 0 \quad \Rightarrow \quad x_0 + c = 0 \quad \Rightarrow \quad x_0 = -c.$$

⚠ **Caution:** For $c > 0$ the curve is shifted to the left by x_0.

⚠ **Caution:** For $c < 0$ the curve is shifted to the right by x_0.

Example 4.43. $y = \sin(x + \pi/2)$: the curve is shifted to the left by $\tfrac{\pi}{2}$.

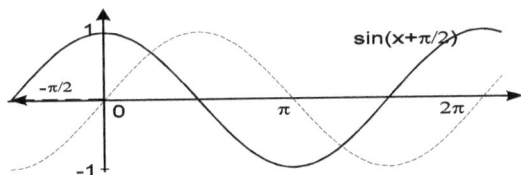

Figure 4.26. Shift along the x-axis

(4) The Meaning of d: The constant d in

$$y(x) = \sin(x) + d$$

results in a **shift** of the sine function **along the y-axis** by the value d.

Example 4.44. $y(x) = \sin(x) + 2 \Rightarrow$ Shift by 2 in the direction of y:

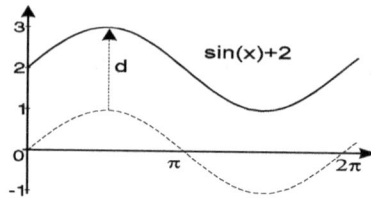

Figure 4.27. Shift along the y-axis

So we have discussed the parameters that occur in the general sine function separately and described their meaning. Finally, we will discuss the situation where all the parameters occur at the same time:

(5) Discussion of the General Sine Function:

$$y(x) = a \sin(bx + c) = a \sin\left(b\left(x + \tfrac{c}{b}\right)\right)$$

The amplitude a is taken directly from the expression of the general sine function. The period p is calculated from b using $p = \frac{2\pi}{b}$. The zero phase c can be read from the first zero, while the phase shift $-\frac{c}{b}$ can be identified from the second zero. In summary, the following table is obtained

Period:	$p = \frac{2\pi}{b}$
1. Zero:	$x_0 = -\frac{c}{b}$
2. Range:	$-a \le y(x) \le a$
3. (Zero) Phase:	c
4. Shift:	$-\frac{c}{b}$

together with the graph of the general sine function which is shown in Fig. 4.28.

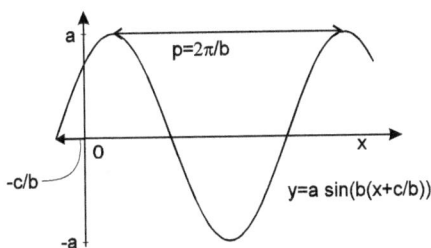

Figure 4.28. General sine function

Example 4.45. $u(t) = 2 \sin\left(\frac{1}{2}t + \frac{1}{2}\pi\right)$. Based on the sine function $\sin(t)$, we double the amplitude: $2 \sin(t)$. Then we change the period to $p = \frac{2\pi}{\frac{1}{2}} = 4\pi$ and get $2 \sin\left(\frac{1}{2}t\right)$. We look at the phase shift of $-\pi$, since $2 \sin\left(\frac{1}{2}t + \frac{1}{2}\pi\right) = 2 \sin\frac{1}{2}(t + \pi)$.

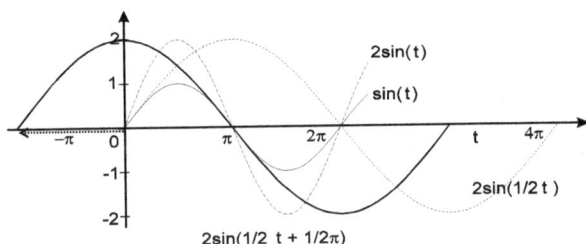

Application Example 4.46 (**Harmonic Resonance**).

Air columns of length L can be excited into harmonic oscillations (*resonance oscillations*) under suitable conditions. The shape of the oscillation is determined by the boundary conditions (open end or fixed end). We determine the parameter k for the case of fixed/fixed ends

$$f(x) = A \sin(kx + \varphi).$$

So we are looking for oscillations with the boundary conditions $f(0) = 0$ and $f(L) = 0$. From $f(0) = 0$ follows $\sin(\varphi) = 0 \to \varphi = 0$ or $\varphi = \pi$. This gives the possible oscillations

$$f_+(x) = A \sin(kx) \quad \text{and} \quad f_-(x) = A \sin(kx + \pi) = -A \sin(kx).$$

From $f(L) = 0$ we get $\sin(kL) = 0 \to kL = n\pi \quad \Rightarrow \quad k = n\frac{\pi}{L}, \, n \in \mathbb{N}_0$. All parameters k are allowed that are multiples of $\frac{\pi}{L}$.   □

4.6.3 Tangent and Cotangent Functions

Based on the sine and cosine functions, we will use the geometric interpretation of the tangent and cotangent functions as the quotient of sine and cosine or of cosine and sine, analogous to the geometric interpretation. Note, however, that the zeros of the denominator must be excluded from the domain.

Definition:

$\tan : \mathbb{R} \setminus \{\frac{\pi}{2} + k \cdot \pi, \quad k \in \mathbb{Z}\} \to \mathbb{R}$ with $x \mapsto \tan(x) := \dfrac{\sin(x)}{\cos(x)}$

is called **Tangent Function**.

$\cot : \mathbb{R} \setminus \{k \cdot \pi, \quad k \in \mathbb{Z}\} \to \mathbb{R}$ with $x \mapsto \cot(x) := \dfrac{\cos(x)}{\sin(x)}$

is called **Cotangent Function**.

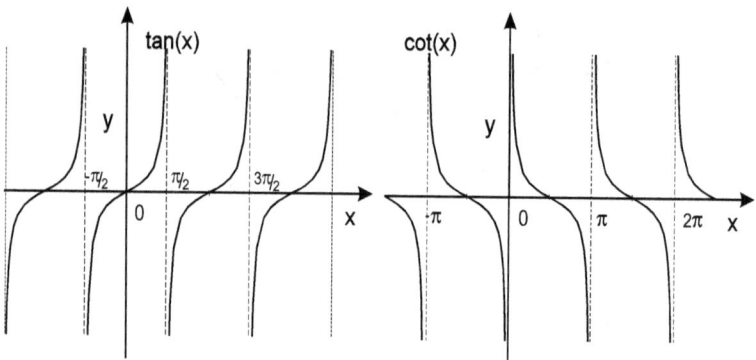

Figure 4.29. Tangent and cotangent functions

Properties of the Tangent and Cotangent Functions:

	$f(x) = \tan(x)$	$f(x) = \cot(x)$
Domain	$\mathbb{R} \setminus \{\frac{\pi}{2} + k \cdot \pi, k \in \mathbb{Z}\}$	$\mathbb{R} \setminus \{k \cdot \pi, k \in \mathbb{Z}\}$
Range	$\mathbb{R}$	$\mathbb{R}$
Period	π	π
Symmetry	odd	odd
Zeros	$x_n = n \cdot \pi$	$x_n = \frac{\pi}{2} + n \cdot \pi$
Pole	$x_n = \frac{\pi}{2} + n\pi$	$x_n = n \cdot \pi$

$n \in \mathbb{Z}$

4.6.4 Important Trigonometric Formulas

The following important relationships apply to the trigonometric functions $(x, x_1, x_2 \in \mathbb{R})$:

(1) Symmetry behavior

$$\sin(-x) = -\sin(x) \qquad\qquad \cos(-x) = \cos(x)$$

$$\tan(-x) = -\tan(x) \qquad\qquad \cot(-x) = -\cot(x)$$

(2) Shift identities

$$\sin(x) = \cos\left(x - \tfrac{\pi}{2}\right) \qquad\qquad \cos(x) = \sin\left(x + \tfrac{\pi}{2}\right)$$

(3) According to the theorem of Pythagoras

$$\sin^2(x) + \cos^2(x) = 1$$

(4) The following addition theorems apply

$$\sin(x_1 \pm x_2) = \sin x_1 \cos x_2 \pm \cos x_1 \sin x_2$$

$$\cos(x_1 \pm x_2) = \cos x_1 \cos x_2 \mp \sin x_1 \sin x_2$$

$$\tan(x_1 \pm x_2) = \frac{\tan x_1 \pm \tan x_2}{1 \mp \tan x_1 \tan x_2}$$

(5) The addition theorems lead to other commonly used formulas

$$\sin(2x) = 2 \cdot \sin(x) \cos(x) \qquad\qquad \sin(3x) = 3\sin(x) - 4\sin^3(x)$$

$$\cos(2x) = \cos^2(x) - \sin^2(x) \qquad\qquad \cos(3x) = -3\cos(x) + 4\cos^3(x)$$

(6) and

$$\sin^2(x) = \tfrac{1}{2}\left(1 - \cos(2x)\right) = \frac{\tan^2(x)}{1 + \tan^2(x)}$$

$$\cos^2(x) = \tfrac{1}{2}\left(1 + \cos(2x)\right) = \frac{1}{1 + \tan^2(x)}.$$

The formulas (4) are fundamental. All other relationships are consequences of these addition theorems.

4.7 Arc Functions

The trigonometric functions assign to each x exactly one function value y. However, the inverse problem often arises: The function value of a trigonometric function is given and the corresponding argument (angle) has to be found.

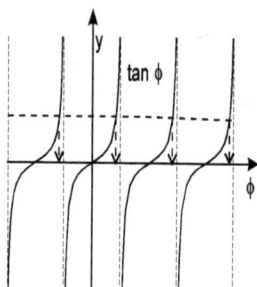

Example 4.47. In an RL circuit, the phase φ between the applied input voltage

$$U_E(t) = U_0 \sin(\omega t)$$

and the output voltage

$$U_A(t) = \frac{R U_0}{\sqrt{R^2 + (\omega L)^2}} \sin(\omega t + \varphi)$$

measured on an ohmic resistor is given by

$$\tan \varphi = -\frac{\omega L}{R}.$$

We are looking for the phase φ. The inverse of the tangent function is not unique because the tangent, like all other trigonometric functions, is periodic. Therefore, the domain must be restricted so that a monotone function is obtained on the restricted domain. The inverse functions of the trigonometric functions are called **Arc functions.**

⚠ **Attention:** In general, the trigonometric functions cannot be inverted because they are periodic functions. However, if the domain is restricted to an interval in which the functions behave strictly monotonously, the trigonometric functions are invertible on this restricted interval. The domain restrictions chosen below give the so-called *principal value* of the functions.

Note: Since the arc functions are the inverse functions of the trigonometric functions, the graph of these arc functions is obtained by **mirroring the corresponding trigonometric function on the identity line** $y = x$. The domain and the range of the arc functions are given by the range and the domain of the corresponding trigonometric functions.

4.7.1 Arcsine Function

The restricted sine function

$$\sin : \left[-\frac{\pi}{2}, \frac{\pi}{2}\right] \to [-1, 1] \quad \text{with} \quad x \mapsto \sin x$$

is strictly increasing in the interval $\left[-\frac{\pi}{2}, \frac{\pi}{2}\right]$. So the inverse function **arcsine** exists.

$$\arcsin : [-1, 1] \to \left[-\frac{\pi}{2}, \frac{\pi}{2}\right] \quad \text{with} \quad x \mapsto \arcsin(x).$$

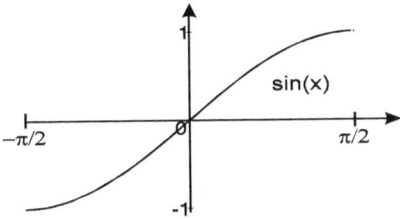

Sine function

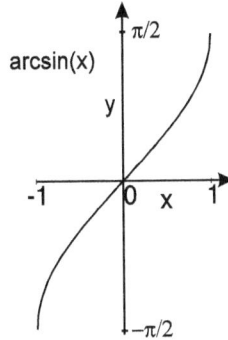

Arcsine function

Properties of the Arcsine Function:

	$\sin(x)$	$\arcsin(x)$
Domain	$\left[-\frac{\pi}{2}, \frac{\pi}{2}\right]$	$[-1, 1]$
Range	$[-1, 1]$	$\left[-\frac{\pi}{2}, \frac{\pi}{2}\right]$
Zeros	$x_0 = 0$	$x_0 = 0$
Symmetry	odd	odd
Monotony	Strictly increasing	Strictly increasing

Examples 4.48:

① $\arcsin 0 = 0, \quad \arcsin\left(\frac{1}{2}\right) = \frac{\pi}{6}, \quad \arcsin\left(\frac{1}{2}\sqrt{2}\right) = \frac{\pi}{4}.$

② Find all solutions to the equation $\sin x = 0.5$.
From $\sin x = 0.5$ we get $x = \arcsin(0.5) = \frac{\pi}{6}$. Due to $\sin(\pi - x) = \sin x$ we get $\pi - x = \frac{\pi}{6}$, so $x = \frac{5}{6}\pi$ is also a solution of the equation. Because of the periodicity of $\sin x = \sin(x + 2\pi)$ we get
$$\mathbb{L} = \{x \in \mathbb{R} : x = \frac{\pi}{6} + k \cdot 2\pi \text{ or } x = \frac{5}{6}\pi + k \cdot 2\pi \quad \text{with } k \in \mathbb{Z}\}.$$

③　Sine and arcsine cancel each other out: Because

$$\arcsin(\sin(x)) = x$$

for all $x \in \left[-\frac{\pi}{2}, \frac{\pi}{2}\right]$ and $\sin(\arcsin(y)) = y$ for $y \in [-1, 1]$!

4.7.2 Arccosine Function

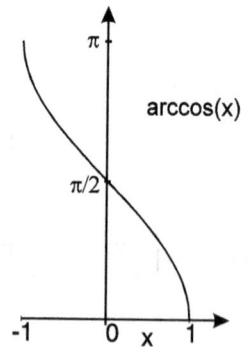

The restricted cosine function

$$\cos : [0, \pi] \to [-1, 1] \quad \text{with} \quad x \mapsto \cos x$$

is strictly decreasing in the interval $[0, \pi]$. Therefore, the inverse function **Arccosine** exists.

The range of the cosine function becomes the domain of the arccosine; the domain of the cosine function becomes the range of the arccosine:

$$\arccos : [-1, 1] \to [0, \pi] \quad \text{with} \quad x \mapsto \arccos(x).$$

The graph of the arccosine is obtained by mirroring the cosine function on the identity line:

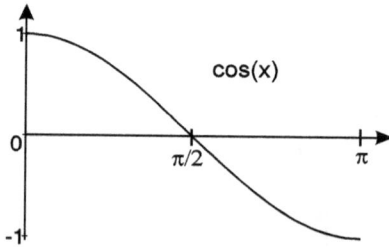

cos(x)

arccos(x)

Cosine function

Arccosine function

Properties of the Arccosine Function:

	$\cos(x)$	$\arccos(x)$
Domain	$[0, \pi]$	$[-1, 1]$
Ranges	$[-1, 1]$	$[0, \pi]$
Zeros	$\frac{\pi}{2}$	1
Monotony	Strictly decreasing	Strictly decreasing

Cosine and arccosine also cancel each other out: Because

$$\arccos(\cos(x)) = x$$

for all $x \in [0, \pi]$ and $\cos(\arccos(y)) = y$ for all $y \in [-1, 1]$!

Example 4.49. $\arccos(1) = 0, \quad \arccos(\frac{1}{2}) = \frac{\pi}{3}, \quad \arccos(-0.237) = 1.8101.$

4.7.3 Arctangent Function

The restricted tangent function

$$\tan : \left(-\frac{\pi}{2}, \frac{\pi}{2}\right) \to \mathbb{R} \quad \text{with} \quad x \mapsto \tan x$$

is strictly increasing in the interval $\left(-\frac{\pi}{2}, \frac{\pi}{2}\right)$ and has the range $\mathbb{R}$. Therefore, the inverse function **Arctangent** exists.

$$\arctan : \mathbb{R} \to \left(-\frac{\pi}{2}, \frac{\pi}{2}\right) \quad \text{with} \quad x \mapsto \arctan(x).$$

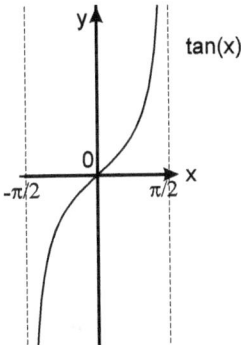

Tangent function

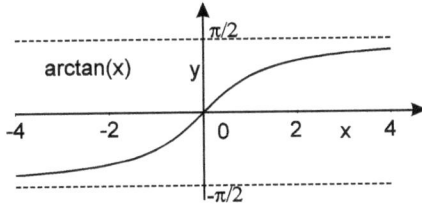

Arctangent function

Properties of the Arctangent function:

	$\tan(x)$	$\arctan(x)$
Domain	$\left(-\frac{\pi}{2}, \frac{\pi}{2}\right)$	$\mathbb{R}$
Range	$\mathbb{R}$	$\left(-\frac{\pi}{2}, \frac{\pi}{2}\right)$
Zeros	$x_0 = 0$	$x_0 = 0$
Symmetry	odd	odd
Monotony	Strictly increasing	Strictly increasing
Asymptotes	$x = \pm\frac{\pi}{2}$	$y = \pm\frac{\pi}{2}$

The graph of the arctangent function is obtained by mirroring the graph of the tangent function on the identity line. Thus, the poles at $x = \pm\frac{\pi}{2}$ are transformed from tangents to asymptotes at $y = -\frac{\pi}{2}$ for $x \to -\infty$ and $y = \frac{\pi}{2}$ for $x \to \infty$ of the arctangent.

Examples 4.50:

① $\arctan(1) = \frac{\pi}{4}$, $\arctan(-3\pi) = -1,4651$.

② Find all solutions of $\tan x = \sqrt{3}$. With $x = \arctan\sqrt{3} = \frac{\pi}{3}$ because of the periodicity of the tangent:

$$\mathbb{L} = \{x \in \mathbb{R} : x = \frac{\pi}{3} + k \cdot \pi, \quad k \in \mathbb{Z}\}.$$

4.7.4 Arccotangent Function
The restricted cotangent function

$$\cot : (0, \pi) \to \mathbb{R} \text{ with } x \mapsto \cot(x)$$

is strictly decreasing in the interval $(0, \pi)$ and has the range $\mathbb{R}$. So the inverse function **Arccotangent** exists.

$$\operatorname{arccot} : \mathbb{R} \to (0, \pi) \text{ with } x \mapsto \operatorname{arccot}(x).$$

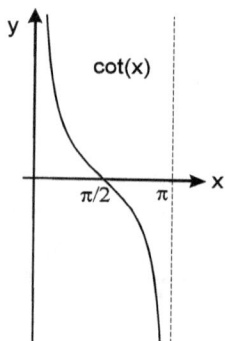

Cotangent function

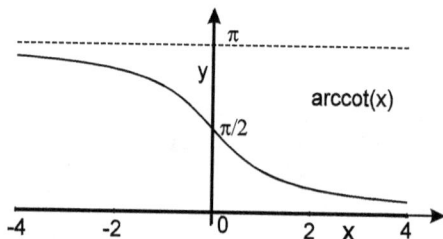

Arccotangent function

The graph of the arccotangent function is obtained by mirroring the graph of the cotangent function on the identity line, so the poles at $x = 0$ and $x = \pi$ are transformed into asymptotes at $y = \pi$ for $x \to -\infty$ and $y = 0$ for $x \to \infty$ of the arccotangent. The zero point of the cotangent at $x = \frac{1}{2}\pi$ becomes the intersection with the y-axis at $x = 0$.

Properties of the Arccotangent Function:

	$\cot(x)$	$\text{arccot}(x)$
Domain	$(0, \pi)$	$\mathbb{R}$
Range	$\mathbb{R}$	$(0, \pi)$
Zeros	$x_0 = \frac{\pi}{2}$	$-$
Monotony	Strictly decreasing	Strictly decreasing
Asymptotes	$x = 0$	$y = 0$
	$x = \pi$	$y = \pi$

Example 4.51. $\text{arccot}(0) = \frac{\pi}{2}$, $\quad \text{arccot}\, 1 = \frac{\pi}{4}$.

Remarks:

(1) On pocket calculators the elementary functions arcsin, arccos, arctan, arccot do not appear as tabs, but symbols such as INV and SIN can be used for the arcsin, or equivalent tab combinations for the other inverse trigonometric functions.

(2) The arccotangent as an independent function does not play a role in practice. It can simply be calculated using the arctangent, since

$$\text{arccot}(x) = \frac{\pi}{2} - \arctan(x)$$

(3) The following formulas apply to arc functions

$$\arccos(x) = \frac{\pi}{2} - \arcsin(x)$$

$$\arcsin(x) = \arctan \frac{x}{\sqrt{1 - x^2}} \qquad \text{for} \; -1 < x < 1.$$

(4) In addition to the functions sin, cos, tan and cot, there are infinitely many intervals in which the corresponding functions are strictly increasing. The inverse functions of the trigonometric functions are also called *cyclometric functions*.

(5) The relationship between the arc functions is as follows:

$$
\begin{aligned}
\arcsin(x) &= -\arcsin(-x) &= \frac{\pi}{2} - \arccos(x) &= \arctan \frac{x}{\sqrt{1-x^2}} \\
\arccos(x) &= \pi - \arccos(-x) &= \frac{\pi}{2} - \arcsin(x) &= \text{arccot} \frac{x}{\sqrt{1-x^2}} \\
\arctan(x) &= -\arctan(-x) &= \frac{\pi}{2} - \text{arccot}(x) &= \arcsin \frac{x}{\sqrt{1+x^2}} \\
\text{arccot}(x) &= \pi - \text{arccot}(-x) &= \frac{\pi}{2} - \arctan(x) &= \arccos \frac{x}{\sqrt{1+x^2}}
\end{aligned}
$$

4.8 Problems on Basic Functions

4.1   Determine domain and range of the following functions

a) $f(x) = \sqrt{x^2 - 1}$ b) $y = \ln |x|$ c) $f(x) = \dfrac{x^2}{4x^2 - 16}$

d) $f(x) = \dfrac{x-1}{x+1}$ e) $y = e^{|x|}$ f) $f(x) = \dfrac{x}{x^2 + 1}$

4.2   Determine the symmetry behavior and the domain of

a) $f(x) = 4x^2 - 16$ b) $f(x) = \dfrac{x^3}{x^2 + 1}$ c) $f(x) = \sin x \cdot \cos x$

d) $f(x) = |x^2 - 16|$ e) $f(x) = \dfrac{x^2 - 1}{1 + x^2}$ f) $f(x) = \dfrac{1}{x - 1}$

4.3   Check for monotony for

a) $y = x^4$ b) $y = \sqrt{x - 1}$ for $x \geq 1$ c) $y = x^3 + 2x$ d) $y = e^{2x}$

4.4   What is the inverse function of

a) $f : \mathbb{R}_{>0} \quad \to \quad ?$ b) $f : \mathbb{R}_{\geq 0} \quad \to \quad ?$

$\qquad x \quad \mapsto \quad y = \frac{1}{(2x)}$ $\qquad x \quad \mapsto \quad y = \sqrt{3x}$

c) $f : \mathbb{R} \quad \to \quad ?$ d) $f : \mathbb{R}_{>-1} \quad \to \quad ?$

$\qquad x \quad \mapsto \quad y = 2e^{x - \frac{1}{2}}$ $\qquad x \quad \mapsto \quad y = \frac{x-1}{x+1}$

4.5   Determine the polynomial of smallest degree that matches the points $(-3, 11), (-1, 7), (0, 5), (4, -3)$.

4.6   Determine the zeros of the functions:
a) $f(x) = x^3 + 2x^2 - 13x + 10$
b) $f(x) = x^3 - x^2 + 2$
c) $f(x) = x^4 - 2x^3 - 25x^2 + 50x$

4.7   Use Horner's scheme to evaluate the function value of function $f(x)$ at point x_0 for
a) $f(x) = x^3 - 2x^2 - 3x + 1$; $x_0 = 2$
b) $f(x) = 0.1x^4 + x^3 + 2x^2 - 4$; $x_0 = 3$.

4.8   Are there polynomials that do not have zeros at all?

4.9   Calculate with Newton's interpolation scheme the coefficients of the polynomial defined by the values
$(0, 1)$; $(1, 0)$; $(2, 5)$; $(-1, 2)$.

4.10   Determine the zeros of the function
a) $f(x) = 3x^3 + 3x^2 - 3x - 3$
b) $f(x) = x^4 - 13x^2 + 36$

4.11   Which polynomial of smallest degree interpolates the pairs
$(-1, 0)$; $(0, 1)$; $(1, 2)$; $(2, 6)$?

4.12 Factorize the polynomial with MAPLE.
$$2x^6 + 3x^5 - 63x^4 - 55x^3 + 657x^2 + 216x - 2160$$

4.13 Determine with MAPLE all zeros of
a) $7x^4 - 59x^3 + 19x^2 + 166x - 1008$
b) $x^3 - x^2 - 100x + 310$

4.14 Use MAPLE to evaluate the polynomials from task 4.13 at point $x_0 = 4$, either by converting the respective expression into a function with **unapply** or by inserting this point into the polynomial with the **subs** command.

4.15 Draw the function $x^3 - x^2 - 100x + 310$ with MAPLE and extract the extreme values from the diagram.

4.16 Use MAPLE to factorize the polynomial as much as possible. Create Horner's scheme for this polynomial. Determine the degree of the polynomial.
$$-17x^6 + 11x^4 - 20x^3 + 13x^2 - 3x + 56x^7 + 4x^5 - 15x^8 + 35x^9$$

4.17 Where are the function's zeros and singularities?
a) $y = \dfrac{x^2 + x - 2}{x - 2}$
b) $y = \dfrac{x^3 - 5x^2 - 2x + 24}{x^3 + 3x^2 + 2x}$
c) $y = \dfrac{x^2 - 2x + 1}{x^2 - 1}$
d) $y = \dfrac{(x - 1)}{(x - 1)^2\,(x + 1)}$

4.18 Determine for the following rational functions: Zeros, poles, asymptotes in infinity. Draw the function graph as well as the asymptotes.
a) $y = \dfrac{x^2-4}{x^2+1}$
b) $y = \dfrac{x^3-6x^2+12x-8}{x^2-4}$
c) $y = \dfrac{x^3-5x^2+8x-4}{x^3-6x^2+12x-8}$
d) $y = \dfrac{(x-1)^2}{(x+1)^2}$

4.19 Determine the zero and pole points of the function
$$h(x) = \frac{x^3 - 6x^2 - 12x + 49}{(x - 2)(x - 7)}$$
by decomposing the numerator and denominator into linear factors. Determine the asymptotes and draw the function together with their asymptotes in a diagram. Draw the function near the zeros and poles.

4.20 Given is the rational expression $\dfrac{x^4+x^3-4x^2-4x}{x^4+x^3-x^2-x}$. Convert this function into the following expressions:
a) $\dfrac{(x+2)(x+1)(x-2)}{x^3+x^2-x-1}$
b) $\dfrac{x^4+x^3-4x^2-4x}{x(x-1)(x+1)^2}$
c) $\dfrac{(x+2)(x-2)}{(x-1)(x+1)}$
d) $\dfrac{x^2}{(x-1)(x+1)} - 4\dfrac{1}{(x-1)(x+1)}$

4.21 If a capacitor with capacitance C is discharged via an ohmic resistor R, its charge Q decreases exponentially with time:
$$\boxed{Q = Q_0\, e^{-\frac{1}{RC}t}.}$$
At what point does the charge drop below 10 % of its initial value Q_0?

4.22 Circuit with inductance L and ohmic resistance R. When a DC voltage source is switched on, the current reaches the final value i_∞ as expected after some time.

$$i\,(t) = i_\infty \left(1 - e^{-\frac{R}{L}\,t}\right).$$

Calculate for $i_0 = 4\,A$, $R = 5\,\Omega$, $L = 2.5\,H$ the time at which the current reaches 95 % of the final state. Sketch the current-time function.

4.23 Determine the parameters a and b of the function $y = a\,e^{-b\,x} + 2$ so that the points $A = (0,\,10)$ and $B = (5,\,3)$ lie on the curve.

4.24 Solve the following exponential equations
a) $e^{x^2 - 2\,x} = 2$ b) $e^x + 2\,e^{-x} = 3$

4.25 Which solution has the logarithmic equation
$\ln \sqrt{x} + 1.5 \cdot \ln\,(\,x\,) = \ln(2x)$?

4.26 Given is a damped oscillation $x\,(t) = A\,e^{-\gamma\,t} \cdot \sin\,(\omega t + \varphi)$. By measuring the amplitudes of two consecutive oscillations, the damping γ can be determined. If $T = \frac{1}{100}\,s$ is the period of the damped oscillation, $x\,(t_0) = 200$ and $x\,(t_0 + T) = 100$ is the amplitude of two consecutive oscillations, then γ is to be calculated.

4.27 Convert from degrees to radians or from radians to degrees:

Degree	$40, 36°$		$278, 19°$
Radians		1.4171 −5.6213	

4.28 Derive from the addition theorem of the cosine function the formula

$$\sin^2 x + \cos^2 x = 1.$$

4.29 Draw the function $f\,(x) = 2 \cdot \cos\,(2\,x - \pi)$.

4.30 Determine the function's amplitude A, period p and phase shift φ:
a) $y = 2 \cdot \sin\left(3\,x - \frac{\pi}{6}\right)$ b) $y = 5 \cdot \cos\,(2\,x + 4.2)$
c) $y = 10 \cdot \sin\,(\pi\,x - 3\,\pi)$ d) $y = 2.4 \cdot \cos\left(4\,x - \frac{\pi}{2}\right)$

4.31 Sketch the function of the harmonic oscillation:
$f\,(t) = 2 \cdot \sin\,(2\,t - 4)$

4.32 Calculate
a) $\arcsin(1)$ b) $\arcsin(\frac{1}{2}\sqrt{2})$ c) $\arcsin(-\frac{1}{2}\sqrt{3})$ d) $\arcsin(0.481)$
e) $\arccos(\frac{1}{2})$ f) $\arccos(\frac{1}{2}\sqrt{3})$ g) $\arccos(-1)$ h) $\arccos(0.8531)$
i) $\arctan(1)$ k) $\arctan(-\sqrt{3})$ l) $\text{arccot}(-\frac{1}{\sqrt{3}})$ m) $\text{arccot}(\frac{1}{3}\sqrt{3})$

4.33 What is the value of x?
a) $\arcsin x = \pi/4$ b) $\arctan x = 0.7749$ c) $\arccos x = 1.021$
d) $\text{arccot}\,x = 2.9208$ e) $(\arccos x)^2 = 0.25$

4.34   Prove $\sin(\arccos(x)) = \sqrt{1 - x^2}$. (Instruction: Set $y = \arccos(x)$.)

4.35   Show within the domain that
   a) $\arcsin(a) = \arccos\sqrt{1 - a^2}$
   b) $\arccos(a) = \arcsin\sqrt{1 - a^2}$
   c) $\text{arccot}(a) = \arctan(\frac{1}{a})$
   d) $\arcsin(a) = \arctan\dfrac{a}{\sqrt{1-a^2}}$

4.36   Simplify   a) $\sin(\arcsin(x))$ b) $\cos(\arccos(x))$ c) $\sin(\arccos(x))$
                   d) $\cos(\arcsin(x))$ e) $\sin(\arctan(x))$ f) $\tan(\arccos(x))$

4.37   Determine the domain of the following functions
   a) $y = x + \arccos(x)$
   b) $y = \sqrt{x} + \arcsin(x)$
   c) $y = \frac{\pi}{2} + \arcsin(x - 1)$
   Plot the functions and determine their range.

Chapter 5
Complex Numbers

5

Complex numbers are an indispensable way of describing electrical circuits. Almost every textbook on circuit design has an introductory chapter that introduces complex numbers. One of the reasons for this is that when complex resistors are introduced, simple rules from DC networks are transferred to AC circuits.

The sum, difference, product and ratio of two complex numbers are not automatically defined by the construction of the complex numbers. It is necessary to declare these operations; but, of course, in such a way that for the special case that the imaginary part is zero, the operations and rules already defined in $\mathbb{R}$ must be valid.

5

5 Complex Numbers

Complex numbers are an indispensable way of describing electrical circuits. Almost every textbook on circuit design has an introductory chapter that introduces complex numbers. One of the reasons for this is that when complex resistors are introduced, simple rules of DC networks are transferred to AC circuits.

First, we discuss the basics of complex numbers within mathematics and start with a mathematical problem: As we saw already in Section 4.2 about polynomials, every polynomial of degree n has at most n different zeros. But already the quadratic polynomial $p(x) = x^2 + 1$ has *no* zeros in $\mathbb{R}$. If the equation $x^2 + 1 = 0$ is solved formally for x, the result would be

$$x_{1/2} = \pm\sqrt{-1} \quad \notin \mathbb{R}.$$

It is extraordinarily successful to extend the real numbers by introducing $\sqrt{-1}$ as a new unit:

$$i := \sqrt{-1} \qquad \textbf{(Imaginary Unit)}.$$

The term *Imaginary Unit* comes from the fact that the square root of any negative real number can be expressed as a real multiple of this unit:

$$\sqrt{-5} = \sqrt{-1 \cdot 5} = \sqrt{-1} \cdot \sqrt{5} = \sqrt{5}\,i.$$

All real multiples of i are called *Imaginary Numbers*. The combination of real and imaginary numbers results in the complex numbers:

Definition: (Complex Numbers). *Expressions of the form*

$$c := a + ib \quad \text{with} \ a, b \in \mathbb{R}$$

are called **Complex Numbers** *and* $\mathbb{C} := \{c = a + ib;\ a, b \in \mathbb{R}\}$ *the* **set of Complex Numbers.**

For $b = 0$ the number $c = a + 0\,i = a \in \mathbb{R}$. So the real numbers are included in the complex numbers. The mathematical meaning of the complex numbers is that every polynomial of degree n has exactly n zeros ($\rightarrow$ 5.2.7 *Fundamental Theorem of Algebra*).

5.1 Representation of Complex Numbers

Each **real** number corresponds to a point on the number line:

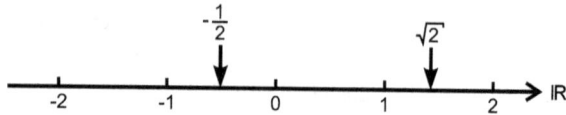

By defining a complex number as a "pair" $c = a + ib$, it has two "components": a purely real component a and an imaginary component ib. To represent complex numbers, the system enters the number plane.

5.1.1 Algebraic Form (Cartesian Form)

Complex numbers

$$c := a + ib \quad \text{with} \ a, b \in \mathbb{R}$$

can be represented by **two** numbers in a plane (Fig. 5.1): If a coordinate system with abscissa a (multiple of unit 1) and ordinate ib (multiple of unit i) is chosen, then every complex number is a point of this plane, the so-called *Gaussian Number Plane*.

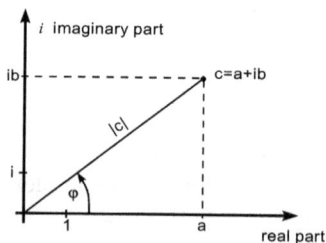

Figure 5.1. Representation of complex numbers $c = a + ib$.

We call

$$a = \operatorname{Re}(c) \text{ the real part of } c$$
$$b = \operatorname{Im}(c) \text{ the imaginary part of } c.$$

⚠ **Caution:** Both the real and imaginary parts of a complex number are real numbers. Note therefore: The imaginary part of a complex number $c = a + ib$ is **not** ib, but only the real value $\operatorname{Im}(c) = b$.

This representation of the complex number is called

$$c = a + ib \qquad \textbf{Algebraic Form (Cartesian Form)}$$

with real part and imaginary parts. As **Magnitude** of a complex number we define the distance to the origin

$$|c| := \sqrt{a^2 + b^2} = \sqrt{(\operatorname{Re}(c))^2 + (\operatorname{Im}(c))^2} \qquad (\textbf{Magnitude of } c).$$

Examples 5.1:

① $\quad c_1 = 4 + 3i \qquad \hookrightarrow \quad |c_1| = 5.$

② $\quad c_2 = -\sqrt{2} + 2i \quad \hookrightarrow \quad |c_2| = \sqrt{6}.$

③ $\quad c_3 = -\frac{3}{2} - 3i \qquad \hookrightarrow \quad |c_3| = \sqrt{\frac{45}{4}}.$

④ $\quad c_4 = 1 - 3i \qquad \hookrightarrow \quad |c_4| = \sqrt{10}.$

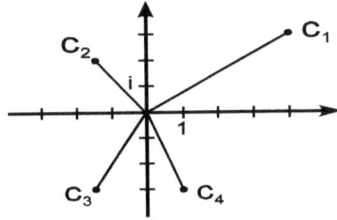

Remarks:

(1) Two complex numbers $c_1 = a_1 + ib_1$ and $c_2 = a_2 + ib_2$ are equal if $a_1 = a_2$ and $b_1 = b_2$. The real part and the imaginary part are thus two individual representatives of a complex number.

(2) So a complex number is nothing more than a point on the complex number plane.

(3) It is also common to assign a *pointer* (location vector) pointing from the origin O to the point c.

5.1.2 Trigonometric Form

If the angle φ between the complex pointer c and the positive axis $\mathbb{R}$ is introduced, according to Fig. 5.1 the following holds

$$\cos\varphi = \frac{a}{|c|} \quad \text{and} \quad \sin\varphi = \frac{b}{|c|}.$$

If we replace $a = |c| \cos \varphi$ and $b = |c| \sin \varphi$ in the algebraic form, we get for any complex number

$$c = a + ib = |c| \cos \varphi + i \, |c| \sin \varphi$$

$$c = |c| \, (\cos \varphi + i \sin \varphi) \qquad \textbf{(Trigonometric Form)}.$$

This representation is called the *Trigonometric Form*, where
- $|c|$ is the *magnitude* of the complex number c and
- φ is the *angle* (argument, phase) of c.

For $c = 0$, φ is not declared! The phase of a complex number is not unique, because the phase changes by 2π or by $360°$ with each full rotation.

Examples 5.2:

① $c_5 = 3 \, (\cos 45° + i \sin 45°)$.

② $c_6 = 4 \, (\cos 150° + i \sin 150°)$.

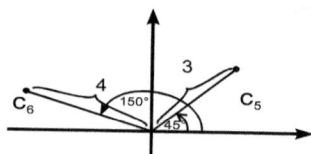

5.1.3 Exponential Form

If we replace $(\cos \varphi + i \sin \varphi)$ in the trigonometric form with the abbreviation

$$e^{i\varphi} := \cos \varphi + i \sin \varphi \qquad \textbf{(Euler's formula)}$$

introduced by Euler (1707-1783), then any complex number can be written as

$$c = |c| \, e^{i\varphi} \qquad \textbf{(Exponential Form)}.$$

At first, we see Euler's formula as just an abbreviation. By convention, the argument φ is given in radians.

Examples 5.3:

① Exponential form of c_5: $\varphi = 45° \hat{=} \frac{\pi}{4}$ $\quad \hookrightarrow c_5 = 3 \, e^{i\frac{\pi}{4}}$.
② Exponential form of c_6: $\varphi = 150° \hat{=} \frac{5}{6}\pi$ $\quad \hookrightarrow c_6 = 4 \, e^{i\frac{5}{6}\pi}$. $\qquad \square$

Special Complex Numbers

$$e^{i\frac{\pi}{2}} = i; \quad e^{i\pi} = -1; \quad e^{i\frac{3}{2}\pi} = -i; \quad e^{2\pi i} = 1.$$

Note: All these complex numbers have the magnitude 1, so they are on the unit circle. $\frac{1}{2}\pi$, π, $\frac{3}{2}\pi$ and 2π indicate the angles with respect to the real axis corresponding to $90°$, $180°$, $270°$ and $360°$.

5.1.4 Converting from one Form to Another

The calculation steps for the transformation of each form will be discussed. We will then choose an appropriate normal form for the complex arithmetic operations.

⊘ **Exponential Representation ⇌ Trigonometric Form:**

If a complex number c is given in the exponential form $c = |c|\, e^{i\varphi}$, the trigonometric form follows directly with Euler's formula

$$c = |c|\, (\cos\varphi + i\sin\varphi).$$

If the complex number $c = |c|\, (\cos\varphi + i\sin\varphi)$ is in trigonometric form, Euler's formula is $c = |c|\, e^{i\varphi}$. If necessary, φ is converted from degrees to radians.

Examples 5.4:

① $\quad c_7 = 5\, e^{i\frac{3}{4}\pi} \quad \hookrightarrow \varphi = \frac{3}{4}\pi \hat{=} 135°. \quad \Rightarrow \quad c_7 = 5\, (\cos 135° + i\sin 135°).$

② $\quad c_8 = \sqrt{2}\, (\cos 60° + i\sin 60°) \quad \hookrightarrow \varphi = 60° \hat{=} \frac{\pi}{3}. \quad \Rightarrow \quad c_8 = \sqrt{2}\, e^{i\frac{\pi}{3}}. \quad \square$

⊘ **Trigonometric Form ⇌ Algebraic Form:**

If the complex number $c = |c|\, (\cos\varphi + i\sin\varphi)$ is given in trigonometric form, then the algebraic form follows by expanding and evaluating the trigonometric functions

$$c = |c|\cos\varphi + i\, |c|\sin\varphi$$

with the real part $|c|\cos\varphi$ and the imaginary part $|c|\sin\varphi$.

If the complex number is given in the algebraic form $c = a + ib$, the trigonometric form follows by determining the magnitude $|c|$ and the angle

φ. To calculate the magnitude $|c|$ we use the theorem of Pythagoras: The square of the real part plus the square of the imaginary part equals the square of the magnitude. The imaginary part divided by the real part gives the tangent of the angle φ.

$$|c| = \sqrt{a^2 + b^2}$$

$$\tan\varphi = \frac{b}{a} \quad \Rightarrow \varphi.$$

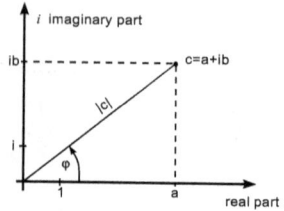

⚠ **Caution:** When calculating the angle $\tan\varphi = \frac{b}{a}$ using the inverse function $arctan$, note that the angle is only given in the range $[-\frac{\pi}{2}, \frac{\pi}{2}]$ (see Section 4.7). The correct angle φ in the range $[0, 2\pi]$ must then be given with an additional diagram.

Examples 5.5:

① $c_9 = 5\left(\cos 135° + i\sin 135°\right) = 5\left(-\frac{1}{2}\sqrt{2}\right) + i\, 5\frac{1}{2}\sqrt{2} = -\frac{5}{2}\sqrt{2} + i\frac{5}{2}\sqrt{2}.$

② $c_{10} = 4\sqrt{2} + i\,4\sqrt{2}.$
$\hookrightarrow |c_{10}| = \sqrt{16\cdot 2 + 16\cdot 2} = \sqrt{64} = 8,$
$\quad \&\ \tan\varphi = \frac{4\sqrt{2}}{4\sqrt{2}} = 1 \quad \hookrightarrow \varphi = 45° \,\hat{=}\, \frac{\pi}{4}.$
$\qquad \Rightarrow \quad c_{10} = 8\left(\cos 45° + i\sin 45°\right) = 8\,e^{i\frac{\pi}{4}}.$

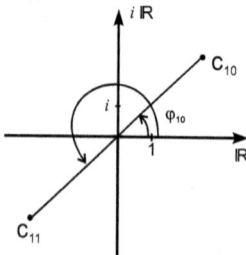

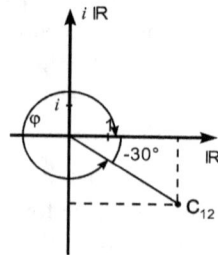

③ $c_{11} = -4\sqrt{2} - i\,4\sqrt{2}.$
$\hookrightarrow |c_{11}| = \sqrt{16\cdot 2 + 16\cdot 2} = \sqrt{64} = 8,$
$\quad \&\ \tan\varphi = \frac{-4\sqrt{2}}{-4\sqrt{2}} = 1 \quad \hookrightarrow \varphi = 45° + 180° = 225° \,\hat{=}\, \frac{5}{4}\pi.$
$\qquad \Rightarrow \quad c_{11} = 8\left(\cos 225° + i\sin 225°\right) = 8\,e^{i\frac{5}{4}\pi}.$

④ $c_{12} = \sqrt{3} - i.$

$\hookrightarrow |c_{12}| = \sqrt{3+1} = 2,$

$\&\ \tan\varphi = \frac{-1}{\sqrt{3}} = -\frac{1}{3}\sqrt{3} \quad \hookrightarrow \varphi = 360° - 30° = 330° \hat{=} \frac{11}{6}\pi.$

$\Rightarrow \quad c_{12} = 2\left(\cos 330° + i\sin 330°\right) = 2\,e^{i\frac{11}{6}\pi}.$ □

⊙ The Conjugated Number (Conjugates)

To perform the division of two complex numbers, we introduce to each complex number c the **Conjugate (complex conjugated number)** c^* or $\bar{c}$, which is the result of mirroring c on the real axis:

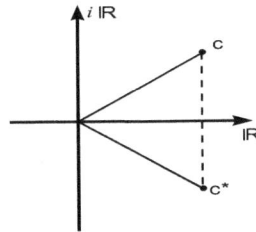

Figure 5.2. c and c*

Definition: (Complex Conjugate).
To each complex number $c = a + ib$ *we introduce the* **complex conjugated number** $c^* := a - ib$.

From the definition of conjugates it follows that

Complex Conjugate

$c = a + ib \qquad\qquad \Rightarrow \quad c^* = a - ib.$

$c = |c|\left(\cos\varphi + i\sin\varphi\right) \quad \Rightarrow \quad c^* = |c|\left(\cos\varphi - i\sin\varphi\right).$

$c = |c|\,e^{i\varphi} \qquad\qquad \Rightarrow \quad c^* = |c|\,e^{-i\varphi}.$

Hint: We get the complex conjugated number to each c very easily by formally replacing i with $-i$. And, of course $(c^*)^* = c$.

5.2 Complex Operations

The sum, difference, product and ratio of two complex numbers are not automatically defined by the construction of the complex numbers. It is necessary to declare these operations; but of course in such a way that for the special case that the imaginary part is zero, the operations and rules already defined in $\mathbb{R}$ must be valid. Let $c_1 = a_1 + i\,b_1$ and $c_2 = a_2 + i\,b_2$ be any two complex numbers. Then we define:

5.2.1 Addition

$$c_1 + c_2 := (a_1 + a_2) + i\,(b_1 + b_2).$$

The addition of two complex numbers means the addition of the real parts and the addition of the imaginary parts. The addition is done in the algebraic form.

Examples 5.6:

① $c_1 = 9 - 2\,i,\ c_2 = 4 + i.$
$\qquad c_1 + c_2 = (9 + 4) + i\,(-2 + 1) = 13 - i.$

② $c_1 = 3(\cos 30° + i\sin 30°),\ c_2 = 4 + i.$ To add c_1 and c_2, first we convert the number c_1 into the algebraic form:
$\qquad c_1 = 3\cos 30° + i\,3\sin 30° = 2.598 + 1.5\,i.$
$\qquad \Rightarrow c_1 + c_2 = (2.598 + 1.5\,i) + (4 + i) = 6.598 + 2.5\,i.$ $\qquad\qquad\square$

5.2.2 Subtraction

$$c_1 - c_2 := (a_1 - a_2) + i\,(b_1 - b_2).$$

Subtracting two complex numbers means subtracting the real and subtracting the imaginary parts. The subtraction is done in the algebraic form.

Examples 5.7:

① $c_1 = 9 - 2\,i,\ c_2 = 4 + i.$
$\qquad c_1 - c_2 = (9 - 2\,i) - (4 + i) = 9 - 4 + i\,(-2 - 1) = 5 - 3\,i.$

② $c_1 = 2\,e^{\frac{\pi}{4}i}$, $c_2 = 4 - 2\,i$. To subtract c_1 and c_2, we first converted c_1 into the algebraic form:

$$\varphi = \frac{\pi}{4} \hat{=} 45° \hookrightarrow c_1 = 2\,e^{\frac{\pi}{4}i} = 2\,(\cos 45° + i \sin 45°)$$

$$= 2\frac{1}{2}\sqrt{2} + i\,2\frac{1}{2}\sqrt{2} = 1.414 + i\,1.414.$$

$$\Rightarrow c_1 - c_2 = (1.414 + i\,1.414) - (4 - 2\,i) = -2.586 + 3.414\,i. \qquad \square$$

Geometric Interpretation. Since the addition and subtraction of two complex numbers are performed by analogy with the corresponding rules of vector calculus (i.e. component by component), the graphical representation of the arithmetic operations corresponds to the force parallelogram, i.e. the vector addition or subtraction.

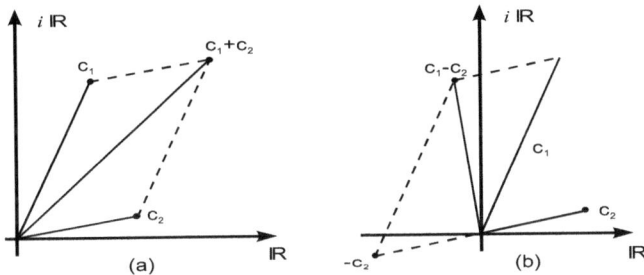

Figure 5.3. Addition (a) and subtraction (b) of complex numbers

Remark: Although a complex number represents only one point in the complex number plane, a complex number is often identified with its *pointer (location vector)* because of the above interpretation of "vector addition".

5.2.3 Multiplication

$$c_1 \cdot c_2 := (a_1 a_2 - b_1 b_2) + i\,(a_1 b_2 + b_1 a_2).$$

This multiplication formula is obtained by multiplying $(a_1 + i\,b_1) \cdot (a_2 + i\,b_2)$ term by term according to the distributive law for real numbers and using the property $i^2 = -1$:

$$c_1 \cdot c_2 = (a_1 + i\,b_1) \cdot (a_2 + i\,b_2)$$
$$= (a_1\,a_2 + a_1\,i\,b_2 + i\,b_1\,a_2 + i\,b_1\,i\,b_2)$$
$$= a_1\,a_2 + i^2\,b_1\,b_2 + i\,a_1\,b_2 + i\,b_1\,a_2$$
$$= (a_1\,a_2 - b_1\,b_2) + i\,(a_1\,b_2 + b_1\,a_2)\,.$$

Examples 5.8:

① $c_1 = 9 - 2\,i,\ c_2 = 4 + i.$
$$c_1 \cdot c_2 = (9 - 2\,i)\,(4 + i) = (36 + 2) + i\,(9 - 8) = 38 + i.$$

② For the product of $c = a + i\,b$ with its conjugate $c^* = a - i\,b$ we have

$$c \cdot c^* = (a + i\,b)\,(a - ib) = a^2 + b^2 = |c|^2\,.$$

This results in the following important formula for $|c|$:

$$\boxed{|c| = \sqrt{a^2 + b^2} = \sqrt{c \cdot c^*}}$$

$\square$

Geometric Interpretation: For the geometric interpretation we perform the multiplication again, but now we start with the trigonometric form of

$$c_1 = |c_1|\,(\cos\varphi_1 + i\sin\varphi_1) \quad \text{and} \quad c_2 = |c_2|\,(\cos\varphi_2 + i\sin\varphi_2)\,.$$

Multiplying the terms gives

$$c_1 \cdot c_2 = |c_1|\,(\cos\varphi_1 + i\sin\varphi_1) \cdot |c_2|\,(\cos\varphi_2 + i\sin\varphi_2)$$

$$= |c_1|\,|c_2|\,\{[\cos\varphi_1\,\cos\varphi_2 - \sin\varphi_1\,\sin\varphi_2] + i\,[\sin\varphi_1\,\cos\varphi_2 + \cos\varphi_1\,\sin\varphi_2]\}.$$

Now let us use the addition theorems for $\cos(\varphi_1 + \varphi_2)$ and $\sin(\varphi_1 + \varphi_2)$ from Section 4.6.4:

$$\cos(\varphi_1 + \varphi_2) = \cos\varphi_1\,\cos\varphi_2 - \sin\varphi_1\,\sin\varphi_2$$
$$\sin(\varphi_1 + \varphi_2) = \sin\varphi_1\,\cos\varphi_2 + \cos\varphi_1\,\sin\varphi_2,$$

we get the product

$$c_1 \cdot c_2 = |c_1| \cdot |c_2| \cdot (\cos(\varphi_1 + \varphi_2) + i\sin(\varphi_1 + \varphi_2))\,.$$

Hint: The multiplication of two complex numbers means **multiplication of magnitudes** and **addition of angles**, which makes it easy to construct $c_1 \cdot c_2$ in the Gaussian number plane.

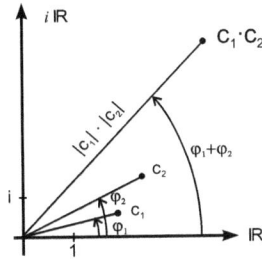

Figure 5.4. Multiplying two complex numbers

For the exponential representation, the formula is simplified to

Complex Multiplication

$$c_1 \cdot c_2 = |c_1| \, e^{i\varphi_1} \cdot |c_2| \, e^{i\varphi_2} = |c_1| \, |c_2| \, e^{i(\varphi_1 + \varphi_2)}.$$

5.2.4 Division

Complex Division

$$\frac{c_1}{c_2} := \frac{a_1 \, a_2 + b_1 \, b_2}{(a_2)^2 + (b_2)^2} + i \, \frac{b_1 a_2 - a_1 b_2}{(a_2)^2 + (b_2)^2} \qquad \text{for} \quad c_2 \neq 0.$$

This division formula is obtained by formally extending $\frac{c_1}{c_2}$ with c_2^*, e.g. by multiplying the upper and lower by the conjugate. Then we expand the numerator and denominator separately:

$$\frac{c_1}{c_2} = \frac{c_1}{c_2} \cdot \frac{c_2^*}{c_2^*} = \frac{a_1 + i \, b_1}{a_2 + i \, b_2} \cdot \frac{a_2 - i \, b_2}{a_2 - i \, b_2} = \frac{(a_1 + i \, b_1)(a_2 - i \, b_2)}{(a_2 + i \, b_2)(a_2 - i \, b_2)}$$

$$= \frac{(a_1 \, a_2 + b_1 \, b_2) + i \, (b_1 \, a_2 - a_1 \, b_2)}{(a_2)^2 + (b_2)^2}.$$

⚠ **Caution:** Also in $\mathbb{C}$ the division by $0 = 0 + i \, 0$ is **not** allowed!

Geometric Interpretation: We do the division in trigonometric form, using the trigonometric formulas for $\cos(\varphi_1 - \varphi_2)$ and $\sin(\varphi_1 - \varphi_2)$. Analogous to the procedure described in Section 5.2.3, we get

$$\frac{c_1}{c_2} = \frac{|c_1|(\cos\varphi_1 + i\sin\varphi_1)}{|c_2|(\cos\varphi_2 + i\sin\varphi_2)} = \frac{|c_1|}{|c_2|}(\cos(\varphi_1 - \varphi_2) + i\sin(\varphi_1 - \varphi_2))$$

and

Complex Division

$$\frac{c_1}{c_2} = \frac{|c_1|\,e^{i\varphi_1}}{|c_2|\,e^{i\varphi_2}} = \frac{|c_1|}{|c_2|}\,e^{i(\varphi_1 - \varphi_2)}.$$

Hint: When taking the quotient of two complex numbers, the **magnitudes are divided** and the **angles are subtracted**. So $\frac{c_1}{c_2}$ can also be constructed geometrically in the Gaussian number plane.

Examples 5.9:

① $c_1 = 9 - 2i$, $c_2 = 4 + i$: To calculate $\frac{c_1}{c_2}$, we extend the quotient with c_2^* and expand the numerator and denominator:

$$\frac{c_1}{c_2} = \frac{9 - 2i}{4 + i} \cdot \frac{4 - i}{4 - i} = \frac{(9 \cdot 4 - 2 \cdot 1) + i(-2 \cdot 4 - 9 \cdot 1)}{17} = 2 - i.$$

② $c_1 = 8\,e^{i\frac{4}{3}\pi}$, $c_2 = 4(\cos 60° + i\sin 60°)$: To calculate $\frac{c_1}{c_2}$, we represent c_2 in the exponential form. Since $60° \triangleq \frac{\pi}{3}$, this is

$$c_2 = 4(\cos 60° + i\sin 60°) = 4\,e^{i\frac{\pi}{3}}.$$

$$\frac{c_1}{c_2} = \frac{8\,e^{i\frac{4}{3}\pi}}{4\,e^{i\frac{\pi}{3}}} = 2\,e^{i\left(\frac{4}{3}\pi - \frac{\pi}{3}\right)} = 2\,e^{i\pi} = -2. \qquad \square$$

Example 5.10. Given are $c_1 = 1 = 1 + i\sqrt{3}$ and $c_2 = -\sqrt{3} + 3i$. Calculate (1) $c_1 \cdot c_2$, (2) $\frac{c_1}{c_2}$, (3) determine the exponential form of the numbers and repeat (4) the multiplication and (5) the division.

(1) $c_1 \cdot c_2 = \left(1 + i\sqrt{3}\right)\left(-\sqrt{3} + 3i\right) = \left(-\sqrt{3} - 3\sqrt{3}\right) + i\,(3 - 3) = -4\sqrt{3}.$

(2) $\dfrac{c_1}{c_2} = \dfrac{1 + i\sqrt{3}}{-\sqrt{3} + 3i} \cdot \dfrac{-\sqrt{3} - 3i}{-\sqrt{3} - 3i} = \dfrac{-\sqrt{3} + 3\sqrt{3} - 3i - 3i}{3 + 9} = \dfrac{\sqrt{3}}{6} - \dfrac{1}{2}i.$

(3) Representation of c_1 and c_2 in exponential form

$$|c_1| = \sqrt{1+3} = 2;\ \tan\varphi = \frac{\sqrt{3}}{1} = \sqrt{3} \Rightarrow \varphi = 60° \hat{=} \frac{\pi}{3} \Rightarrow c_1 = 2\,e^{i\frac{\pi}{3}}.$$

$$|c_2| = 2\sqrt{3};\ \tan\varphi = -\frac{3}{\sqrt{3}} \Rightarrow \varphi = \pi - \frac{\pi}{3} = \frac{2}{3}\pi \Rightarrow c_2 = 2\sqrt{3}\,e^{i\frac{2}{3}\pi}.$$

(4) $c_1 \cdot c_2 = 2\,e^{i\frac{\pi}{3}} \cdot 2\sqrt{3}\,e^{i\frac{2}{3}\pi} = 4\sqrt{3}\,e^{i\pi} = -4\sqrt{3}.$

(5) $\dfrac{c_1}{c_2} = \dfrac{2}{2\sqrt{3}}\,e^{i\left(\frac{\pi}{3} - \frac{2}{3}\pi\right)} = \frac{1}{3}\sqrt{3}\,e^{-i\frac{\pi}{3}}.$ □

Hint: Multiplication and division can be done very easily in trigonometric or exponential form: Multiplication involves multiplying the
magnitudes and adding the angles, while division involves dividing the
magnitudes and subtracting the angles.

5.2.5 Power c^n

Taking the complex number c to the power of n $(n \in \mathbb{N})$ is easy if we
consider the complex number c to be in exponential form: $c = |c|\,e^{i\varphi}$. Then
the following applies

$$c^2 = c \cdot c = |c|\,e^{i\varphi}\,|\cdot c|\,e^{i\varphi} = |c|^2\,e^{i\,2\,\varphi}$$
$$c^3 = c^2 \cdot c = |c|^2\,e^{i\,2\,\varphi} \cdot |c|\,e^{i\varphi} = |c|^3\,e^{i\,3\,\varphi}$$

and so on.

Mathematical induction provides

Power of Complex Numbers

$$c = |c|\,e^{i\varphi} \qquad \Rightarrow \qquad c^n = |c|^n\,e^{i\,n\,\varphi}.$$

Examples 5.11:

① Find $\left(2\sqrt{2} + i\,2\sqrt{2}\right)^5$.

To take the complex number $c = 2\sqrt{2} + i\,2\sqrt{2}$ to the power of 5, we first
have to represent it in its exponential form. To do this, we calculate
the magnitude $|c|$ and the angle φ:

$$\left.\begin{array}{l} |c| = \sqrt{4\cdot 2 + 4\cdot 2} = \sqrt{16} = 4 \\ \tan\varphi = \frac{2\sqrt{2}}{2\sqrt{2}} = 1 \hookrightarrow \varphi = \frac{\pi}{4} \end{array}\right\} \Rightarrow c = 4\,e^{i\frac{\pi}{4}}.$$

Then, we use the power formula: $\Rightarrow c^5 = 4^5\,e^{i\frac{\pi}{4}\cdot 5} = 1024\,e^{i\frac{5}{4}\pi}.$

② Find $\left(\sqrt{3}-i\right)^6$.

According to Example 5.5 ④ $c = \sqrt{3} - i = 2\,e^{i\frac{11}{6}\pi}$.

According to the power formula we get

$$c^6 = \left(2\,e^{i\frac{11}{6}\pi}\right)^6 = 2^6\,e^{i\frac{11}{6}\pi\cdot 6} = 64\,e^{i\,11\pi} = -64. \qquad \square$$

5.2.6 Roots

Complex Roots

For any complex number $c = |c|\,(\cos\varphi + i\sin\varphi) = |c|\,e^{i\varphi}$ we get the **n-th root** $(n \in \mathbb{N})$ by

$$c^{\frac{1}{n}} = \left\{ \sqrt[n]{|c|}\left(\cos\left(\frac{\varphi+k\cdot 360°}{n}\right) + i\sin\left(\frac{\varphi+k\cdot 360°}{n}\right)\right) \right.$$
$$\left. k = 0, 1, \dots, n-1 \right\}$$

$$= \left\{ \sqrt[n]{|c|}\,e^{i\frac{\varphi+k\cdot 2\pi}{n}};\ k = 0, 1, 2, \dots, n-1 \right\}, \qquad (*)$$

where $\sqrt[n]{|c|}$ is the n-th root of $|c|$ in the real numbers.

Proof: To show that the complex numbers

$$W_k := \sqrt[n]{|c|}\,e^{i\frac{\varphi+k\,2\pi}{n}};\ k = 0, 1, 2, \dots, n-1$$

are the n-th roots of c, it is sufficient to show that $(W_k)^n = c$. Since the n-th root of a complex number has the universal property that it must give exactly c when we take its n-th power! Of course, this is due to the calculation rules for complex powers:

$$(W_k)^n = \left(\sqrt[n]{|c|}\right)^n e^{i\frac{\varphi+k\,2\pi}{n}\cdot n} = |c|\,e^{i(\varphi+k\,2\pi)} = |c|\,e^{i\varphi},$$

because $e^{i(\varphi+k\,2\pi)} = e^{i\varphi}$ for $k \in \mathbb{N}$. $\qquad \square$

The n-th roots W_k are different for $k = 0, \dots, n-1$, but are repeated for $k \geq n$. Note that the n-th power of a complex number is **unambiguous**, but the n-th roots are multiple **ambiguous**.

Examples 5.12:

① $\left(4\sqrt{2}+i\,4\sqrt{2}\right)^{\frac{1}{3}} = \left(8\,e^{i\frac{\pi}{4}}\right)^{\frac{1}{3}} = \{\sqrt[3]{8}\,e^{i\frac{\frac{\pi}{4}+k\cdot2\pi}{3}} \ ; k = 0,1,2\}$
$$= \{2\,e^{i\frac{\pi}{12}}\,,\ 2\,e^{i\frac{9}{12}\pi}\,,\ 2\,e^{i\frac{17}{12}\pi}\}.$$

② $(-1)^{\frac{1}{5}} = \left(1\,e^{i\pi}\right)^{\frac{1}{5}} = \{\sqrt[5]{1}\,e^{i\frac{\pi+k\cdot2\pi}{5}} \ ; k = 0,1,2,3,4\}$
$$= \{e^{i\frac{\pi}{5}}\,,\ e^{i\frac{3}{5}\pi}\,,\ e^{i\pi}\,,\ e^{i\frac{7}{5}\pi}\,,\ e^{i\frac{9}{5}\pi}\}. \qquad\qquad \square$$

Special Case: n-th roots of 1: Each complex solution of $z^n = 1$ is called the n-th *root of unity*. With $c = 1$ we obtain applying formula $(*)$:

$$1^{\frac{1}{n}} = \left(1\,e^{i0}\right)^{\frac{1}{n}} = \{1\,,\ e^{i\frac{2\pi}{n}}\,,\ e^{i\frac{4\pi}{n}}\,,\dots,\ e^{i\frac{2\pi\,(n-1)}{n}}\}.$$

The magnitude of these numbers is always 1, i.e. the n-th unit roots are on the unit circle. The difference of the angles is $\frac{2\pi}{n}$, so that they are obtained successively by rotation about $\frac{2\pi}{n}$ starting from the complex number 1.

Example 5.13. Find all 9-th roots of unity:

$$(1)^{\frac{1}{9}} = \left(1\,e^{i0}\right)^{\frac{1}{9}}$$
$$= \{\sqrt[9]{1}\,e^{i\frac{0+k\,2\pi}{9}} \ ; k = 0,\dots,8\}$$
$$= \{1\,e^0\,,\ e^{i\frac{2\pi}{9}}\,,\ e^{i\frac{4\pi}{9}}\,,\dots,\ e^{i\frac{16}{9}\pi}\}.$$

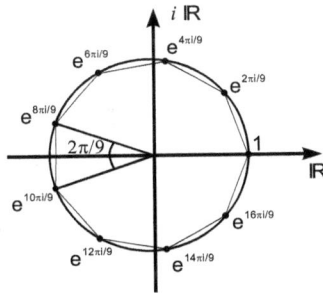

Figure 5.5. 9th roots of unity $c = 1$

Theorem: For $n > 1$ the following holds: $\displaystyle\sum_{k=0}^{n-1} e^{i\frac{\varphi+k\cdot2\pi}{n}} = 0.$

Proof: This theorem is obvious because of its geometric property, since $e^{i\frac{\varphi+k\cdot 2\pi}{n}}$ represents the n-th root of unity of the complex number $e^{i\varphi}$.

If all n roots of unity are added (vector addition), the sum is zero. Formally, this statement is obtained using the geometric sum formula (see Section 1.2.3)

$$\sum_{k=0}^{n-1} e^{i\frac{\varphi+k\cdot 2\pi}{n}} = \sum_{k=0}^{n-1} e^{i\frac{\varphi}{n}} e^{i k\frac{2\pi}{n}}$$

$$= e^{i\frac{\varphi}{n}} \sum_{k=0}^{n-1} \left(e^{i\frac{2\pi}{n}}\right)^k = e^{i\frac{\varphi}{n}} \frac{1-\left(e^{i\frac{2\pi}{n}}\right)^n}{1-e^{i\frac{2\pi}{n}}} = 0,$$

because for the numerator of the last fraction it is

$$1 - \left(e^{i\frac{2\pi}{n}}\right)^n = 1 - e^{i2\pi} = 1 - 1 = 0. \qquad \square$$

5.2.7 Fundamental Theorem of Algebra

We interpret the ambiguity of the n-th root as follows: Every polynomial of n-th degree of the form $p(z) = z^n - a$ ($n \in \mathbb{N}$, $a \in \mathbb{C}$) has exactly n zeros, namely the n-th roots of a. This property generalizes to any complex polynomial of degree n:

Fundamental Theorem of Algebra

An arbitrary complex polynomial of degree n

$$p(z) = a_n z^n + a_{n-1} z^{n-1} + \ldots + a_1 z + a_0$$

$$(a_k \in \mathbb{C}, \, a_n \neq 0, \, z \in \mathbb{C})$$

has **exactly** n zeros.

Appendum

If the coefficients of $p(z)$ are real (i.e. $a_k \in \mathbb{R}$), the zeros are real or occur in pairs with complex and conjugate values.

Note: The appendix becomes important for applications when we solve differential equations and use the characteristic polynomial. Each root of this characteristic polynomial gives us a solution. So if we have a real problem, the coefficients are real. So for each complex root we have not only a solution but also the complex conjugate.

While the fundamental theorem ensures that every polynomial of degree n has n zeros, it says nothing about how to find these zeros. There is also no general formula for calculating zeros in the complex domain, except in simple special cases. So we either have to guess the zeros and to reduce the degree by polynomial division, or we have to find them numerically.

Example 5.14. Find the zeros of

$$p(z) = z^3 - 2z - 4.$$

The fundamental theorem says that there are exactly 3 zeros. To get a zero we check that $z = 0, \pm 1, \pm 2 : \; \hookrightarrow z = 2$ is a zero. By polynomial division we get:

$$
\begin{array}{llll}
(z^3 & -2z & -4) & : \quad (z-2) \quad = \quad z^2 + 2z + 2. \\
\underline{-(z^3 \quad -2z^2)} & & & \\
\quad\quad 2z^2 & -2z & & \\
\quad\underline{-(2z^2 \quad -4z)} & & & \\
\quad\quad\quad 2z & -4 & & \\
\quad\quad\underline{-(2z \quad -4\,)} & & & \\
\quad\quad\quad\quad 0 & & & \\
\end{array}
$$

$$\Rightarrow (z^3 - 2z - 4) = (z-2)(z^2 + 2z + 2).$$

The quadratic formula gives $z_{2/3} = -1 \pm \sqrt{1-2} = -1 \pm i$. So the zeros of the polynomial are: $2, \, -1 + i, \, -1 - i$. □

Remark: The appendix to the Fundamental Theorem can be proved directly: If $p(z) = a_n z^n + a_{n-1} z^{n-1} + \cdots + a_1 z + a_0$ is a real polynomial and z_0 is a zero of p, then z_0^* is also a zero of p:

$$
\begin{aligned}
p(z_0^*) &= a_n (z_0^*)^n + a_{n-1}(z_0^*)^{n-1} + \cdots + a_1 z_0^* + a_0 \\
&= a_n (z_0^n)^* + a_{n-1} (z_0^{n-1})^* + \cdots + a_1 (z_0)^* + a_0 \\
&= (a_n z_0^n + a_{n-1} z_0^{n-1} + \cdots + a_1 z_0 + a_0)^* \\
&= (p(z_0))^* = 0^* = 0.
\end{aligned}
$$
□

5.3 Applications

5.3.1 Description of Harmonic Oscillations with Complex Numbers

The spring pendulum known from mechanics has the property that for an undamped oscillation, the deflection from the rest position $s(t)$ has the time behavior $s(t) = A\cos(\omega t + \varphi)$. The system oscillates at the frequency $\omega = \sqrt{\frac{D}{m}}$, where D is the spring constant and m is the mass. This function has a maximum amplitude A and a zero phase φ. The period of the oscillation is $T = \frac{2\pi}{\omega}$. A periodic motion with frequency ω and amplitude A is called a **harmonic oscillation**. The voltage behavior of an LC alternating current circuit is also $U(t) = U_0 \cos(\omega t + \varphi)$ with frequency $\omega = \sqrt{\frac{1}{LC}}$ and period $T = 2\pi\sqrt{LC}$.

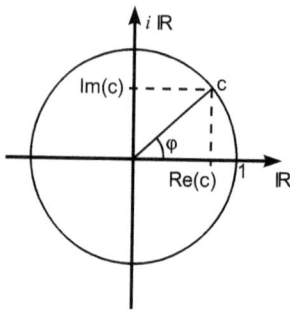

To describe harmonic oscillations within complexes, we consider the complex number

$$c = \cos\varphi + i\sin\varphi = e^{i\varphi}.$$

Since $|c| = 1$, c is a number on the unit circle. If the point c is projected onto the real axis, we get the real part of c: $\text{Re}(c) = \cos\varphi$. If the point c is projected onto the imaginary axis, we get the imaginary part of c: $\text{Im}(c) = \sin\varphi$.

If the angle φ varies as a function of time $\varphi = \omega \cdot t$ with a constant angular frequency $\omega = \frac{2\pi}{T}$, then $\boxed{e^{i\varphi} = e^{i\omega t}}$ for $0 \le t \le T$ passes through the complete unit circle in the complex plane.

$\cos(\omega t)$ and $\sin(\omega t)$ are the projections of the complex vector $e^{i\omega t}$ on the real and imaginary axes, respectively.

Visualization: This is an animation showing the projections of $e^{i\omega t}$ on the x- and y-axis. The vector $e^{i\omega t}$ running on the unit circle is animated together with its real and imaginary parts by varying the time t from 0 to $\frac{2\pi}{T}$ (see Fig. 5.6).

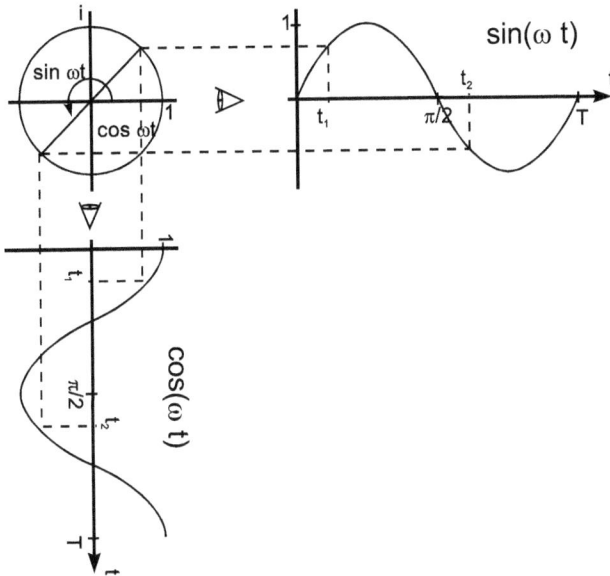

Figure 5.6. Sine and cosine as real and imaginary parts of $e^{i\omega t}$

A harmonic oscillation with amplitude A and zero phase φ_0 can thus be written as

$$A\cos(\omega t + \varphi_0) = \text{Re}\left(A\,e^{i(\omega t + \varphi_0)}\right) = \text{Re}\left(A\,e^{i\varphi_0}\,e^{i\omega t}\right)$$
$$A\sin(\omega t + \varphi_0) = \text{Im}\left(A\,e^{i(\omega t + \varphi_0)}\right) = \text{Im}\left(A\,e^{i\varphi_0}\,e^{i\omega t}\right).$$

The complex description of a harmonic oscillation is therefore

$$\hat{y}(t) = A\,e^{i\varphi_0}\,e^{i\omega t}$$

with the *complex amplitude* $A\,e^{i\varphi_0}$ and the pure time behavior $e^{i\omega t}$.

5.3.2 Superposition of Equal-Frequency Oscillations

We will now calculate the *superposition* of two **equal-frequency** harmonic oscillations in the complex domain. We take advantage of the fact that the complex amplitude includes both the amplitude A and the phase φ of the oscillation. Given are the two oscillations

$$u_1(t) = u_1 \sin(\omega t + \varphi_1)$$
$$u_2(t) = u_2 \sin(\omega t + \varphi_2).$$

We look for the amplitude A and the phase φ of the superposition

$$u(t) = u_1(t) + u_2(t) = A \sin(\omega t + \varphi).$$

To calculate the superposition, we interpret $u_1(t)$ as the imaginary part of the complex oscillation $\hat{u}_1(t) = u_1 e^{i(\omega t + \varphi_1)}$ and $u_2(t)$ as the imaginary part of $\hat{u}_2(t) = u_2 e^{i(\omega t + \varphi_2)}$ and perform the superposition in the complex domain. Then we take the imaginary part, which gives the result $u_1(t) + u_2(t)$ in the real numbers

$$u_1(t) + u_2(t) = \operatorname{Im} \hat{u}_1(t) + \operatorname{Im} \hat{u}_2(t)$$
$$= \operatorname{Im}(\hat{u}_1(t) + \hat{u}_2(t)) = \operatorname{Im} \hat{u}(t) = u(t).$$

$$
\begin{array}{lll}
\textbf{Transition} & \hat{u}_1(t) & = \quad u_1 e^{i(\omega t + \varphi_1)} = u_1 e^{i\varphi_1} e^{i\omega t}, \\
\textbf{to Complexes} & \hat{u}_2(t) & = \quad u_2 e^{i(\omega t + \varphi_2)} = u_2 e^{i\varphi_2} e^{i\omega t}.
\end{array}
$$

$$
\begin{array}{lll}
\textbf{Complex} & \hat{u}(t) & = \quad \hat{u}_1(t) + \hat{u}_2(t) \\
\textbf{Addition} & & = \quad u_1 e^{i\varphi_1} e^{i\omega t} + u_2 e^{i\varphi_2} e^{i\omega t} \\
& & = \quad \left(u_1 e^{i\varphi_1} + u_2 e^{i\varphi_2}\right) e^{i\omega t} \\
& & = \quad A e^{i\varphi} e^{i\omega t} = A e^{i(\omega t + \varphi)}.
\end{array}
$$

The complex amplitude of the superposition $A e^{i\varphi}$ is the sum of the two single amplitudes $u_1 e^{i\varphi_1}$ and $u_2 e^{i\varphi_2}$. The superposition of two oscillations of the same frequency corresponds to the vector addition of these complex amplitudes. The addition can be done mathematically as well as graphically (see Fig. 5.7).

When the complex addition is performed, A and φ are given by

$$A e^{i\varphi} = u_1 e^{i\varphi_1} + u_2 e^{i\varphi_2}$$
$$= u_1 (\cos(\varphi_1) + i \sin(\varphi_1)) + u_2 (\cos(\varphi_2) + i \sin(\varphi_2))$$
$$= (u_1 \cos(\varphi_1) + u_2 \cos(\varphi_2)) + i (u_1 \sin(\varphi_1) + u_2 \sin(\varphi_2)).$$

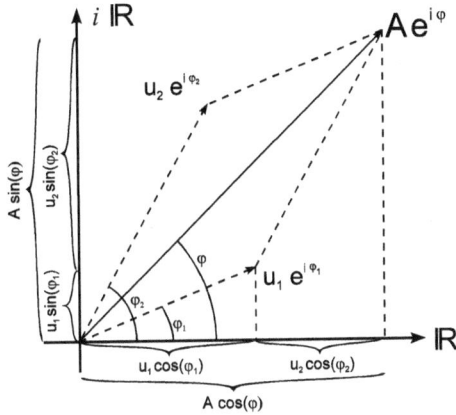

Figure 5.7. Graphical addition of complex amplitudes

To obtain the results in the form of $A = \sqrt{\mathrm{Re}^2 + \mathrm{Im}^2}$ and $\tan(\varphi) = \frac{\mathrm{Im}}{\mathrm{Re}}$, we use the identity $\cos^2\alpha + \sin^2\alpha = 1$ and the addition theorem for the cosine

$$A^2 = u_1^2 \cos^2(\varphi_1) + u_2^2 \cos^2(\varphi_2) + 2u_1 u_2 \cos(\varphi_1)\cos(\varphi_2) +$$
$$u_1^2 \sin^2(\varphi_1) + u_2^2 \sin^2(\varphi_2) + 2u_1 u_2 \sin(\varphi_1)\sin(\varphi_2)$$
$$= u_1^2 + u_2^2 + 2u_1 u_2 \cos(\varphi_1 - \varphi_2)$$

and

$$\tan(\varphi) = \frac{u_1 \sin(\varphi_1) + u_2 \sin(\varphi_2)}{u_1 \cos(\varphi_1) + u_2 \cos(\varphi_2)}.$$

Transition to Real → $u(t) = u_1(t) + u_2(t) = \mathrm{Im}\left(A\, e^{i(\omega t + \varphi)}\right) = A\sin(\omega t + \varphi).$

Summary: If two harmonic oscillations $u_1(t) = u_1 \sin(\omega t + \varphi_1)$ and $u_2(t) = u_2 \sin(\omega t + \varphi_2)$ have the **same** frequency ω, then the superposition is again a harmonic oscillation

$$u(t) = u_1(t) + u_2(t) = A\sin(\omega t + \varphi)$$

with amplitude $A = \sqrt{u_1^2 + u_2^2 + 2u_1 u_2 \cos(\varphi_1 - \varphi_2)}$

and phase $\tan(\varphi) = \dfrac{u_1 \sin(\varphi_1) + u_2 \sin(\varphi_2)}{u_1 \cos(\varphi_1) + u_2 \cos(\varphi_2)}.$

Remarks:

(1) In the same way, the superposition of two cosine oscillations $u_1(t) = u_1 \cos(\omega t + \varphi_1)$, $u_2(t) = u_2 \cos(\omega t + \varphi_2)$ is obtained by interpreting these oscillations as the *real part* of the corresponding complex oscillations. $u(t) = \mathrm{Re}\left(A\, e^{i(\omega t + \varphi)}\right)$ then gives the cosine of the superposition.

(2) If one oscillation is represented in cosine $u_1(t) = a_1 \cos(\omega t + \varphi_1)$ and the other one in sine $u_2(t) = a_2 \sin(\omega t + \varphi_2)$, a common form of representation must be chosen. Either we write

$$u_1(t) = a_1 \cos(\omega t + \varphi_1) = a_1 \sin\left(\omega t + \varphi_1 + \frac{\pi}{2}\right)$$

and perform the superposition in the sine form, or we write

$$u_2(t) = a_2 \sin(\omega t + \varphi_2) = a_2 \cos\left(\omega t + \varphi_2 - \frac{\pi}{2}\right)$$

and perform the superposition in the cosine form.

(3) ⚠ **Caution:** The superposition of two harmonic oscillations of **different** frequencies does not generally result in a periodic function any more. Only if the ratio of the frequencies is a rational number, we get a periodic function again, but even then not a harmonic function.

Example 5.15. Find the superposition of the two alternating voltages

$$u_1(t) = 4\sin(2t) \quad \text{and} \quad u_2(t) = 3\cos\left(2t - \frac{\pi}{6}\right).$$

Before superimposing these two harmonic functions, it is necessary to represent e.g. $u_2(t)$ as a sine function:

$$u_2(t) = 3\cos\left(2t - \frac{\pi}{6}\right) = 3\sin\left(2t - \frac{\pi}{6} + \frac{\pi}{2}\right) = 3\sin\left(2t + \frac{\pi}{3}\right).$$

Transition
to Complexes $\longrightarrow$

$$\hat{u}_1(t) = 4\,e^{i2t}$$
$$\hat{u}_2(t) = 3\,e^{i\left(2t + \frac{\pi}{3}\right)} = 3\,e^{i\frac{\pi}{3}}\,e^{i2t}.$$

Complex
Addition $\longrightarrow$

$$\hat{u}(t) = \hat{u}_1(t) + \hat{u}_2(t)$$
$$= 4\,e^{i2t} + 3\,e^{i\frac{\pi}{3}}\,e^{i2t} = \left(4 + 3\,e^{i\frac{\pi}{3}}\right)e^{i2t}.$$

Addition of complex amplitudes

$$c = 4 + 3\,e^{i\frac{\pi}{3}} = 4 + 3\left(\cos\frac{\pi}{3} + i\sin\frac{\pi}{3}\right)$$
$$= 4 + 3\left(\frac{1}{2} + i\frac{1}{2}\sqrt{3}\right) = 5.5 + i\,2.6,$$

Representation of c in exponential form

$|c| = \sqrt{5.5^2 + 2.6^2} = 6.08$,

$\tan\varphi = \frac{\text{Im}\, c}{\text{Re}\, c} = \frac{2.6}{5.5} \quad \hookrightarrow \varphi = 25.28° \hat{=} 0.44$,

$\Rightarrow c = A\, e^{i\varphi}$ with $A = 6.08 \quad \varphi = 0.44$

$\Rightarrow \hat{u}(t) = 6.08\, e^{i\,0.44}\, e^{i\,2\,t} = 6.08\, e^{i\,(2\,t+0.44)}$.

Transition $\longrightarrow$
to Real

$\qquad u(t) = \text{Im}\,\hat{u}(t) = 6.08\ \sin(2\,t + 0.44)$.

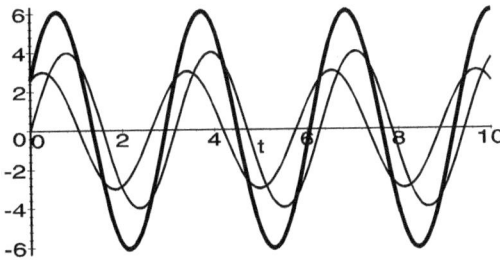

Figure 5.8. Superposition of the two equal-frequency oscillations

This method can easily be applied to the case of the superposition of more than two harmonic oscillations of the same frequency.   □

5.3.3 Description of AC Circuits

We consider electrical networks consisting of ohmic resistors, capacitors and inductors. In AC circuits, the voltages $U(t)$ and the currents $I(t)$ are sinusoidal or cosinusoidal:

$$U(t) = U_0 \cos(\omega t + \varphi_1), \quad I(t) = I_0 \cos(\omega t + \varphi_2).$$

We switch to the complex formulation and interpret them as real parts of the complex functions

$$\hat{U}(t) = U_0\, e^{i(\omega t+\varphi_1)} = U_0\, e^{i\varphi_1}\, e^{i\omega t} = \hat{U}_0\, e^{i\omega t}$$
$$\hat{I}(t) = I_0\, e^{i(\omega t+\varphi_2)} = I_0\, e^{i\varphi_2}\, e^{i\omega t} = \hat{I}_0\, e^{i\omega t}.$$

We show that Ohm's law is transferred to the inductive and capacitive elements when this complex formulation is chosen.

(1) Ohmic Resistance R. For an ohmic resistance, the relationship between voltage and current is given by $U(t)=R\,I(t)$. This law also applies to a complex alternating current $\hat{I}(t) = \hat{I}_0\, e^{i\omega t}$.

$$\hookrightarrow \hat{U}(t) = R\,\hat{I}_0\, e^{i\omega t} = R\,\hat{I}(t).$$

An *ohmic resistance* is described by the real resistance R. The current and voltage are in phase.

(2) Capacitance C. For a capacitor with capacity C, the relationship between the charge Q and the applied voltage U is:

$$Q = C \cdot U \quad \hookrightarrow \quad I(t) = \frac{d}{dt}Q(t) = C \cdot \dot{U}(t).$$

Especially for $\hat{U}(t) = \hat{U}_0\, e^{i\omega t}$

$$\hat{I}(t) = C \cdot \left(\hat{U}_0\, e^{i\omega t}\right)' = C \cdot \hat{U}_0\, e^{i\omega t}\, i\omega = C \cdot i\omega\, \hat{U}(t).$$

So we introduce the *complex resistance*

$$\hat{R}_C := \frac{\hat{U}(t)}{\hat{I}(t)} = \frac{1}{i\omega C} = -i\frac{1}{\omega C}.$$

The complex resistance $\hat{R}_C = \frac{1}{i\omega C}$ is associated with a capacitor. Voltage and current are shifted by $-90°$. The formula used $\left(e^{i\omega t}\right)' = i\omega\, e^{i\omega t}$ will be discussed in Volume II: Chapter Taylor Series.

(3) Inductance L. For a coil with inductance L, the relationship between current and induced voltage is given by the **induction law**

$$U(t) = L\frac{d\,I(t)}{dt}.$$

Specifically for $\hat{I}(t) = \hat{I}_0\, e^{i\omega t}$ we have

$$\hat{U}(t) = L\left(\hat{I}_0\, e^{i\omega t}\right)' = L\,\hat{I}_0\, e^{i\omega t}\, i\omega = i\omega L\,\hat{I}(t).$$

The *complex resistance* $\hat{R}_L$ is associated with a coil of inductance L.

$$\hat{R}_L := \frac{\hat{U}(t)}{\hat{I}(t)} = i\omega L.$$

$i\omega L$ is on the positive imaginary axis. The phase between the voltage and the current is $+90°$; the voltage leads the current by $90°$.

Summary: Modelling RCL Networks

For RCL networks with alternating voltages $\hat{U}(t) = \hat{U}_0 e^{i\omega t}$ or alternating currents $\hat{I}(t) = \hat{I}_0 e^{i\omega t}$ ohmic laws apply in the form of $\hat{U}(t) = \hat{R}\hat{I}(t)$ when assigning **complex resistances (impedances)**:

$$
\begin{aligned}
\text{Ohmic resistance } R \quad & \hat{R}_\Omega & = & \quad R \\
\text{Capacitance } C \quad & \hat{R}_C & = & \quad \tfrac{1}{i\omega C} \\
\text{Inductance } L \quad & \hat{R}_L & = & \quad i\omega L
\end{aligned}
$$

Conclusion: Using Kirchhoff's laws, the equivalent circuit of two complex resistors $\hat{R}_1$ and $\hat{R}_2$ yields a **complex total resistance (= equivalent resistance)** $\hat{R}$:

(a)   Series circuit $\quad \hat{R} \;=\; \hat{R}_1 + \hat{R}_2.$

(b)   Parallel circuit $\quad \dfrac{1}{\hat{R}} \;=\; \dfrac{1}{\hat{R}_1} + \dfrac{1}{\hat{R}_2} \;$ or $\; \hat{R} = \dfrac{\hat{R}_1 \hat{R}_2}{\hat{R}_1 + \hat{R}_2}.$

$\operatorname{Re}\hat{R}$ is called the *active resistance*, $\operatorname{Im}\hat{R}$ the *reactance* and $\left|\hat{R}\right|$ the *real impedance.*

In AC circuits, therefore, the well-known rules for the equivalent circuit of resistances can be used, as in the DC circuits, when capacitance and inductance change to complex values! The complex total resistance of elements connected in series is the sum of the complex individual resistances. The inverse total resistance $\frac{1}{\hat{R}}$ of elements connected in parallel is the sum of the inverse complex individual resistances.

5.3.4 Examples for AC Circuits

Example 5.16 (RLC Series Circuit, with MAPLE-Worksheet):
Fig. 5.9 shows a series circuit consisting of an ohmic resistor R_Ω, a capacitance C and an inductance L. The complex values are added to give the complex total resistance

$$\hat{R} = R_\Omega + \hat{R}_C + \hat{R}_L = R_\Omega + \frac{1}{i\omega C} + i\omega L$$
$$\hat{R} = R_\Omega + i\left(\omega L - \tfrac{1}{\omega C}\right).$$

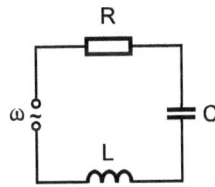

Figure 5.9. RLC circuit

The addition is represented graphically by the pointer diagrams.

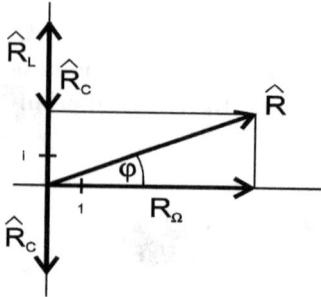

Impedance diagram   Voltage diagram

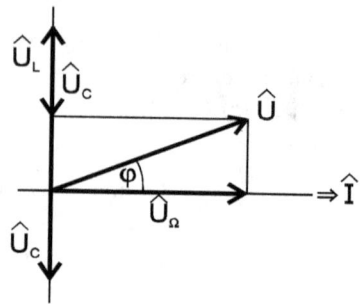

The *reactance* is $\operatorname{Im} \hat{R} = \omega L - \frac{1}{\omega C}$, the *active resistance* is $\operatorname{Re} \hat{R} = R_\Omega$ and the **real impedance** is

$$R = \left|\hat{R}\right| = \sqrt{R_\Omega^2 + \left(\omega L - \frac{1}{\omega C}\right)^2}.$$

The **phase** between voltage and current is given by

$$\tan \varphi = \frac{\operatorname{Im} \hat{R}}{\operatorname{Re} \hat{R}} = \frac{\omega L - \frac{1}{\omega C}}{R_\Omega}.$$

Discussion: Multiplying the resistances by $\hat{I}$ gives the voltage diagram.

(1) U_Ω decreases at the ohmic resistance. It is in phase with the current I.

(2) U_L decreases at the inductance. U_L leads the current by 90°.

(3) U_C decreases at the capacity. U_C lags the current by 90°.

For $R = 1$, $L = 1$ and $C = 1$, the following curves are obtained for the equivalent resistance $R(\omega)$ and the phase $\varphi(\omega)$:

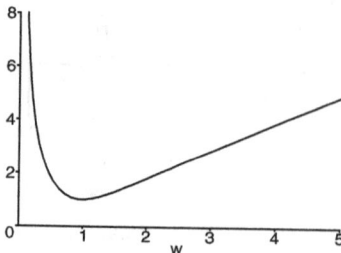

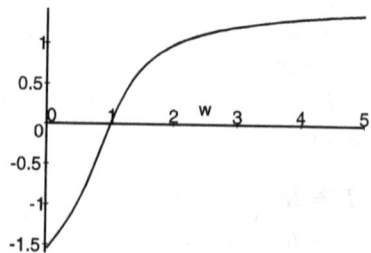

Equivalent resistance $|R(\omega)|$ Phase $\varphi(\omega)$

Example 5.17 (LC-Parallel Circuit, with MAPLE-Worksheet):

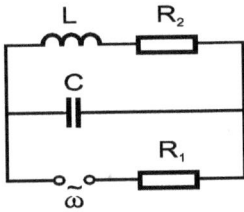

Figure 5.10. LC circuit

For the circuit shown in Fig. 5.10, the complex equivalent resistance is calculated by first replacing L and R_2 with the series equivalent resistance $R_s = R_2 + i\omega L$. R_s is parallel to C, so that the coupling coefficients add up

$$Y_p = i\,\omega\,C + \frac{1}{i\,\omega\,L + R_2}.$$

The total complex resistance is now the sum of R_1 and $R_p = \frac{1}{Y_p}$:

$$R_{total} = R_1 + \frac{1}{Y_p} = R_1 + \frac{1}{i\,\omega\,C + \frac{1}{i\,\omega\,L + R_2}}$$

$$R_{total} = \frac{R_1\,\omega^2\,C\,L - R_1\,\omega\,C\,R_2\,i - R_1 - \omega\,L\,i - R_2}{\omega^2\,C\,L - \omega\,C\,R_2\,i - 1}.$$

We can see from this expression that the total resistance is a complex rational function in ω, and that 2 is the highest exponent that occurs. This reflects the fact that the circuit has two energy stores, C and L. For the values $C = 20 \cdot 10^{-6}, L = 20 \cdot 10^{-3}, R_1 = 50, R_2 = 500$ the total resistance is given as a function in ω

$$R_{total} = \frac{0.8000\,10^{-11}\,\omega^4 + 0.004960\,\omega^2 + 550}{0.1600\,10^{-12}\,\omega^4 + 0.00009920\,\omega^2 + 1}$$
$$+ \frac{(-0.8000\,10^{-8}\,\omega^3 - 4.980\,\omega)\,i}{0.1600\,10^{-12}\,\omega^4 + 0.00009920\,\omega^2 + 1}.$$

The curves of total resistance and phase as a function of ω are given by

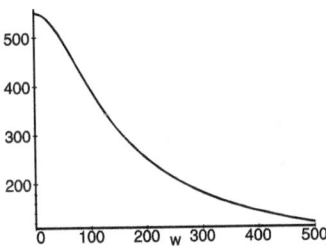

Total resistance $|R(\omega)|$

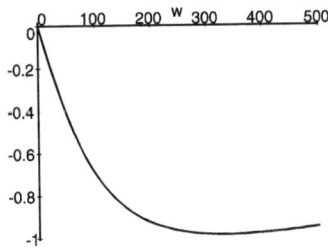

Phase $\varphi(\omega)$ ☐

5.4 Problems on Complex Numbers

5.1 Specify the exponential form of the following complex numbers
  a) $3\sqrt{3} + 3i$ b) $-2 - 2i$ c) $1 - \sqrt{3}i$ d) 5 e) $-5i$ f) -1

5.2 What are the trigonometric and algebraic normal form of the numbers
  a) $3\sqrt{2}\,e^{i\frac{\pi}{4}}$ b) $2\,e^{i\frac{2\pi}{3}}$ c) $e^{i\pi}$ d) $4\,e^{i\frac{4\pi}{3}}$

5.3 Which are the corresponding complex conjugated numbers of
  a) $3 + \sqrt{2}i$ b) $4\,(\cos 125° + i\sin 125°)$ c) $5\,e^{i\frac{3}{2}\pi}$ d) $\sqrt{3}\,e^{i\,0.734}$

5.4 Determine the trigonometric normal form of
  a) $-1 + \sqrt{3}i$ b) $-1 + i$ c) $\sqrt{2} + \sqrt{2}i$ d) $-3 - 4i$

5.5 Calculate
  a) $2\,(5 - 3i) - 3\,(-2 + i) + 5\,(i - 3)$ b) $(3 - 2i)^3$ c) $\frac{5}{3 - 4i} + \frac{10}{4 + 3i}$
  d) $\left(\dfrac{1 - i}{1 + i}\right)^{10}$ e) $\left|\dfrac{2 - 4i}{5 - 7i}\right|^2$ f) $\dfrac{(1 + i)\,(2 + 3i)\,(4 - 2i)}{(1 + 2i)^2\,(1 - i)}$

5.6 Let $z_1 = 1 - i$, $z_2 = -2 + 4i$, $z_3 = \sqrt{3} - 2i$. What is the algebraic normal form of
  a) $z_1^2 + 2z_1 - 3$ b) $|2z_2 - 3z_1|^2$ c) $(z_3 - z_3^*)^5$
  d) $|z_1 z_2^* + z_2 z_1^*|$ e) $\left|\frac{z_1 + z_2 + 1}{z_1 - z_2 + i}\right|$ f) $\frac{1}{2}\left(\frac{z_3}{z_3^*} + \frac{z_3^*}{z_3}\right)$
  g) $((z_2 + z_3)\,(z_1 - z_3))^*$ h) $\left|z_1^2 + z_2^{*\,2}\right|^2 + \left|z_3^{*\,2} - z_2^2\right|^2$ i) $\mathrm{Im}\left\{\frac{z_1 z_2}{z_3}\right\}$

5.7 Calculate
  a) $\left(-1 + \sqrt{3}i\right)^{10}$ b) $[2\,(\cos 45° + i\sin 45°)]^3$ c) $\left(3\sqrt{3} + 3i\right)^6$ d) $\left(2\,e^{i\frac{5}{3}\pi}\right)^7$

5.8 Specify all solutions in the complex from
  a) $z^4 + 81 = 0$
  b) $z^6 + 1 = \sqrt{3}i$

5.9 Identify all complex solutions
  a) $z^5 - 2z^4 - z^3 + 6z - 4 = 0$
  b) $4x^4 + 4x^3 - 7x^2 + x - 2 = 0$

5.10 Solve problems 5.1 - 5.9 with MAPLE.

5.11 What are the real and imaginary parts of the following complex numbers?
  a) $\dfrac{-2 + 7i}{15i}$ b) $\dfrac{1 + i}{1 - i}$ c) $\dfrac{1 - i}{1 + 2i} - \dfrac{1 + 3i}{1 - 2i}$ d) $\dfrac{2\,e^{i\frac{\pi}{4}}}{(1 + i)\,(2 + i)}$
  e) $2\,e^{i\,120°}$ f) $3\,e^{i\frac{5\pi}{6}}$ g) $5\,e^{-i\frac{\pi}{2}}$ h) $7\,e^{i\pi}$ i) $\dfrac{2 - i}{2 + i} \cdot e^{-i\frac{\pi}{3}}$
  Determine magnitude and angle.

5.12 What is the exponential form of the following complex numbers?
  a) $-1 - i$ b) $-1 + i$ c) $3 + 4i$ d) $-3 - 4i$ e) $2i$ f) -2 g) $1 - 2i$

5.13 Let $z = x + iy$ and z^* be the complex number conjugated to z. Find out expanded expressions for:

a) $a = \left|\frac{z}{z^*}\right|$ b) $b = \mathrm{Re}\left\{z^{-2}\right\}$ c) $c = \mathrm{Im}\left\{z^{*\,3}\right\}$ d) $d = \mathrm{Im}\left\{\left(z^3\right)^*\right\}$

5.14 Calculate

a) $\left(\frac{3+4i}{5}\right)^{10}$ b) $\left(i + \frac{1}{1+i}\right)^6$ c) $\left[(1+i)\cdot e^{-i\frac{\pi}{6}}\right]^9$

5.15 Calculate all real and complex solutions of the equations

a) $z^3 = i$

b) $z^2 = -1 + i\sqrt{3}$

c) $32\,z^5 - 243 = 0$

d) $z^3 + \frac{4}{1+i} = 0$

e) $z^4 + \frac{1+2e^{i\frac{\pi}{2}}}{2+e^{-i\frac{\pi}{2}}} = 0$

f) $z^2 - 2iz + 3 = 0$

5.16 Determine all zeros of the function $z^4 - 3z^3 + 2z^2 + 2z - 4$.

5.17 a) Calculate the complex and real impedance for the circuit sketched in Fig. 1a ($R = 100\Omega$, $C = 20\mu F$, $L = 0.2H$, $\omega = 10^6\,\frac{1}{s}$).
b) Determine the complex and real impedance for the parallel circuit sketched in Fig. 1b ($R = 100\Omega$, $L = 0.5H$, $\omega = 500\frac{1}{s}$).

Fig. 1a

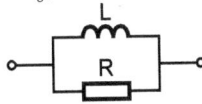

Fig. 1b

5.18 a) Calculate the complex impedance of the circuit shown in Fig. 2a as a function of ω.
b) Calculate the complex impedance of the circuit shown in Fig. 2b at an angular frequency $\omega = 300\,s^{-1}$ for the parameters $R_1 = 50\Omega$, $L_1 = 1H$, $R_2 = 300\Omega$, $C_1 = 10\mu F$, $R_3 = 20\Omega$, $L_2 = 1.5H$.

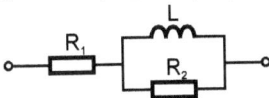

Fig. 2a

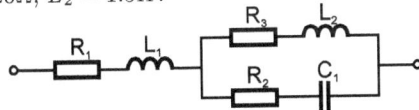

Fig. 2b

5.19 Given are the two alternating voltages $u_1(t)$ and $u_2(t)$:
$u_1(t) = 100\,V \cdot \sin(\omega t)$ $u_2(t) = 150\,V \cdot \cos\left(\omega t - \frac{\pi}{4}\right)$
Determine the alternating voltage resulting from the superposition $u_1(t) + u_2(t)$ and draw all three graphs ($\omega = 314\frac{1}{s}$) in one diagram.

5.20 The mechanical oscillations $y_1(t) = 20\,cm \cdot \sin\left(\pi t + \frac{\pi}{10}\right)$ and $y_2(t) = 15\,cm \cdot \cos\left(\pi t + \frac{\pi}{6}\right)$ are superimposed in an undisturbed manner. What is the resulting oscillation? (Calculate in the cosine representation!)

5.21 Show graphically that $3\cos\left(\omega t + \frac{\pi}{6}\right) + 2\cos\left(\omega t + \frac{\pi}{4}\right) = A\cos(\omega t + \varphi)$ with $A \approx 5$, $\varphi \approx 36°$.

Chapter 6
Limits and Continuity

6

The limit of a sequence is the fundamental concept in this chapter. It is the basis for constructing the limit of a function, which in turn is necessary for the concept of continuity. For these continuous functions, we will introduce the method of bisection to numerically determine the zeros of expressions.

6

6 Limits and Continuity

The limit of a sequence is the fundamental concept in this chapter. It is the basis for constructing the limit of a function, which in turn is necessary for the concept of continuity. For continuous functions, we will introduce the method of bisection to numerically determine the zeros of expressions. In the next chapter, we extend the evaluation of limits to calculus.

6.1 Sequences of Real Numbers

We start with a simple problem: Given is

$$-\frac{1}{2}, \frac{1}{4}, -\frac{1}{6}, \frac{1}{8}, \dots$$

Let's specify four more elements of this sequence. This means, of course $-\frac{1}{10}, \frac{1}{12}, -\frac{1}{14}, \frac{1}{16}$. Somewhat more difficult is the task of finding a regularity to determine the 100-th element of the sequence. Here, we need the formula $(-1)^n \frac{1}{2n}$, in which we substitute $n = 100$. And, we are also looking for the value of the elements for very large n. In other words, we are looking for the limit value of the sequence. We define:

> **Definition: (Sequence of Numbers).** *A **sequence of real numbers** is defined as an ordered set of real numbers indexed by $n \in \mathbb{N}$.*
> **Notation:** $(a_n)_n = a_1, a_2, a_3, \dots, a_n, \dots$.
>
> The numbers $a_1, a_2, \dots$ are called the *elements* of the sequence, a_n is the *n-th element* or the *general element* of the sequence *(sequence rule)*.

Examples 6.1:

① $(a_n)_n = 1, 2, 3, 4, \dots, n, \dots$; $\qquad a_n = n$.

② $(a_n)_n = 1, \frac{1}{2}, \frac{1}{3}, \frac{1}{4}, \dots, \frac{1}{n}, \dots$; $\qquad a_n = \frac{1}{n}$.

③ $(a_n)_n = -1, +1, -1, +1, -1, +1, \dots$; $\quad a_n = (-1)^n$.

④ $(a_n)_n = -\frac{1}{2}, \frac{1}{4}, -\frac{1}{6}, \frac{1}{8}, -\frac{1}{10}, \dots$; $\quad a_n = (-1)^n \frac{1}{2n}$.

⑤ $(a_n)_n = 0.1, 0.11, 0.111, 0.1111, \dots$; $\quad a_1 = 0.1$ and
$$a_n = a_{n-1} + 10^{-n}; \ (n \geq 2).$$

A sequence of numbers can be interpreted as a **discrete function** $F : \mathbb{N} \to \mathbb{R}$ with $n \longmapsto a(n) = a_n$. The function F assigns to each natural number n exactly one real number $F(n) = a_n$. Here the variable n appears as an index; the function values are numbered.

Representation of a number sequence: The elements of a sequence can be displayed on the real *number line*. For example, for $a_n = \frac{1}{2^{n-1}}$ we obtain the sequence

$$(a_n)_n = 1, \frac{1}{2}, \frac{1}{4}, \frac{1}{8}, \frac{1}{16}, \frac{1}{32}, \ldots$$

which is shown on the left side. According to the interpretation as a discrete function, sequences can also be represented by the function graph (see right figure). Since the function is only defined for $n \in \mathbb{N}$, the points must not be connected!

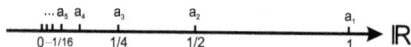

Sequence as ordered real numbers   Sequence as a discrete function

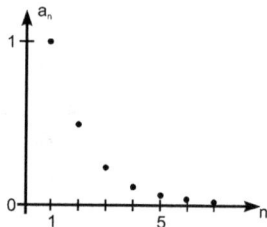

> **Limit of a Sequence**

We discuss the behavior of the sequence $a_n = 1 - \frac{1}{n}$, $(n \in \mathbb{N})$, especially for large n. So we create a table of values

n	1	2	3	4	$\cdots$	10	$\cdots$	100	$\cdots$	1000	$\cdots$	10000
a_n	0	$\frac{1}{2}$	$\frac{2}{3}$	$\frac{3}{4}$	$\cdots$	0.9	$\cdots$	0.99	$\cdots$	0.999	$\cdots$	0.9999

If we look at large n, we see that the elements of the sequence approach 1. We observe that the difference between 1 and the elements of the sequence becomes smaller as n increases. Alternatively, we can plot the sequence as a discrete function (see Fig. 6.1).

The property of this sequence is that all elements a_n are less than 1 and that as n increases, a_n approaches the number 1. So the distance between a_n and the value 1 becomes smaller as n increases:

$$|a_n - 1| \to 0 \quad \text{for} \quad n \to \infty.$$

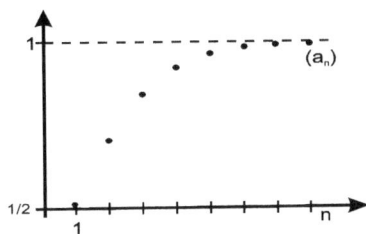

Figure 6.1. Limit of the sequence $1 - \frac{1}{n}$

The number 1 is called the *limit* of the sequence $a_n = 1 - \frac{1}{n}$.

Visualization: On the homepage there is a worksheet where sequences are specified, and if they have a limit then the convergence process is shown in the form of an animation.

Definition: (Limit).

(1) *A real number a is the* **limit** *of the sequence* $(a_n)_{n \in \mathbb{N}}$ *if for every* $\varepsilon > 0$ *there exists a natural number* n_0 *such that for all* $n \geq n_0$

$$|a_n - a| < \varepsilon.$$

(2) *A sequence is called* **convergent** *if it has a limit. Then we use the notation*

$$a_n \overset{n \to \infty}{\longrightarrow} a \quad \text{or} \quad \lim_{n \to \infty} a_n = a.$$

(3) *A sequence is called* **divergent** *if it has no limit.*

Definition (1) says that a is the limit of a sequence, if the distance between the sequence and the limit, $|a_n - a|$, becomes as small as a desired ε. All elements a_n starting from n_0 have a smaller distance to the limit value a than ε. Descriptively, this means:

> **Conclusion:** A sequence a_n converges if there exists a number (=limit) a such that the distance
>
> $$d = |a_n - a| \to 0 \quad \text{for} \quad n \to \infty.$$

Remarks:

(1) If a sequence converges to a limit a, then the index n_0 depends on the choice of the distance ε.

(2) If a sequence diverges, $a_n \to \pm\infty$ is not necessarily true.

(3) We will learn theorems that allow us to compute the limit of a sequence.

> **Conclusion:** The limit of a sequence $(a_n)_n$ is unique.

Proof: If there would be two limits a and b of the sequence $(a_n)_n$, then by definition it is $|a - a_n| \to 0$ for $n \to \infty$ and $|b - a_n| \to 0$ for $n \to \infty$. Then

$$|a - b| = |a - a_n + a_n - b| \le |a - a_n| + |a_n - b| \to 0 \text{ for } n \to \infty.$$

Hence, $|a - b| = 0$ and $a = b$. $\qquad\qquad\square$

Examples 6.2:

① The sequence

$$(a_n)_n = \left(\frac{1}{n}\right)_n = 1, \frac{1}{2}, \frac{1}{3}, \frac{1}{4}, \ldots$$

converges to the limit 0 : $\qquad \lim_{n\to\infty} a_n = \lim_{n\to\infty} \frac{1}{n} = 0.$

Sequences converging to zero are called **zero sequences**.

② The sequence

$$(a_n)_n = \left(1 + \frac{1}{2^n}\right)_n = 1.5, 1.25, 1.125, 1.0625, 1.03125, \ldots$$

converges to 1, because for the distance from a_n to 1 we have

$$d = |a_n - a| = \left|1 + \frac{1}{2^n} - 1\right| = \frac{1}{2^n} \xrightarrow{n\to\infty} 0.$$

③ The sequence

$$(a_n)_n = (n)_n = 1, 2, 3, 4, \ldots$$

is increasing to infinity and therefore divergent.

④ ⚠ The sequence

$$(a_n)_n = ((-1)^n)_n = -1, 1, -1, 1, -1, \ldots$$

has two so-called *accumulation points*, namely 1 and -1. Thus, it does not converge to **a** limit. Therefore, it is divergent.

⑤ $x \in \mathbb{R}$ is fixed.

$$(a_n)_n = (x^n)_n = x, x^2, x^3, x^4, x^5, \ldots, x^n, \ldots.$$

For x-values with $-1 < x < 1$ the sequence converges to zero, for $x = 1$ it converges to 1, for all other x-values the sequence diverges. □

To prove that a sequence converges, the limit must already be known according to the definition of convergence, since the distance $d = |a_n - a|$ must be determined. The following *monotonicity criterion* makes a statement about the convergence of a sequence without the limit being known. It says that a monotonically increasing sequence that is bounded at the top always has a limit. A monotonically decreasing sequence bounded at the bottom also has a limit.

Monotonicity Criterion

(1) Let $(a_n)_n$ be a sequence with the properties

(i) $a_n \leq a_{n+1}$ (monotonic), (ii) $a_n \leq A$ (bounded),

then the sequence converges to a limit $a \leq A$.

(2) Let $(a_n)_n$ be a sequence with the properties

(i) $a_n \geq a_{n+1}$ (monotonic), (ii) $a_n \geq A$ (bounded),

then the sequence converges to a limit $a \geq A$.

Example 6.3 (Exponential Sequence, with MAPLE-Worksheet): The sequence

$$a_n = \left(1 + \frac{1}{n}\right)^n$$

is convergent, because it is a strictly increasing sequence bounded (without proof) by e.g. 3:

n	1	10^1	10^2	10^3	10^4	10^5
a_n	2	2.59374	2.70481	2.71692	2.71814	2.71826

The limit value e is called *Euler's number*.

$$e = 2.71828\ 18284\ 59045\ 23536\ 0287\ldots.$$

We graphically represent the limit value along with an ε-environment as a function plot for the sequence $(1 + \frac{1}{n})^n$.

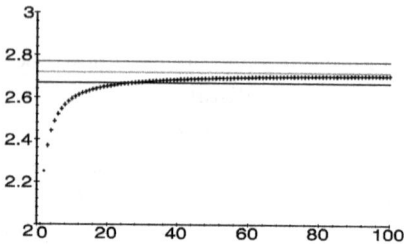

Figure 6.2. Limit of $(1 + \frac{1}{n})^n$

Euler's number e appears in many scientific applications. Mathematically, it is one of the most important real numbers. The exponential function is based on e. In the chapter Taylor Series (Volume II) we will learn an alternative method to approach the number e by a faster convergent sequence:

$$e = \sum_{k=0}^{\infty} \frac{1}{k!} = 1 + 1 + \frac{1}{2!} + \frac{1}{3!} + \frac{1}{4!} + \cdots + \frac{1}{n!} + \cdots. \qquad \square$$

Example 6.4 (Babylonian Root Calculation, with Worksheet): The *recursively* defined sequence

$$b_0 = a \quad , \quad b_{n+1} = \tfrac{1}{2}\left(b_n + \frac{a}{b_n}\right) \qquad (*)$$

is a strictly decreasing sequence for all $a > 0$ bounded by $\sqrt{a}$.

(1.) We show that $b_n^2 - a \geq 0$ to prove that b_n is bounded by $\sqrt{a}$:

$$b_n^2 - a = \frac{1}{4}\left(b_{n-1} + \frac{a}{b_{n-1}}\right)^2 - a = \frac{1}{4}\left(b_{n-1} - \frac{a}{b_{n-1}}\right)^2 \geq 0\,.$$

(2.) We show that $b_{n+1} - b_n \leq 0$ to prove that b_n is decreasing:

$$b_{n+1} - b_n = \frac{1}{2}\left(b_n + \frac{a}{b_n}\right) - b_n = \frac{a}{2b_n} - \frac{b_n}{2} = \frac{a - b_n^2}{2b_n} \leq 0\,,$$

because according to (1) $b_n^2 - a \geq 0$ and therefore $a - b_n^2 \leq 0$.

(3.) The limit of the sequence is determined by the definition equation of b_n (∗) by forming $n \to \infty$ on both sides of the equation. If the limit of the sequence is written as $b := \lim\limits_{n\to\infty} b_n = \lim\limits_{n\to\infty} b_{n+1}$, then according to the limit calculation rules

$$\lim_{n\to\infty} b_{n+1} = \lim_{n\to\infty} \frac{1}{2}\left(b_n + \frac{a}{b_n}\right)$$

$$\Rightarrow b = \frac{1}{2}\left(b + \frac{a}{b}\right)\,.$$

We solve this equation with respect to b and get

$$b^2 = a \quad \Rightarrow \quad \boxed{b = \sqrt{a}.}$$

The above sequence is therefore an approximate method of calculating square roots, which was already known to the Babylonians. In fact, it is a special case of Newton's method, which will be introduced in Section 7.8.

(4.) The following table illustrates the fast convergence of the sequence for $a = 2$:

n	1	2	3	4	5
b_n	1.5	1.41666666	1.41421568	1.41421356	1.41421356

After 4 iterations we get $\sqrt{2} \approx 1.414213562$ with an accuracy of 9 digits! □

The monotonicity criterion ensures the convergence of a sequence, but it does not provide the limit value. The limit calculation rules for sequences provide a way to compute the limit of a sequence for many, but not all, cases.

Calculation Rules for Limits

Let $(a_n)_n$ and $(b_n)_n$ be convergent sequences with $\lim\limits_{n\to\infty} a_n = a$ and $\lim\limits_{n\to\infty} b_n = b$. Let $c \in \mathbb{R}$. Then it holds

$(L1) \qquad \lim\limits_{n\to\infty} c\,a_n \qquad = c \lim\limits_{n\to\infty} a_n \qquad\qquad = c \cdot a$

$(L2) \qquad \lim\limits_{n\to\infty} (a_n \pm b_n) = \lim\limits_{n\to\infty} a_n \pm \lim\limits_{n\to\infty} b_n = a \pm b$

$(L3) \qquad \lim\limits_{n\to\infty} (a_n \cdot b_n) = \lim\limits_{n\to\infty} a_n \cdot \lim\limits_{n\to\infty} b_n = a \cdot b$

$(L4) \qquad \lim\limits_{n\to\infty} \left(\dfrac{a_n}{b_n}\right) = \lim\limits_{n\to\infty} a_n \big/ \lim\limits_{n\to\infty} b_n = \dfrac{a}{b}, \quad (b_n, b \neq 0).$

Examples 6.5 (Limit of number sequences):

① $\quad a_n = \dfrac{4n^3 - 6}{6n^3 + 2n^2} = \dfrac{4n^3 - 6}{6n^3 + 2n^2} \cdot \dfrac{\frac{1}{n^3}}{\frac{1}{n^3}} = \dfrac{4 - \frac{6}{n^3}}{6 + \frac{2}{n}} \xrightarrow{n\to\infty} \dfrac{4}{6} = \dfrac{2}{3}.$

② $\quad a_n = \dfrac{n - 1}{2n^2 + 1} = \dfrac{n - 1}{2n^2 + 1} \cdot \dfrac{\frac{1}{n^2}}{\frac{1}{n^2}} = \dfrac{\frac{1}{n} - \frac{1}{n^2}}{2 + \frac{1}{n^2}} \xrightarrow{n\to\infty} \dfrac{0}{2} = 0.$

③ $\quad a_n = \dfrac{3^{n+1} + 2^n}{3^n + 1} = \dfrac{3^{n+1} + 2^n}{3^n + 1} \cdot \dfrac{\frac{1}{3^n}}{\frac{1}{3^n}} = \dfrac{3 + \left(\frac{2}{3}\right)^n}{1 + \left(\frac{1}{3}\right)^n} \xrightarrow{n\to\infty} \dfrac{3}{1} = 3,$

here $\left(\frac{2}{3}\right)^n \to 0$ and $\left(\frac{1}{3}\right)^n \to 0$ for $n \to \infty$ according to Example 6.2 ⑤.

Tip: In the examples, the quotient is extended with the reciprocal of the leading term. This manipulation is necessary to be able to apply the limit rules for **convergent** sequences. □

6.2 Function Limits

In Section 6.1 we examined the limits of number sequences $(x_n)_{n\in\mathbb{N}}$. This procedure is now extended directly to *function limits* by considering sequences of the form $(f(x_n))_{n\in\mathbb{N}}$. As an introduction, we will examine the behavior of the function $f(x) = x^2$ at the point $x_0 = 2$. To do this, we will use the sequence

$$(x_n)_n = 1.9,\ 1.99,\ 1.999,\ 1.9999, \ldots \xrightarrow{n\to\infty} 2$$

and calculate the value of the function for each element of the sequence

$$(f(x_n))_n = 3.61, 3.9601, 3.996, 3.9996, \ldots \xrightarrow{n\to\infty} 4.$$

The **sequence of function values** converges to the value 4.

To obtain the function limit for a given function f at a position x_0, we choose a sequence of numbers $x_n \xrightarrow{n\to\infty} x_0$ from the domain of f and apply the function f to x_n. Then the convergence of the sequence $(f(x_n))_n$ (= *limit of the function at the position x_0*) is checked. In our example, for any other sequence $(x_n)_n$ that converges to the value 2, $f(x_n) \xrightarrow{n\to\infty} 4$ is also valid:

$$\lim_{\substack{n\to\infty}} f(x_n) = \lim_{\substack{x\to 2 \\ (x<2)}} f(x) = \lim_{\substack{x\to 2 \\ (x<2)}} x^2 = 4.$$

Since the elements of the sequence are $x < 2$, this limit is called the *left-sided limit* of $f(x) = x^2$ at the position $x_0 = 2$.

Example 6.6 (With MAPLE-Worksheet). This behavior can be illustrated by plotting both the sequence $(x_n)_n$ and the function sequence $(f(x_n))_n$ on a graph. For a clearer presentation, we now choose the sequence $x_n = 2 - \frac{1}{n^2} \xrightarrow{n\to\infty} 2$:

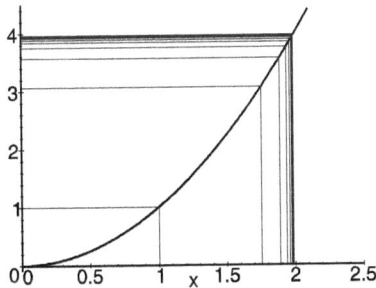

Figure 6.3. Left limit of the function at $x_0 = 2$

The x_n values approach the number 2 from the left; the $f(x_n)$ function values approach the y value 4. Similarly, the *right-sided* limit of the function at $x_0 = 2$ is obtained by using the following sequence of numbers, for example

$$(x_n)_n = 2.1, 2.01, 2.001, 2.0001, \ldots \to 2.$$

The corresponding function sequence is

$$(f(x_n))_n = 4.41, 4.041, 4.004, 4.0004, \ldots \to 4.$$

Again, the function limit is independent of the selected sequence of numbers x_n. For the right-sided limit, we get

$$\lim_{\substack{n \to \infty}} f(x_n) = \lim_{\substack{x \to 2 \\ (x>2)}} f(x) = \lim_{\substack{x \to 2 \\ (x>2)}} x^2 = 4.$$

So for the function $f(x) = x^2$ with $x_0 = 2$ both the left and the right limits of the function exist and are both equal to 4.   □

Definition: (Function Limit). *Let f be a function defined in an environment of x_0. If for* **any** *sequence $(x_n)_n$ within the domain of f converging to x_0, it is*

$$\lim_{n \to \infty} f(x_n) = g \in \mathbb{R},$$

then g is called the **limit** *of $f(x)$ for $x_n \overset{n \to \infty}{\longrightarrow} x_0$.*

Notation: $\lim_{n \to \infty} f(x_n) = \lim_{x \to x_0} f(x) = g$, *when $x_n \overset{n \to \infty}{\longrightarrow} x_0$.*

Remarks:

(1) It is **not** required that x_0 is an element of the function's domain.

(2) $x \to x_0$ means that x comes as close as possible to the position x_0, **without** reaching the value x_0!

(3) It can happen that, although $x_0 \notin \mathbb{D}$, the function limit (i.e. the left limit equals the right limit) exists.

(4) The *left-sided limit* is often also written as

$$g_l := \lim_{\substack{x \to x_0 \\ (x<x_0)}} f(x) = \lim_{h \to 0} f(x_0 - h)$$

and the *right-sided limit*

$$g_r := \lim_{\substack{x \to x_0 \\ (x>x_0)}} f(x) = \lim_{h \to 0} f(x_0 + h).$$

Examples 6.7:

① The **Heaviside Function:**

$$f : \mathbb{R} \to \mathbb{R} \text{ with } f(x) = \begin{cases} 0 & \text{for } x < 0 \\ 1 & \text{for } x \geq 0 \end{cases}$$

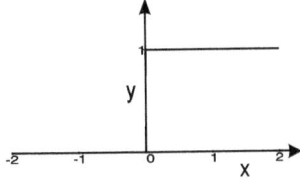

The Heaviside function is zero for negative values of x and 1 for positive values of x. In applications, it is often referred to as a step function $S(x)$ or a power-on function. The Heaviside function is not continuous at $x_0 = 0$ because the right limit is not equal to the left limit:

$$g_l = \lim_{h \to 0} f(x_0 - h) = \lim_{h \to 0} f(-h) = \lim_{h \to 0} 0 = 0,$$

$$g_r = \lim_{h \to 0} f(x_0 + h) = \lim_{h \to 0} f(h) = \lim_{h \to 0} 1 = 1.$$

② For the function

$$f : \mathbb{R} \setminus \{2\} \to \mathbb{R} \text{ with } f(x) = \frac{x^2 - 2x}{x - 2}$$

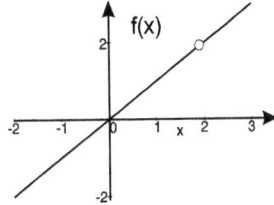

the function limit exists at the point $x_0 = 2$, although $x_0 \notin \mathbb{D}$:

$$g_l = \lim_{\substack{x \to x_0 \\ (x < x_0)}} f(x) = \lim_{\substack{x \to 2 \\ (x < 2)}} \frac{x^2 - 2x}{x - 2} = \lim_{\substack{x \to 2 \\ (x < 2)}} x = 2,$$

$$g_r = \lim_{\substack{x \to x_0 \\ (x > x_0)}} f(x) = \lim_{\substack{x \to 2 \\ (x > 2)}} \frac{x^2 - 2x}{x - 2} = \lim_{\substack{x \to 2 \\ (x > 2)}} x = 2.$$

The factor $(x - 2)$ appears in both the numerator and the denominator, so it can be reduced.

③ The function

$$f : \mathbb{R} \setminus \{0\} \to \mathbb{R} \text{ with } f(x) = \frac{1}{x}$$

has **no** limit at $x_0 = 0$, because

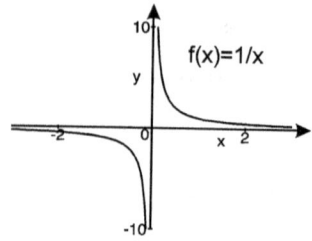

f(x)=1/x

$$g_l = \lim_{h \to 0} f(x_0 - h) = \lim_{h \to 0} -\frac{1}{h} = -\infty,$$

$$g_r = \lim_{h \to 0} f(x_0 + h) = \lim_{h \to 0} \frac{1}{h} = +\infty.$$ □

The question of convergence is not always as straightforward as in the examples above. Then additional geometric considerations are required.

Example 6.8. $\lim_{x \to 0} \dfrac{\sin(x)}{x} = ?$

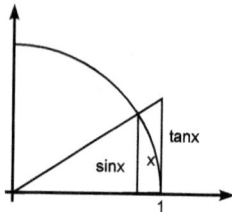

Geometrically, the limit of the function $f(x) = \frac{\sin x}{x}$ at $x_0 = 0$ can be estimated: Looking at the unit circle we see that

$$\tan x > x > \sin x \text{ for } 0 < x < \tfrac{1}{2}\pi.$$

$$\Rightarrow \quad \frac{1}{\cos x} > \frac{x}{\sin x} > 1 \quad \Rightarrow \quad \cos x < \frac{\sin x}{x} < 1$$

When we take the limit, we have to replace $<$ by $\leq$. So

$$1 = \lim_{x \to 0} \cos x \leq \lim_{x \to 0} \frac{\sin x}{x} \leq 1 \quad \Longrightarrow \quad \lim_{x \to 0} \frac{\sin x}{x} = 1.$$ □

Example 6.9. Similar geometric considerations lead to the formula

$$\lim_{x \to 0} \frac{e^x - 1}{x} = 1.$$

Because for small, positive values of x it is

$$1 + x < e^x < 1 + x + x^2.$$

Hence $x < e^x - 1 < x(x+1)$ or $1 < \frac{e^x - 1}{x} < x+1$. For $x \to 0$ we obtain the formula.   □

Behavior of Functions at Infinity ($x \to \infty$):

If for **every** sequence $(x_n)_{n \in \mathbb{N}}$ from the domain of f with $x_n \overset{n \to \infty}{\Longrightarrow} \infty$ also $f(x_n) \overset{n \to \infty}{\Longrightarrow} g$ converges, then g is called *the limit of the function* $x \to \infty$:

$$\lim_{x \to \infty} f(x) = g.$$

Similarly, the function limit is defined at $x \to -\infty$. To calculate the function limits, both for $x_n \to x_0$ and for $x_n \to \pm\infty$, the same calculation rules apply as for sequences of real numbers:

Calculation Rules for Function Limits

Assuming that the limits $\lim\limits_{x \to x_0} f(x)$ and $\lim\limits_{x \to x_0} g(x)$ exist, the following rules apply ($c \in \mathbb{R}$):

(F1) $\lim\limits_{x \to x_0} c f(x) = c \lim\limits_{x \to x_0} f(x).$

(F2) $\lim\limits_{x \to x_0} (f(x) \pm g(x)) = \lim\limits_{x \to x_0} f(x) \pm \lim\limits_{x \to x_0} g(x).$

(F3) $\lim\limits_{x \to x_0} (f(x) \cdot g(x)) = \lim\limits_{x \to x_0} f(x) \cdot \lim\limits_{x \to x_0} g(x).$

(F4) $\lim\limits_{x \to x_0} \dfrac{f(x)}{g(x)} = \dfrac{\lim\limits_{x \to x_0} f(x)}{\lim\limits_{x \to x_0} g(x)},$ if $\lim\limits_{x \to x_0} g(x) \neq 0.$

Remarks:

(1) In order to apply rule ($F4$), we must not only exclude $g(x_0) = 0$, but also guarantee that $g(x) \neq 0$ for the sequence $x \to x_0$.

(2) These rules also apply to the limits of functions for $x \to \pm\infty$, if the limits $\lim\limits_{x \to \pm\infty} f(x)$ and $\lim\limits_{x \to \pm\infty} g(x)$ exist.

(3) For limits of type $\frac{0}{0}$ and $\frac{\infty}{\infty}$ we must apply the *rules of l'Hospital* described in Section 7.7.3.

Examples 6.10:

① $\displaystyle \lim_{x \to 0} \frac{x^2 - 2x + 5}{\cos x} = \frac{\displaystyle\lim_{x \to 0}\left(x^2 - 2x + 5\right)}{\displaystyle\lim_{x \to 0} \cos x} = \frac{5}{1} = 5.$

② $\displaystyle \lim_{x \to \infty} \frac{2x^2 + 4}{x^2 - 1} = \lim_{x \to \infty} \frac{2x^2 + 4}{x^2 - 1} \cdot \frac{\frac{1}{x^2}}{\frac{1}{x^2}} = \lim_{x \to \infty} \frac{2 + \frac{4}{x^2}}{1 - \frac{1}{x^2}} = \frac{2}{1} = 2.$

③ $\displaystyle \lim_{x \to 1} \frac{x - 1}{x^2 - 1} = \lim_{x \to 1} \frac{x - 1}{(x - 1)(x + 1)} = \lim_{x \to 1} \frac{1}{x + 1} = \frac{1}{2}.$

④ $\displaystyle \lim_{x \to \infty} \frac{4 + 2x}{x^2 + 1} = \lim_{x \to \infty} \frac{4 + 2x}{x^2 + 1} \cdot \frac{\frac{1}{x^2}}{\frac{1}{x^2}} = \lim_{x \to \infty} \frac{\frac{4}{x^2} + \frac{2}{x}}{1 + \frac{1}{x^2}} = \frac{0}{1} = 0.$

⑤ $\displaystyle \lim_{x \to 0} \frac{\sqrt{x + 1} - 1}{x} = \lim_{x \to 0} \frac{\left(\sqrt{x + 1} - 1\right)\left(\sqrt{x + 1} + 1\right)}{x\left(\sqrt{x + 1} + 1\right)}$

$\displaystyle \qquad\qquad = \lim_{x \to 0} \frac{(x + 1) - 1}{x\left(\sqrt{x + 1} + 1\right)} = \lim_{x \to 0} \frac{1}{\sqrt{x + 1} + 1} = \frac{1}{2}.$　□

6.3 Continuity of a Function

A function $f : \mathbb{R} \to \mathbb{R}$ is said to be *continuous* if the graph has no jumps. For most applications this illustrative interpretation is sufficient. For example, the function $f : \mathbb{R} \to \mathbb{R}$ with $f(x) = x^2$ is continuous in this sense (see Fig. 6.4, (a)).

However, to classify situations like the function $f : \mathbb{R}\setminus\{\frac{\pi}{2} + k\pi, \, k \in \mathbb{Z}\} \to \mathbb{R}$ with $f(x) = \tan x$ (see Fig. 6.4 (b)), a more precise definition is required.

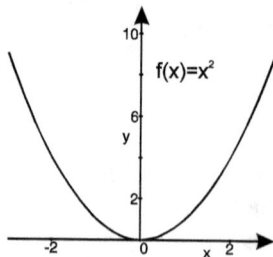

(a) Square function　　　　　　　　　　　(b) Tangent

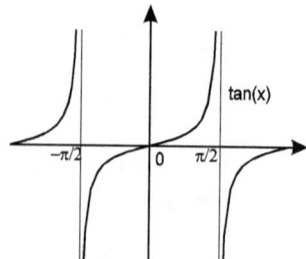

Figure 6.4. Continuous functions

Definition: (Continuity). *Let* $x_0 \in \mathbb{D}$ *and let the function* f *be defined in an environment of* x_0. *The function* f *is said to be* **continuous** *at* x_0 *if the function limit exists at* x_0 *and matches the function value* $f(x_0)$.

In short: f **is continuous** *at* $x_0 \in \mathbb{D}$, *if*

$$\lim_{h \to 0} f(x_0 + h) = \lim_{h \to 0} f(x_0 - h) = f(x_0).$$

Remarks:

(1) The continuity at point x_0 assumes that $x_0 \in \mathbb{D}$. Points where f is not defined are *definition gaps*. There, the continuity is not checked.

(2) If f is continuous at every point $x \in \mathbb{D}$, then f is said to be a *continuous function*.

(3) At a point x_0 the continuity of a function can also be rewritten:

$$\lim_{x \to x_0} f(x) = f\left(\lim_{x \to x_0} x\right) = f(x_0).$$

Examples 6.11:

① A **polynomial** $f : \mathbb{R} \to \mathbb{R}$ with $f(x) = a_0 + a_1 x + \cdots + a_n x^n$ is **continuous** at every point $x \in \mathbb{R}$.

② The **absolute value function**

$f : \mathbb{R} \to \mathbb{R}$ with $f(x) = |x|$

is **continuous** at any point, especially at $x_0 = 0$:

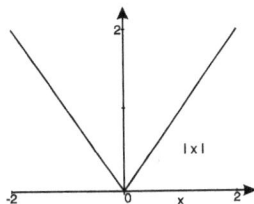

$$\lim_{h \to 0} f(x_0 + h) = \lim_{h \to 0} f(h) = \lim_{h \to 0} h = 0$$
$$\lim_{h \to 0} f(x_0 - h) = \lim_{h \to 0} f(-h) = \lim_{h \to 0} |-h| = 0$$
$$f(0) = 0.$$

③ The **Step function (Heaviside function)**, which is used to describe switch-on processes, $S : \mathbb{R} \to \mathbb{R}$ with

$$S(x) = \begin{cases} 1 & \text{for } x \geq 0 \\ 0 & \text{for } x < 0 \end{cases}$$

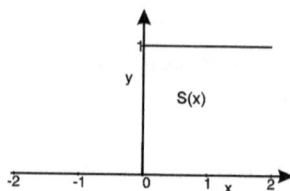

is **not continuous** at $x_0 = 0$, because it has a jump:

$$\lim_{h \to 0} f(x_0 + h) = \lim_{h \to 0} S(h) = 1,$$
$$\lim_{h \to 0} f(x_0 - h) = \lim_{h \to 0} S(-h) = 0.$$

④ The **Sign function (Signum function)** $sign : \mathbb{R} \to \mathbb{R}$ with

$$sign(x) = \begin{cases} 1 & \text{for } x > 0 \\ 0 & \text{for } x = 0 \\ -1 & \text{for } x < 0 \end{cases}$$

is **not continuous** at $x_0 = 0$.

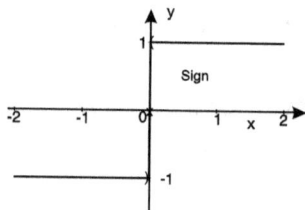

⑤ The function

$$f : \mathbb{R} \setminus \{2\} \to \mathbb{R}$$

with $f(x) = \dfrac{x^2 - 2x}{x - 2}$

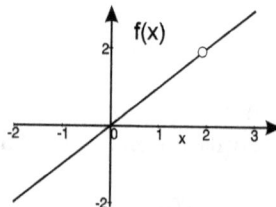

has a definition gap at $x_0 = 2$. According to the Example 6.7 ② it holds: Both the right and the left limit at $x_0 = 2$ exist and are equal. So we define the **continuous extension** of f:

$$\tilde{f} : \mathbb{R} \to \mathbb{R} \quad \text{with} \quad \tilde{f}(x) = \begin{cases} f(x) & \text{for } x \neq 2 \\ 2 & \text{for } x = 2 \end{cases}.$$

The function $\tilde{f}$ is continuous at x_0 and thus for all $x \in \mathbb{R}$. Often the notation $\tilde{f}$ is omitted and the same function name f is used as the name for the continuous extension. ☐

6.4 Interval Halving Method (Bisection Method)

A simple numerical method for finding the zeros of a function is based on the following descriptive proposition: Every continuous function that changes its sign in an interval $[a, b]$ has a zero in that interval (see Fig. 6.5):

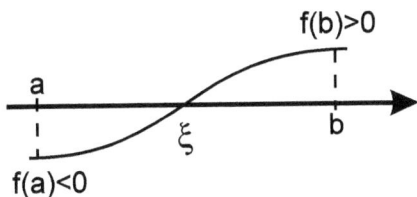

Figure 6.5. Interval Halving Method

Intermediate Value Theorem

Let $f : [a, b] \to \mathbb{R}$ be a continuous function with different signs at $f(a)$ and $f(b)$. This means either $(f(a) < 0$ and $f(b) > 0)$ or $(f(a) > 0$ and $f(b) < 0)$. Then there exists an intermediate point $\xi \in (a, b)$ with the property

$$f(\xi) = 0.$$

The idea of the proof is to calculate the center of the interval m and to compare the function values

$$f(a), f(m), f(b).$$

Replace this interval boundary by m, whose function value has the same sign as $f(m)$. Then repeat the procedure for the halved interval, and so on.

This method directly returns an algorithm (iteration) to calculate a zero in the interval $[a, b]$. The only requirement is that the function values at the interval boundaries have different signs. Since the interval length is always halved, this method is also called the interval halving method.

Visualization: The animation shows that the bisection method converges to the zero contained in the interval. At each iteration step, the interval is halved by replacing the right or left boundary of the interval. □

Remarks:

(1) ⚠ The function f can have multiple zeros in the interval $[a, b]$. However, the bisection method returns only one.

(2) Algorithms that enclose the zero point in a systematically decreasing interval are called *inclusion algorithms*. Starting from an initial interval, this interval is systematically reduced by the same algorithm. This process is called an *iteration*.

(3) The number of iterations required for a given accuracy can be estimated before the calculation, since the distance of the zero point from the left interval boundary is

$$|a_n - \xi| \leq \frac{b-a}{2^n} \quad \text{for } n = 1, 2, 3, \ldots.$$

If f is a continuous function on $[0, 1]$ with a change of sign and a zero point $\xi \in (0, 1)$ is to be determined up to an accuracy of 4 decimal places, then the distance of $|a_n - \xi|$ must not be greater than $5 \cdot 10^{-5}$. n must be chosen so that $\frac{b-a}{2^n} \leq 5 \cdot 10^{-5} \Rightarrow n = 14.$
(With an accuracy of 6 decimals, n is already 21.)

Example 6.12 (With MAPLE-Procedure): Given is the function

$$f(x) = x^3 - \sqrt{x^2 + 1}.$$

Since $f(1) = -0.4142$ and $f(2) = 5.7639$ have different signs, there is a zero in the interval $[1, 2]$. The bisection method applies:

n	a	b	$f(\frac{a+b}{2})$	
	1.0	2.0	1.5722	
1	1.0	1.5	0.3523	
2	1.0	1.25	−0.0813	
3	1.125	1.25	0.1220	
4	1.125	1.1875	0.0171	
5	1.125	1.1562	−0.0329	
6	1.1406	1.1562	−0.0081	
7	1.1484	1.1562	0.0044	
8	1.1484	1.1523	−0.0018	
9	1.1503	1.1523	0.0013	
10	1.1503	1.1513	$-2 \cdot 10^{-4}$	
11	1.1508	1.1513	$5 \cdot 10^{-4}$	
12	1.1508	1.1511	$1 \cdot 10^{-4}$	$\Rightarrow \quad \xi \approx 1.15095 \quad \square$

6.5 Problems on Limits and Continuity

6.1 For which $n \in \mathbb{N}$ apply the next inequalities

a) $\left| \dfrac{1}{n^2} \right| < 10^{-6}$ b) $\left| \dfrac{1}{n^2} + 1 \right| < 1 + 10^{-8}$ c) $\left| \dfrac{1}{n+1} \right| < 10^{-10}$

6.2 Determine - if possible - the limit of the number sequences for $n \to \infty$

a) $a_n = \dfrac{2n+1}{4n}$ b) $a_n = \dfrac{n^2+4}{n}$ c) $a_n = \dfrac{n^2+4n-1}{n^2-3n}$

d) $a_n = \dfrac{5n^3+4n+1}{n^3}$ e) $a_n = \dfrac{1}{n} \left(\sum_{k=1}^{n} k \right) - \dfrac{1}{2}n$ f) $a_n = \dfrac{4(n+1)^4}{3n^4+3n+5}$

g) $a_n = \dfrac{4n+1}{5n-1}$ h) $a_n = \dfrac{3n^2+4n}{\sqrt[3]{n^6+n^4+1}}$ i) $\sin\left(\dfrac{\pi n^3+n^2}{2(n^3+4)} \right)$

6.3 Show that the sequence $a_n = \dfrac{2n+1}{4n}$ converges.

6.4 a) Given is the recursively defined sequence a_n with $a_0 = 0$ and $a_{n+1} = \frac{1}{3}(a_n + 1)$ for $n \geq 0$. Determine the general rule a_n and calculate the limit of the sequence.

b) Show: If $\lim_{n\to\infty} \left| \dfrac{a_{n+1}}{a_n} \right| < 1$, then a_n is a null sequence.

6.5 Prove the limit calculation rules

(L1) $\lim_{n\to\infty} (a_n + b_n) = \lim_{n\to\infty} a_n + \lim_{n\to\infty} b_n$

(L2) $\lim_{n\to\infty} (a_n \cdot b_n) = \left(\lim_{n\to\infty} a_n \right) \cdot \left(\lim_{n\to\infty} b_n \right)$

6.6 Calculate the limits

a) $\lim_{n\to\infty} \dfrac{1}{n} \left(1 + \dfrac{1}{2} + \ldots + \dfrac{1}{n} \right)$ b) $\lim_{n\to\infty} \dfrac{n}{\sqrt[n]{n!}}$

6.7 Calculate the limits of the functions:

a) $\lim_{x\to 1} (x^3 + 5x^2 - 3x + 4)$ b) $\lim_{x\to 0} \dfrac{x^2 - 2x}{x^2 + 3x}$ c) $\lim_{x\to\infty} \dfrac{x}{1 + x^2}$

6.8 Calculate the function limits

a) $\lim_{x\to 1} \dfrac{x^2 - 1}{x^2 + 1}$ b) $\lim_{x\to -3} \dfrac{x^2 - x - 12}{x + 3}$ c) $\lim_{x\to 0} \dfrac{\sin(2x)}{\sin(x)}$

d) $\lim_{x\to 2} \dfrac{(x-2)(3x+1)}{4x-8}$ e) $\lim_{x\to 0} \dfrac{\sqrt{1+x}-1}{x}$ f) $\lim_{x\to\infty} \dfrac{x^2}{x^2 - 4x + 1}$

6.9 Which limit value does the function have $f(x) = \dfrac{1-x}{1-\sqrt{x}}$ for $x \to 1$?

6.10 Show that the function $f(x) = \begin{cases} x & x \leq 0 \\ x - 2 & x > 0 \end{cases}$ is discontinuous at $x_0 = 0$.

6.11 Show that the function $f(x) = \begin{cases} \frac{x^2-1}{x-1} & \text{for } x \neq 1 \\ 2 & \text{for } x = 1 \end{cases}$ is continuous at $x_0 = 1$.

6.12 Can the definition gaps of the function $f(x) = \dfrac{x^2 - x}{x^3 - x^2 + x - 1}$ be continuously extended?

Chapter 7
Differential Calculus

7

One of the most important tasks in applied mathematics is to calculate the derivative of a function. Many physical laws are described by the derivative of a physical quantity. For example, if the *space-time law* $s(t)$ is given for a motion, then the velocity $v(t)$ is the derivative of this space-time law with respect to time t. The concrete determination of the velocity requires that we are able to perform this manipulation on the function $s(t)$.

Whenever physical quantities change non-linearly with time, the derivative is needed to describe the change. Also the determination of the extreme values of a function or optimization tasks lead to the problem of the derivative of a function. The derivative is thus motivated by the theoretical description of physical quantities (velocity, acceleration) and by the mathematical description of curves (tangent, curve discussion).

7

7 Differential Calculus

One of the most important tasks in applied mathematics is to calculate the derivative of a function. Many physical laws are described by the derivative of a physical quantity. For example, if the *space-time law* $s(t)$ is given for a motion, then the velocity $v(t)$ is the derivative of this space-time law with respect to time t. The concrete determination of the velocity requires that we are able to perform this manipulation on the function $s(t)$.

Whenever physical quantities change non-linearly with time, the derivative is needed to describe the change. Also the determination of the extreme values of a function or optimization tasks lead to the problem of the derivative of a function. The derivative is thus motivated by the theoretical description of physical quantities (velocity, acceleration) and by the mathematical description of curves (tangent, curve discussion).

7.1 Introduction

One of the most important methods in applied mathematics is the differentiation of a function. As an example, we consider two simple space-time laws from kinematics.

Application Example 7.1 (Uniform Motion).

Fig. 7.1 shows the space-time law of a uniform motion. If the path changes within $\triangle t$ by $\triangle s$, the body moves with the **velocity**

$$v := \frac{\triangle s}{\triangle t}.$$

The velocity does not depend on the time t or on the interval $\triangle t$.

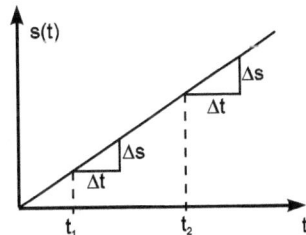

Figure 7.1. Uniform motion

Application Example 7.2 (Uniformly Accelerated Motion).

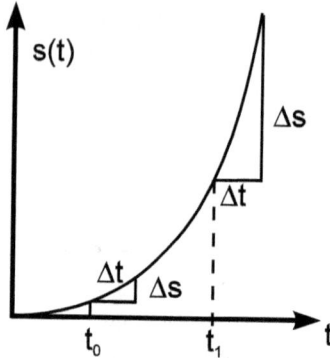

Figure 7.2. Accelerated motion

Fig. 7.2 shows the space-time law of a uniformly accelerated motion.

$$s(t) = \tfrac{1}{2} g t^2.$$

By means of a hypothetical series of measurements, we will determine the value of the velocity at time $t_0 = 1$. It turns out that the "measured" velocity v now depends on the measurement interval $\triangle t$: The smaller $\triangle t$, the more accurate is the value of the velocity at that time. We calculate the ratio $\frac{\triangle s}{\triangle t}$ depending on $\triangle t$:

$$v(t_0) = \frac{\triangle s}{\triangle t} = \frac{\tfrac{1}{2} g (t_0 + \triangle t)^2 - \tfrac{1}{2} g t_0^2}{\triangle t} = \frac{1}{2} g \frac{(t_0 + \triangle t)^2 - t_0^2}{\triangle t}.$$

Measurement interval	Velocity	
$\triangle t = 1$	$v(1) = \tfrac{1}{2} g \frac{4-1}{1}$	$= \tfrac{3}{2} g$
$\triangle t = \tfrac{1}{2}$	$v(1) = \tfrac{1}{2} g \frac{\left(1+\tfrac{1}{2}\right)^2 - 1}{\tfrac{1}{2}}$	$= \tfrac{5}{4} g$
$\triangle t = \tfrac{1}{4}$	$v(1) = \tfrac{1}{2} g \frac{\left(1+\tfrac{1}{4}\right)^2 - 1}{\tfrac{1}{4}}$	$= \tfrac{9}{8} g$
$\triangle t = \tfrac{1}{8}$	$v(1) = \tfrac{1}{2} g \frac{\left(1+\tfrac{1}{8}\right)^2 - 1}{\tfrac{1}{8}}$	$= \tfrac{17}{16} g$
$\triangle t = \tfrac{1}{16}$	$v(1) = \tfrac{1}{2} g \frac{\left(1+\tfrac{1}{16}\right)^2 - 1}{\tfrac{1}{16}}$	$= \tfrac{33}{32} g$

Observation: The smaller the measurement interval $\triangle t$, the closer the velocity approaches the value g. Mathematics provides a tool, called *Differential Calculus*, to make $\triangle t$ equal to zero.

$$v(t_0) := \lim_{\triangle t \to 0} \frac{s(t_0 + \triangle t) - s(t_0)}{\triangle t} = \left.\frac{ds}{dt}\right|_{t_0} \qquad \text{(Instantaneous Velocity)}$$

For each t_0 the following applies

$$v\left(t_0\right) = \lim_{\triangle t \to 0} \frac{s\left(t_0 + \Delta t\right) - s\left(t_0\right)}{\Delta t} = \lim_{\triangle t \to 0} \frac{1}{2}\, g\, \frac{\left(t_0 + \Delta t\right)^2 - t_0^2}{\Delta t}$$

$$= \lim_{\triangle t \to 0} \frac{1}{2}\, g\, \frac{2\,\Delta t\, t_0 + \left(\Delta t\right)^2}{\Delta t} = \lim_{\triangle t \to 0} \frac{1}{2}\, g\, \left(2\, t_0 + \Delta t\right) = g t_0 \qquad \square$$

This procedure can be applied to any function:

> **Definition: (Derivative of a Function).** *A function* $f : \mathbb{D} \to \mathbb{R}$ *is said to be* **differentiable** *at* $x_0 \in \mathbb{D}$ *if there exists the limit*
>
> $$f'\left(x_0\right) := \lim_{\triangle x \to 0} \frac{f\left(x_0 + \triangle x\right) - f\left(x_0\right)}{\triangle x} := \lim_{x \to x_0} \frac{f\left(x\right) - f\left(x_0\right)}{x - x_0}.$$
>
> *It is called the first* **derivative** *of the function* f *at* x_0. *A function is differentiable if it can be differentiated at all points* $x \in \mathbb{D}$ *of its domain.*

Notes:

(1) The derivative of the function $y = f\left(x\right)$ is also represented by the symbols

$$f'\left(x\right),\; y',\; \frac{dy}{dx},\; \frac{d\,f\left(x\right)}{dx}.$$

(2) The calculation of the derivative is called the *Differentiation of* $f\left(x\right)$.

(3) The quotient

$$\frac{f\left(x_0 + \triangle x\right) - f\left(x_0\right)}{\triangle x}$$

is called the *difference quotient* and

$$\lim_{\triangle x \to 0} \frac{f\left(x_0 + \triangle x\right) - f\left(x_0\right)}{\triangle x}$$

is the *differential quotient*.

(4) For $\triangle x \to 0$, both the numerator and the denominator of the difference quotient approach zero. So the case is $\frac{0}{0}$! To determine the derivative for elementary functions, the fraction must be transformed until $\triangle x$ disappears. Then in the remaining term $\triangle x \to 0$ is taken into account (see Examples 7.3).

Geometric Interpretation of the Derivative: Given is a function $f(x)$ and a point $x_0 \in \mathbb{D}$. Then the **gradient of the secant** defined by the points $P(x_0, f(x_0))$ and $Q(x_0 + \triangle x, f(x_0 + \triangle x))$ is given by

$$\frac{f(x_0 + \triangle x) - f(x_0)}{\triangle x} \qquad \text{(Difference quotient)}.$$

Performing $\triangle x \to 0$ keeps the point P fixed, but moves Q along the curve towards P. The slope of the secant changes to the slope of the tangent at the point $P(x_0, f(x_0))$. **The derivative of a function at x_0 is therefore the slope of the tangent at point $(x_0, f(x_0))$.**

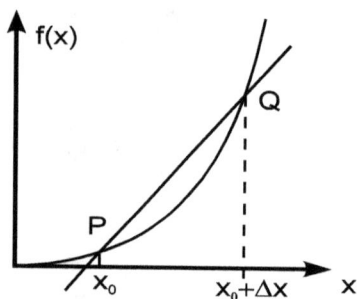

Figure 7.3. Difference quotient as slope of secant

Visualization: The animation shows the movement of point Q along the curve f towards point P. The secant approaches the tangent and the slope of the secant becomes the slope of the tangent.

Numerical Differentiation: The limit $h \to 0$ cannot be calculated numerically, because it would immediately cause an *Overflow*. The fraction would be divided by 0. To calculate the derivative of a function f at position x_0 *approximately*, we replace the derivative of a function f at point x_0 by the secant slope

$$f'(x_0) \approx \frac{f(x_0 + h) - f(x_0)}{h}$$

where $h > 0$. This is called the **right-sided difference formula**. A more accurate approximation is obtained by taking the average of the right and left formulas. This gives the **central difference formula**:

$$f'(x) \approx \frac{1}{2} \frac{f(x_0 + h) - f(x_0 - h)}{h}.$$

Examples 7.3:

① $\boxed{f(x) = c \;\Rightarrow\; f'(x) = 0}$ (Constant function):

$$f'(x) = \lim_{h \to 0} \frac{f(x+h) - f(x)}{h} = \lim_{h \to 0} \frac{c - c}{h} = 0.$$

② $\boxed{f(x) = x \;\Rightarrow\; f'(x) = 1}$ (Identical function):

$$f'(x) = \lim_{h \to 0} \frac{f(x+h) - f(x)}{h} = \lim_{h \to 0} \frac{x + h - x}{h} = 1.$$

③ $\boxed{f(x) = x^2 \;\Rightarrow\; f'(x) = 2x}$ (Quadratic function):

$$f'(x) = \lim_{h \to 0} \frac{f(x+h) - f(x)}{h} = \lim_{h \to 0} \frac{(x+h)^2 - x^2}{h}$$
$$= \lim_{h \to 0} \frac{2hx + h^2}{h} = \lim_{h \to 0} (2x + h) = 2x.$$

④ $\boxed{f(x) = \dfrac{1}{x} \;\Rightarrow\; f'(x) = -\dfrac{1}{x^2}}$ ($x \neq 0$):

$$f'(x) = \lim_{h \to 0} \frac{1}{h} (f(x+h) - f(x)) = \lim_{h \to 0} \frac{1}{h} \left(\frac{1}{x+h} - \frac{1}{x} \right)$$
$$= \lim_{h \to 0} \frac{1}{h} \frac{-h}{x(x+h)} = -\frac{1}{x^2}.$$

⑤ $\boxed{f(x) = e^x \;\Rightarrow\; f'(x) = e^x}$ (Exponential function):

$$f'(x) = \lim_{h \to 0} \frac{1}{h} \left(e^{x+h} - e^x \right) = \lim_{h \to 0} \frac{1}{h} e^x \left(e^h - 1 \right)$$
$$= e^x \lim_{h \to 0} \frac{1}{h} \left(e^h - 1 \right).$$

According to the Example 6.9
$$\lim_{h \to 0} \tfrac{1}{h} \left(e^h - 1 \right) = 1, \text{ so that } \boxed{(e^x)' = e^x}.$$

Table 7.1: Derivatives of Elementary Functions

	$f(x)$	$f'(x)$
Power function	x^n	$n\,x^{n-1}$
Trigonometric	$\sin(x)$	$\cos(x)$
funtions	$\cos(x)$	$-\sin(x)$
	$\tan(x)$	$\dfrac{1}{\cos^2(x)}$
	$\cot(x)$	$-\dfrac{1}{\sin^2(x)}$
Arc functions	$\arcsin(x)$	$\dfrac{1}{\sqrt{1-x^2}}$
	$\arccos(x)$	$-\dfrac{1}{\sqrt{1-x^2}}$
	$\arctan(x)$	$\dfrac{1}{1+x^2}$
	$\operatorname{arccot}(x)$	$-\dfrac{1}{1+x^2}$
Exponential function	e^x	e^x
	a^x	$a^x\,\ln a$
Logarithmic function	$\ln(x)$	$\dfrac{1}{x}$
	$\log_a(x)$	$\dfrac{1}{\ln(a)\cdot x}$
Hyperbolic functions	$\sinh(x)$	$\cosh(x)$
	$\cosh(x)$	$\sinh(x)$
	$\tanh(x)$	$\dfrac{1}{\cosh^2(x)}$
	$\coth(x)$	$-\dfrac{1}{\sinh^2(x)}$
Area functions	$ar\sinh(x)$	$\dfrac{1}{\sqrt{x^2+1}}$
	$ar\cosh(x)$	$\dfrac{1}{\sqrt{x^2-1}}$
	$ar\tanh(x)$	$\dfrac{1}{1-x^2}$ for $|x|<1$
	$ar\coth(x)$	$\dfrac{1}{1-x^2}$ for $|x|>1$

⑥ $f(x) = \sin(x)$ $\Rightarrow$ $f'(x) = \cos(x)$ (Sine function):

$$f'(x) = \lim_{h \to 0} \frac{1}{h}(\sin(x+h) - \sin(x)) = \lim_{h \to 0} \frac{1}{h}\left(2\cos\left(\frac{2x+h}{2}\right)\sin\frac{h}{2}\right)$$

$$= \left\{\lim_{h \to 0} \cos\left(\frac{2x+h}{2}\right)\right\} \cdot \left\{\lim_{h \to 0} \frac{1}{\frac{h}{2}}\sin\frac{h}{2}\right\} = \cos(x),$$

because according to the Example 6.8 $\lim\limits_{h \to 0} \frac{\sin h}{h} = 1$.

Higher Derivatives of f: If the derivative $f': \mathbb{D} \to \mathbb{R}$ exists and if $f'(x)$ can be differentiated again, its derivative is called the *second derivative* $f''(x)$ of f. $f''(x)$ is the derivative of the function $f'(x)$, i.e. $f'' = (f')'$. Repeated differentiation finally leads to higher order derivatives, which are written:

1.   Derivative $y' = f'(x) = \frac{d}{dx} f(x)$.

2.   Derivative $y'' = f''(x) = \frac{d^2}{dx^2} f(x)$.

$\vdots$

n-th   Derivative $y^{(n)} = f^{(n)}(x) = \frac{d^n}{dx^n} f(x)$.

7.2 Rules for Differentiation

To differentiate complicated expressions and functions, rules for differentiation are needed. In the following, the most important rules are introduced and illustrated with examples. We assume that all functions occurring in the formulas are differentiable.

7.2.1 Constant Factor Rule

Constant Factor Rule

A constant factor is used as a common factor when differentiating:

$$y = c\,f(x) \quad \Rightarrow \quad y' = c\,f'(x).$$

Examples 7.4:

① $x(t) = 220 \sin(t)$ $\Rightarrow$ $x'(t) = 220 (\sin(t))' = 220 \cos(t)$.

② $f(x) = 9 x^5$ $\Rightarrow$ $f'(x) = 9 (x^5)' = 45 x^4$.

③ $s(t) = -5 e^t$ $\Rightarrow$ $s'(t) = -5 (e^t)' = -5 e^t$.   □

7.2.2 Sum Rule

Sum Rule

A sum of functions is differentiated term by term:

$$y = f_1(x) + f_2(x) \quad \Rightarrow \quad y' = f_1'(x) + f_2'(x) \,.$$

Examples 7.5:

① $y = \cos x - \sin x$ $\Rightarrow y' = (\cos x)' - (\sin x)' = -\sin x - \cos x$.

② $y = 4 + 2x - 3x^2 + 9x^5$ $\Rightarrow y' = 2 - 6x + 45x^4$.

③ $y = 10 \ln x + 5 \tan x$ $\Rightarrow y' = 10 \dfrac{1}{x} + 5 \dfrac{1}{\cos^2 x}$.   □

7.2.3 Product Rule

Product Rule

The derivative of a function y, which can be written as the product of two functions $u(x) \cdot v(x)$, is calculated as follows

$$y = u(x) \cdot v(x) \quad \Rightarrow \quad y' = u'(x) \cdot v(x) + u(x) \cdot v'(x) \,.$$

Examples 7.6:

① $y(x) = \ln x \cdot e^x$ $\Rightarrow$ $y'(x) = (\ln x)' e^x + \ln x \cdot (e^x)'$
$$= \frac{1}{x} e^x + \ln x \, e^x.$$

② $a(z) = \sin z \cdot \cos z$ $\Rightarrow$ $a'(z) = (\sin z)' \cos z + \sin z \, (\cos z)'$
$$= \cos^2 z - \sin^2 z.$$

③ $y(x) = x^n \cosh x$ $\Rightarrow$ $y'(x) = (x^n)' \cosh x + x^n (\cosh x)'$
$$= n \, x^{n-1} \cosh x + x^n \sinh x.$$ □

Example 7.7 (Power Rule)

$$y = x^n \quad \Rightarrow \quad y' = n\,x^{n-1} \qquad (n \geq 0).$$

The **power rule** is verified by using mathematical induction and the product rule:

$$n = 0: \qquad y = x^0 = 1 \quad \Rightarrow \quad y' = 0 = 0 \cdot x^{-1}$$

$$n \to n+1: \quad y = x^{n+1} \quad \Rightarrow \quad y' = \left(x^{n+1}\right)' = \left(x^n \cdot x\right)'$$
$$= \left(x^n\right)' \cdot x + x^n \cdot \left(x\right)'$$

With the induction assumption $\left(x^n\right)' = n\,x^{n-1}$ we have

$$y' = n\,x^{n-1} \cdot x + x^n \cdot 1 = n\,x^n + x^n = (n+1)\,x^n. \qquad \square$$

Notes:

(1) The product rule can also be applied to functions consisting of more than two factors:

$$y = u \cdot v \cdot w \quad \Rightarrow \quad y' = (u \cdot v)' \cdot w + u \cdot v \cdot w'$$
$$= (u' \cdot v + u \cdot v') \cdot w + u \cdot v \cdot w'$$
$$= u' \cdot v \cdot w + u \cdot v' \cdot w + u \cdot v \cdot w'.$$

(2) A generalization of the product rule is the *Leibniz rule* for the n-th derivative of a product:

$$y = u \cdot v \Rightarrow y^{(n)} = \sum_{i=0}^{n} \binom{n}{i} u^i v^{n-i}.$$

7.2.4 Quotient Rule

Quotient Rule

The derivative of a function y, which can be written as the quotient of two functions $\frac{u(x)}{v(x)}$, is calculated by:

$$y = \frac{u(x)}{v(x)} \quad \Rightarrow \quad y' = \frac{u'(x) \cdot v(x) - u(x) \cdot v'(x)}{v^2(x)}.$$

Examples 7.8:

① $y = \dfrac{4x^2 + 3x + 5}{2x^6 + 2x^7}$:

$$y' = \frac{\left(4x^2 + 3x + 5\right)' \cdot \left(2x^6 + 2x^7\right) - \left(4x^2 + 3x + 5\right) \cdot \left(2x^6 + 2x^7\right)'}{\left(2x^6 + 2x^7\right)^2}$$

$$= \frac{(8x + 3) \cdot \left(2x^6 + 2x^7\right) - \left(4x^2 + 3x + 5\right) \cdot \left(12x^5 + 14x^6\right)}{\left(2x^6\right)^2 (1 + x)^2}$$

$$= \frac{(8x + 3)\, x \cdot (1 + x) - \left(4x^2 + 3x + 5\right) \cdot (6 + 7x)}{2x^7\, (1 + x)^2}.$$

② $y = \tan x = \dfrac{\sin x}{\cos x}$:

$$y' = \frac{(\sin x)' \cos x - \sin x\, (\cos x)'}{\cos^2 x}$$

$$= \frac{\cos^2 x + \sin^2 x}{\cos^2 x} = \frac{1}{\cos^2 x} = 1 + \tan^2 x\,.$$

③ $y = \cot x = \dfrac{\cos x}{\sin x}$:

$$y' = \frac{(\cos x)' \sin x - \cos x\, (\sin x)'}{\sin^2 x}$$

$$= \frac{-\sin^2 x - \cos^2 x}{\sin^2 x} = \frac{-1}{\sin^2 x} = -1 - \cot^2 x\,.$$

④ $y = \dfrac{1}{x^n} = x^{-n}$.

$$\Rightarrow \quad y' = \frac{-n\, x^{n-1}}{x^{2n}} = -n\, \frac{1}{x^{n+1}} = -n\, x^{-n-1} \quad (n > 0)\,. \qquad \square$$

Remark: From the Examples 7.7 and 7.8 ④ we generalize the power rule from $n \in \mathbb{N}$ to $n \in \mathbb{Z}$:

$$\boxed{y = x^n \quad \Rightarrow \quad y' = n\, x^{n-1}} \qquad \text{for } n \in \mathbb{Z}.$$

7.2.5 Chain Rule

So far we have differentiated elementary functions such as $\sin(t)$, $\cos(t)$, $e^x \cdot \sin(x)$ etc. In most applications, however, functions of the form $f(\omega t + \varphi)$ occur. During evaluation, the inner function $\omega t + \varphi$ is calculated first and then the outer function f is applied to the result. The chain rule makes a statement about how nested (= chained) functions $y = f(g(x))$ are differentiated.

> **Chain Rule**
>
> The derivative of a nested function $y = f(g(x))$ is obtained from the product of the outer and inner derivatives:
>
> $$y = f(g(x)) \quad \Rightarrow \quad y' = f'(g(x)) \cdot g'(x).$$

In simple terms, the chain rule says that nested functions $y = f(g(x))$ are differentiated by first taking the derivative of the outer function f' and evaluating it at $g(x)$ (= outer derivative), and then multiplying the result by the derivative of the inner function $g'(x)$ (= inner derivative).

Examples 7.9:

① $\quad s(t) = (t^2 - 3)^4$.
$\quad s'(t) = 4(t^2 - 3)^3 \cdot (t^2 - 3)' = 4(t^2 - 3)^3 \cdot 2t = 8t(t^2 - 3)^3$.

② $\quad y = e^{5x+4}$.
$\quad y' = e^{5x+4} \cdot (5x + 4)' = e^{5x+4} \cdot 5$.

③ $\quad f(x) = \ln(x^2 + 4x + 2)$.
$\quad f'(x) = \dfrac{1}{x^2 + 4x + 2} \cdot (x^2 + 4x + 2)' = \dfrac{2x + 4}{x^2 + 4x + 2}$.

④ $\quad x(t) = x_0 \sin(\omega t + \varphi)$.
$\quad x'(t) = x_0 \cos(\omega t + \varphi) \cdot (\omega t + \varphi)' = x_0 \cos(\omega t + \varphi) \cdot \omega$. ☐

Sometimes the chain rule is executed with a substitution: If $y = f(g(x))$ then we set $u = g(x)$ (inner function) and $y = f(u)$ (outer function). The derivative is then

$$y' = f'(u) \cdot u'(x) \qquad \text{or} \qquad \frac{dy}{dx} = \frac{dy}{du} \cdot \frac{du}{dx}.$$

Examples 7.10:

① $y = \ln\left(1 + x^2\right)$. We set $u = 1 + x^2$ (inner function) and $f(u) = \ln u$ (outer function). Because of $f'(u) = \frac{1}{u}$ and $u'(x) = 2x$ we get

$$y' = f'(u) \cdot u'(x) = \frac{1}{u} \cdot 2x = \frac{1}{1 + x^2} \cdot 2x.$$

② $y = \sqrt[3]{\left(x^2 + 4x + 10\right)^2} = \left(x^2 + 4x + 10\right)^{\frac{2}{3}}$. We define $u = x^2 + 4x + 10$ (inner function) and $f(u) = u^{\frac{2}{3}}$ (outer function). So we get $f'(u) = \frac{2}{3}u^{-\frac{1}{3}}$ and $u'(x) = 2x + 4$ and subsequently:

$$y' = f'(u) \cdot u'(x) = \frac{2}{3} u^{-\frac{1}{3}} (2x + 4) = \frac{2}{3} \left(x^2 + 4x + 10\right)^{-\frac{1}{3}} (2x + 4).$$

③ $y = e^{x \cdot \sin x}$. With $u = x \cdot \sin x$ and $f(u) = e^u$ we get $f'(u) = e^u$ and $u'(x) = 1 \cdot \sin x + x \cos x$. Then

$$y' = f'(u) \cdot u'(x) = e^u \cdot (\sin x + x \cos x) = e^{x \sin x} (\sin x + x \cos x).$$

The chain rule always means differentiating from the outside in. It can also be used for multiple nested functions.

Examples 7.11:

① $y = \ln\left(\sin\left(2x - 3\right)\right)$. Repeated use of the chain rule gives

$$y' = \frac{1}{\sin\left(2x - 3\right)} \cdot \left(\sin\left(2x - 3\right)\right)'$$
$$= \frac{1}{\sin\left(2x - 3\right)} \cdot \cos\left(2x - 3\right) \cdot 2 = 2 \cot\left(2x - 3\right).$$

② $y = \exp\left(\sin\left(4x + 2x^2\right)\right)$. Repeated use of the chain rule gives

$$y' = \exp\left(\sin\left(4x + 2x^2\right)\right) \cdot \left(\sin\left(4x + 2x^2\right)\right)'$$
$$= \exp\left(\sin\left(4x + 2x^2\right)\right) \cdot \cos\left(4x + 2x^2\right) \cdot \left(4x + 2x^2\right)'$$
$$= \exp\left(\sin\left(4x + 2x^2\right)\right) \cdot \cos\left(4x + 2x^2\right) \cdot \left(4 + 4x\right). \qquad \square$$

7.2.6 Proofs for the Formulas 7.2.1 - 7.2.5

To prove the formulas, we use the definition of the derivative of a function as the differential quotient

$$f'(x) = \lim_{h \to 0} \frac{f(x+h) - f(x)}{h}.$$

(1) The constant factor rule is derived from

$$\frac{1}{h}(cf(x+h) - cf(x)) = c\frac{1}{h}(f(x+h) - f(x))$$

by taking the limit and applying the limit calculation rule $L1$.

(2) The summation rule is derived from

$$\frac{1}{h}([f_1(x+h) + f_2(x+h)] - [f_1(x) + f_2(x)])$$

$$= \frac{1}{h}(f_1(x+h) - f_1(x)) + \frac{1}{h}(f_2(x+h) - f_2(x))$$

by taking the limit and applying the limit calculation rule $L2$.

(3) The product rule is derived from

$$\frac{1}{h}(u(x+h)v(x+h) - u(x)v(x))$$

$$= \frac{1}{h}(u(x+h)v(x+h) - u(x)v(x+h)$$
$$+ u(x)v(x+h) - u(x)v(x))$$
$$= \frac{1}{h}(u(x+h) - u(x))v(x+h) + u(x)\frac{1}{h}(v(x+h) - v(x))$$

by taking the limit and applying the limit calculation rules $(L2), (L3)$.

(4) Because of the identity

$$\frac{1}{h}\left(\frac{1}{v(x+h)} - \frac{1}{v(x)}\right) = -\frac{\frac{1}{h}(v(x+h) - v(x))}{v(x+h)v(x)}$$

we take the limit and apply the limit calculation rule $(L4)$

$$\boxed{\left(\frac{1}{v}\right)' = -\frac{v'}{v^2}.}$$

The product rule then gives the derivative of a quotient

$$\left(\frac{u}{v}\right)' = \left(u \cdot \frac{1}{v}\right)' = u'\frac{1}{v} + u\left(\frac{1}{v}\right)' = \frac{u'}{v} - u\frac{v'}{v^2} = \frac{u'v - uv'}{v^2}.$$

(5) The statement is derived from
$$\frac{1}{h}\left(f\left(g\left(x+h\right)\right) - f\left(g\left(x\right)\right)\right)$$

$$= \frac{f\left(g\left(x+h\right)\right) - f\left(g\left(x\right)\right)}{g\left(x+h\right) - g\left(x\right)} \cdot \frac{1}{h}\left(g\left(x+h\right) - g\left(x\right)\right)$$

by taking the limit and applying the limit calculation rule $(L3)$.   □

7.2.7 Derivative of Inverse Functions

Let's look at the problem: Given is a function $y = f\left(x\right)$ and its derivative $y' = f'\left(x\right)$. We want to find the derivative of the inverse function $g\left(x\right) = f^{-1}\left(x\right)$. To calculate this derivative, we set

$$y = f\left(x\right)$$

and solve this equation for x:

$$x = f^{-1}\left(y\right) = g\left(y\right).$$

Therefore

$$y = f\left(g\left(y\right)\right).$$

By differentiating $\frac{d}{dy}$ and applying the chain rule, we obtain

$$1 = \frac{d}{dy}y = \frac{d}{dy}f\left(g\left(y\right)\right) = f'\left(g\left(y\right)\right) \cdot g'\left(y\right) = f'\left(x\right) \cdot g'\left(y\right).$$

We solve this equation with respect to $g'\left(y\right)$, the derivative of the inverse function is then given by

$$g'\left(y\right) = \frac{1}{f'\left(x\right)} = \frac{1}{f'\left(g\left(y\right)\right)}.$$

Finally, the variables x and y are swapped with each other.

Derivative of the Inverse Function

Given is a function $f(x)$ with its derivative $f'(x)$. g is the inverse function. The derivative of the inverse function $g(y)$ is given by

$$g'(y) = \frac{1}{f'(x)} \qquad (*)$$

The variable x is replaced by $g(y)$, and then the variable x is formally swapped with y on both sides of the equation. We get $g'(x)$.

Examples 7.12:

① **Derivative of** $\ln x$:
   Given: $y = f(x) = e^x$ with $f'(x) = e^x$.
   Find: Derivative of the inverse function $g(x) = \ln x$.
   Formulation: $y = e^x$
   Solve for x: $x = g(y) = \ln y$

$$\hookrightarrow g'(y) = \frac{1}{f'(x)} = \frac{1}{e^x} = \frac{1}{y}$$

   Exchange the variables: $g'(x) = \boxed{(\ln x)' = \dfrac{1}{x}}.$

② **Derivative of** $\arctan x$:
   Given: $y = f(x) = \tan x$ with $f'(x) = \dfrac{1}{\cos^2 x} = \tan^2 x + 1.$
   Find: Derivative of the inverse function $g(x) = \arctan x$.
   Formulate: $y = \tan x$
   Solve for x: $x = g(y) = \arctan y$

$$\hookrightarrow g'(y) = \frac{1}{f'(x)} = \frac{1}{\tan^2 x + 1} = \frac{1}{y^2 + 1}$$

   Exchange the variables: $g'(x) = \boxed{(\arctan x)' = \dfrac{1}{x^2 + 1}}.$ □

Remarks:

(1) It is important to express the derivative of the function f again by the function f! Otherwise, we will not be able to simplify the result directly, as the Example 7.12 ② shows: If we take for f' e.g. $f'(x) = \frac{1}{\cos^2 x}$, we would get the result

$$g'(y) = \frac{1}{f'(x)} = \cos^2 x = \cos^2(\arctan y).$$

(2) In the same way as for arctan the derivatives of arcsin, arccos, arccot are calculated (see problems).

Examples 7.13 (Derivative of Hyperbolic and Area Functions):

① The derivatives of the functions

$\sinh: \mathbb{R} \to \mathbb{R}$ with $\sinh(x) := \frac{1}{2}(e^x - e^{-x})$ **(hyperbolic sine)**

$\cosh: \mathbb{R} \to \mathbb{R}$ with $\cosh(x) := \frac{1}{2}(e^x + e^{-x})$ **(hyperbolic cosine)**

are calculated directly using the derivative of the exponential function. The following applies

$$\sinh'(x) = \cosh(x) \quad \text{and} \quad \cosh'(x) = \sinh(x).$$

② For hyperbolic sine and cosine, the relationship

$$\cosh^2(x) - \sinh^2(x) = 1$$

applies. By inserting the hyperbolic sine and cosine directly into the left side, the equation is checked

$$\cosh^2(x) - \sinh^2(x) = \left(\frac{1}{2}(e^x + e^{-x})\right)^2 - \left(\frac{1}{2}(e^x - e^{-x})\right)^2 = 1.$$

③ The derivative of **tangential hyperbolic**

$$\tanh: \mathbb{R} \to \mathbb{R} \quad \text{with} \quad \tanh(x) := \frac{\sinh(x)}{\cosh(x)}$$

is calculated using the derivatives of sinh and cosh. Then, applying the quotient rule, we get

$$\tanh'(x) = \frac{\cosh^2(x) - \sinh^2(x)}{\cosh^2(x)}.$$

Using the identity ② we obtain the final result

$$\tanh'(x) = \frac{1}{\cosh^2(x)} = 1 - \tanh^2(x).$$

④ Analogous to the tangential hyperbolic, we define the **cotangent hyperbolic**

$$\coth : \mathbb{R} \setminus \{0\} \to \mathbb{R} \quad \text{with} \quad \coth(x) := \frac{\cosh(x)}{\sinh(x)}$$

with its derivative

$$\coth'(x) = -\frac{1}{\sinh^2(x)} = 1 - \coth^2(x).$$

⑤ The **area functions** are the inverse functions of sinh, cosh, tanh and coth. For example, the derivative of $ar\sinh(x)$ follows as the derivative of the inverse function:

Formulation: $y = \sinh(x) \quad \hookrightarrow \quad f'(x) = \cosh(x)$

Solve for x: $x = g(y) = ar\sinh(x)$

$$\hookrightarrow g'(y) = \frac{1}{f'(x)} = \frac{1}{\cosh(x)} = \frac{1}{\sqrt{1+\sinh^2 x}} = \frac{1}{\sqrt{1+y^2}}.$$

Exchange the variables: $ar\sinh'(x) = \dfrac{1}{\sqrt{1+x^2}}.$

7.2.8 Logarithmic Differentiation

The function $f(x) = x^{\cos x}$ cannot be derived directly with the differentiation rules 7.2.1 - 7.2.7. However, if the logarithm

$$\ln f(x) = \ln x^{\cos x} = \cos x \cdot \ln x$$

is first applied to this equation and then the chain rule (left side) and product rule (right side) are used to differentiate

$$\frac{1}{f\left(x\right)} \cdot f'\left(x\right) = -\sin x \cdot \ln x + \cos x \cdot \frac{1}{x}$$

we get

$$f'\left(x\right) = f\left(x\right) \cdot \left[-\sin x \cdot \ln x + \cos x \cdot \frac{1}{x}\right].$$

This procedure is called **Logarithmic Differentiation** because the logarithmic function is applied first, followed by a differentiation.

Logarithmic Differentiation

A function $y(x)$ with

$$y(x) = [u\left(x\right)]^{v\left(x\right)}$$

$\left(u\left(x\right) > 0\right)$ is differentiated logarithmically in three steps:

(1) We take the logarithm of the expression:

$$\ln y(x) = v\left(x\right) \cdot \ln\left(u\left(x\right)\right).$$

(2) We differentiate this equation using the chain rule:

$$\frac{1}{y(x)} \cdot y'\left(x\right) = v'\left(x\right) \cdot \ln\left(u\left(x\right)\right) + v\left(x\right) \cdot \frac{1}{u\left(x\right)} \cdot u'\left(x\right).$$

(3) We solve this equation with respect to $y'(x)$.

⚠ **Caution:** When differentiating the logarithmic equation, the function y depends on x. Therefore, the chain rule must be applied when differentiating the left side:

$$\frac{d}{dx} \ln y\left(x\right) = \frac{1}{y\left(x\right)} \cdot y'\left(x\right).$$

Examples 7.14:

① $y = x^x$.

Logarithm: $\ln y = \ln x^x = x \ln x$.

Differentiate: $\frac{1}{y} y' = 1 \cdot \ln x + x \cdot \frac{1}{x} = \ln x + 1$.

Solve: $y' = y (\ln x + 1) = x^x (\ln x + 1)$.

② Logarithmic differentiation is used to calculate the derivative of the general power function $\boxed{y = x^\alpha}$ ($x > 0$, $\alpha \in \mathbb{R}$ fixed):

Logarithm: $\ln y = \ln x^\alpha = \alpha \ln x$.

Differentiation: $\frac{1}{y} \cdot y' = \alpha \cdot \frac{1}{x}$.

Solve: $y' = y \alpha \frac{1}{x} = x^\alpha \alpha x^{-1} = \alpha x^{\alpha-1}$.

$$\Rightarrow \quad y = x^\alpha \ \Rightarrow \ y' = \alpha x^{\alpha-1} \qquad (\alpha \in \mathbb{R}).$$ □

Tip: In many practical examples, the function to be differentiated consists of several products and quotients. Although the derivative can be calculated using the product and quotient rule, logarithmic differentiation makes the calculation simple and convenient:

Example 7.15. We use the method of logarithmic differentiation to find the derivative of

$$y = \frac{\sin (x - 2) \ e^{2x}}{(x - 1)^3 (x^2 + 3)^5}.$$

Logarithm:

$$\ln y = \ln \left[\sin (x - 2) \ e^{2x} \right] - \ln \left[(x - 1)^3 (x^2 + 3)^5 \right]$$
$$= \ln (\sin (x - 2)) + \ln (e^{2x}) - \ln (x - 1)^3 - \ln (x^2 + 3)^5$$
$$= \ln (\sin (x - 2)) + 2x - 3 \ln (x - 1) - 5 \ln (x^2 + 3).$$

Differentiate:

$$\frac{1}{y} \cdot y' = \frac{1}{\sin (x - 2)} \cdot \cos (x - 2) + 2 - 3 \frac{1}{x - 1} - 5 \frac{1}{x^2 + 3} \cdot 2x.$$

Solve:

$$y' = y \cdot \left[\cot (x - 2) + 2 - \frac{3}{x - 1} - \frac{10x}{x^2 + 3} \right].$$ □

7.2.9 Implicit Differentiation

In some applications, a function f is given in an *implicit form* $F(x, f(x)) = 0$ and the functional expression $y = f(x)$ is difficult or impossible to express explicitly. The derivative of such implicit functions is calculated using the chain rule:

Example 7.16. Given is the **circle equation**

$$F(x, y) = (x - 4)^2 + (y + 5)^2 - 25 = 0$$

with center $(x_0, y_0) = (4, -5)$ and radius 5. We look for the slope at point $(x, y) = (7, -1)$.

We differentiate each term of the equation with respect to x. Note that $y = y(x)$ depends on the variable x! If the function F is zero, then the derivative of F with respect to x is also zero:

$$2(x - 4) + 2(y + 5) \cdot y' - 0 = 0.$$

Solving for y' we get

$$\boxed{y' = -\frac{x - 4}{y + 5}}$$

and inserting the point $(x, y) = (7, -1)$, the slope is

$$y' = -\frac{7 - 4}{-1 + 5} = -\frac{3}{4}.$$ □

Implicit Differentiation

A function $y(x)$ is given *implicitly* by

$$F(x, y(x)) = 0.$$

The derivative of the function y is obtained by differentiating F with respect to x. Each term including y must be differentiated using the chain rule. Then solve the differentiated equation for y'.

Examples 7.17:

① Find the derivative of the function $y(x)$, which is implicitly defined by the equation:

$$e^y - e^{2x} = x \cdot y.$$

From $e^y - e^{2x} = x \cdot y$ it follows $\quad e^y - e^{2x} - x \cdot y = 0.$

Differentiation: $\quad e^y \cdot y' - e^{2x} \cdot 2 - (1 \cdot y + x \cdot y') = 0$

$$\Rightarrow (e^y - x) \, y' - 2 \, e^{2x} - y = 0.$$

Solve for: $\quad y' = \dfrac{2 \, e^{2x} + y}{e^y - x}.$

② Find the derivative of the function $y(x)$, which is implicitly defined by the equation

$$x \cdot \sin 2y = 1 - 3 \, y^2.$$

$x \cdot \sin 2y = 1 - 3 \, y^2 \quad \Rightarrow \quad x \cdot \sin 2y - 1 + 3 \, y^2 = 0.$

Differentiation: $\quad 1 \cdot \sin 2y + x \, \cos 2y \cdot 2y' - 0 + 3 \cdot 2y \cdot y' = 0.$

Solve for: $\quad y' = \dfrac{-\sin 2y}{2x \, \cos 2y + 6y}.$ $\qquad\qquad\qquad\qquad$ □

7.3 Applications in Physics and Engineering

7.3.1 Kinematics

The motion of an object is usually expressed by a **space-time law** $s = s\,(t)$. The instantaneous velocity is the derivative of the space-time law with respect to time t

$$v\,(t) := \dot{s}\,(t) = \frac{d}{dt}\, s\,(t) \qquad\qquad (Velocity)$$

and the acceleration indicates the change in velocity:

$$a\,(t) := \dot{v}\,(t) = \frac{d}{dt}\, v\,(t) = \ddot{s}\,(t) \qquad\qquad (Acceleration).$$

These are the basic kinematic equations that apply to all motion. We discuss the free fall and the damped spring pendulum.

(1) For a **Free Fall** the following formula applies

$$s\left(t\right) = \frac{1}{2} g t^2 + v_0 t + s_0,$$

where s_0 is the initial position and v_0 is the initial velocity. For velocity and acceleration we know

$$v\left(t\right) = \dot{s}\left(t\right) = g t + v_0,$$
$$a\left(t\right) = \dot{v}\left(t\right) = g = const.$$

(2) A **Damped Spring Pendulum** oscillates with

$$x\left(t\right) = x_0 \, e^{-\gamma t} \cos\left(\omega t\right),$$

where x_0 is the initial displacement, γ is the friction coefficient and ω is the oscillation frequency. The velocity and acceleration are

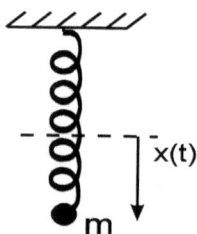

Figure 7.4.
Spring pendulum

$$v\left(t\right) = \dot{x}\left(t\right) = -\gamma x_0 \, e^{-\gamma t} \cos\left(\omega t\right) - \omega x_0 \, e^{-\gamma t} \sin\left(\omega t\right),$$

$$a\left(t\right) = \dot{v}\left(t\right) = \gamma^2 x_0 \, e^{-\gamma t} \cos\left(\omega t\right) + \gamma \omega x_0 \, e^{-\gamma t} \sin\left(\omega t\right)$$
$$+ \gamma \omega x_0 \, e^{-\gamma t} \sin\left(\omega t\right) - \omega^2 x_0 \, e^{-\gamma t} \cos\left(\omega t\right)$$
$$= \left(\gamma^2 - \omega^2\right) x_0 e^{-\gamma t} \cos\left(\omega t\right) + 2\,\gamma \omega x_0 \, e^{-\gamma t} \sin\left(\omega t\right).$$

For the special case without friction $\left(\gamma = 0\right)$, it is

$$x\left(t\right) = x_0 \, \cos\left(\omega t\right)$$
$$\ddot{x}\left(t\right) = a\left(t\right) = -\omega^2 x_0 \, \cos\left(\omega t\right) = -\omega^2 x\left(t\right).$$

So the force of the spring is

$$F = m\,a = m\,\ddot{x}\left(t\right) = -m\,\omega^2 x\left(t\right) \sim x\left(t\right).$$

This is *Hooke's law*, which states that the restoring force is proportional to the displacement $x\left(t\right)$.

7.3.2 Faraday's Law of Induction

Faraday's Law of Induction gives a relationship between the change in the magnetic flux and the induced voltage: A change in the magnetic flux ϕ induces an electric voltage U_i in a conductor consisting of n coils according to

$$U_i(t) = -n \frac{d}{dt} \phi(t).$$

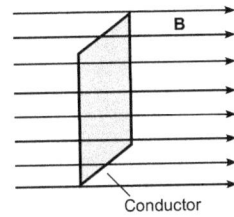

Conductor

Figure 7.5. Magnetic field

The magnetic flux is $\phi = B \cdot A_{eff}$, where B is the applied magnetic field and A_{eff} is the effective area penetrated by the magnetic field.

(1) Applying the law of induction to a coil rotating in a constant magnetic field, the effective area penetrated by the magnetic field is

$$A_{eff} = A \cos \varphi (t) = A \cos (\omega t).$$

According to the law of induction, the alternating voltage is

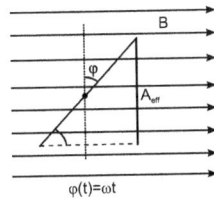

Figure 7.6. Rotating

$$U_i = -n \frac{d}{dt} \phi = n \frac{d}{dt} B A \cos (\omega t)$$
$$= n B A \omega \sin (\omega t)$$

with the induced peak voltage $U_0 = n B A \omega$.

(2) If the conductor is fixed and the magnetic field perpendicular to the conductor varies according to $B = B_0 \cos (\omega t)$ with amplitude B_0 and frequency ω, the voltage U_i is induced in the conductor loop (cross-sectional area A) according to the formula

$$U_i = -n \frac{d}{dt} \phi = -n \frac{d}{dt} A B_0 \cos (\omega t) = n A B_0 \omega \sin (\omega t).$$

7.3.3 Electrostatics

(1) In a *plate capacitor* with anode voltage ϕ_A, cathode voltage ϕ_K and gap distance d the potential $\phi(x)$ is given by

$$\phi(x) = \left(1 - \frac{x}{d}\right) \phi_A + \frac{x}{d} \phi_K$$

$$= \phi_A + \left(\frac{x}{d}\right)(\phi_K - \phi_A).$$

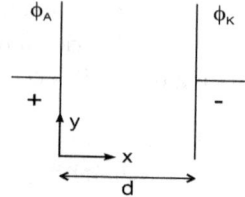

The *electric field* E is defined as

$$E(x) := -\frac{d}{dx}\phi(x) \qquad \text{(electric field)}.$$

Then

$$E(x) = -\frac{d}{dx}\phi(x) = -\frac{1}{d}(\phi_K - \phi_A) = \frac{\phi_A - \phi_K}{d} = const.$$

(2) **Condenser Microphone.** A constant voltage U_0 is applied to the plates of a condenser microphone. The pressure of the sound waves (frequency ω) changes the plate spacing according to the formula

$$d = d_0 + a \sin(\omega t).$$

So the charge Q on the capacitor varies according to

$$Q(t) = C(t) \cdot U_0 = \frac{\varepsilon_0 A}{d(t)} \cdot U_0,$$

as the capacity C changes with time. The current

$$I(t) = \tfrac{d}{dt} Q(t)$$

is therefore given by

$$I(t) = \frac{d}{dt} Q(t) = \varepsilon_0 A U_0 \frac{d}{dt} \frac{1}{d_0 + a \sin(\omega t)}$$

$$= -\varepsilon_0 A U_0 \frac{\omega a \cos(\omega t)}{[d_0 + a \sin(\omega t)]^2}.$$

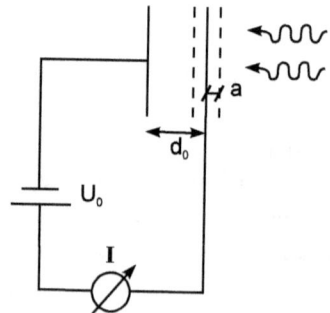

7.4 Differential of a Function

To find the growth of a differentiable function f in the vicinity of a point x_0, we determine the value of the function at the point x_0 and at $x_0 + \triangle x$. When the abscissa value changes by $\triangle x$, the functional value changes by $\triangle y$. For the **growth** $\triangle y$ of the function f, we have

$$\triangle y = f(x_0 + \triangle x) - f(x_0).$$

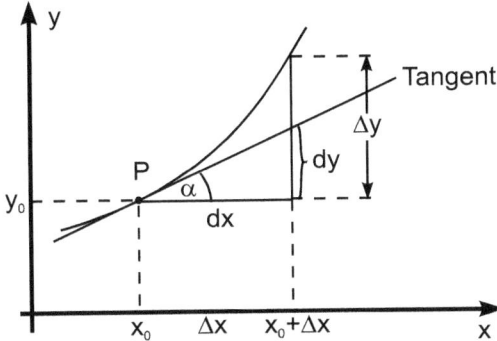

Figure 7.7. Differential of a function

We compare the increase of the function $\triangle y$ with the change of the tangent dy. We call

$$dx \quad : \quad \textit{Independent Differential}$$
$$dy \quad : \quad \textit{Dependent Differential} \quad (= \text{Change in the Tangent})$$

According to Fig. 7.7 we get

$$\tan \alpha = f'(x_0) = \frac{dy}{dx} \quad \Rightarrow \quad dy = f'(x_0) \cdot dx.$$

Definition: (Differential of a Function).
The Differential of a function $y = f(x)$

$$dy = df = f'(x_0) \cdot dx$$

describes the **change along the tangent** at x_0, when the abscissa changes as dx. Sometimes df is used instead of dy to indicate the differential of the function f.

Δy: **Change in the function** when changing the x-value by Δx.
dy: **Change in the tangent** when changing the x-value by dx.

Examples 7.18:

① Find the differential of the function $f(x) = \sqrt{x+1}$ at $x_0 = 0$:

$$f(x) = (x+1)^{\frac{1}{2}} \hookrightarrow f'(x) = \frac{1}{2}(x+1)^{-\frac{1}{2}} \hookrightarrow f'(x_0 = 0) = \frac{1}{2}.$$

Therefore, $df = f'(x_0)\, dx = \frac{1}{2}\, dx$.

② Find the differential of $f(x) = \arctan(x)$ at the position $x_0 = 0$:

$$f(x) = \arctan(x) \hookrightarrow f'(x) = \frac{1}{1+x^2} \hookrightarrow f'(x_0 = 0) = 1.$$

Hence, $df = 1 \cdot dx$.   □

7.4.1 Linearizing Functions

From the differential df of a function we will now draw an important conclusion for applications: For small $\Delta x = dx$ the approximation

$$\Delta y \approx dy$$

applies. For small $\Delta x = dx$ the change of the tangent is comparable to the change of the function. Using the formula for dy, we obtain

$$\Rightarrow \Delta y \approx dy = f'(x_0) \cdot \Delta x.$$

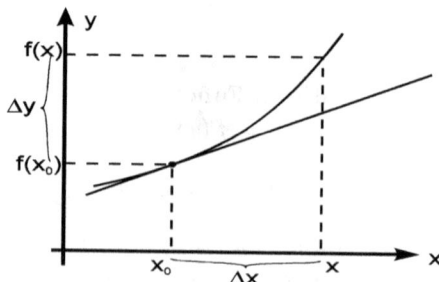

Figure 7.8. Linearizing a function

For small Δx the function f near the point x_0 can be replaced by its tangent. This procedure is called the *linearization of the function f*. Using

the approximation

$$\Delta y = f(x) - f(x_0) \approx dy = f'(x_0)(x - x_0)$$

we conclude for the **linearization** of the function f at the position x_0:

$$f(x) \approx f(x_0) + f'(x_0) \cdot (x - x_0).$$

Examples 7.19:

① Given is the function $f(x) = \dfrac{4x}{x(x+1)}$. Find the linearization of the function at $x_0 = 1$:

Since $f'(x) = \dfrac{4x(x+1) - 4x(2x+1)}{x^2(x+1)^2}$ is $f'(x_0 = 1) = -1$

$$\Rightarrow f(x) \approx f(x_0) + f'(x_0)(x - x_0) = 2 + (-1)(x-1) = 3 - x.$$

② Given is the function $f(\varphi) = \sin\varphi$. Find the linearization of the function at $\varphi_0 = 0$:

Since $f'(\varphi) = \cos\varphi$ is $f'(\varphi_0 = 0) = 1$

$$\Rightarrow f(\varphi) \approx f(\varphi_0) + f'(\varphi_0)(\varphi - \varphi_0) = 0 + 1 \cdot (\varphi - 0) = \varphi$$

$$\Rightarrow \boxed{\sin\varphi \approx \varphi} \quad \text{(for small angles).} \qquad \square$$

Application Example 7.20 (Harmonic Pendulum).

A mass m is attached to a string of the length l. The angle $\varphi(t)$ is to be found as a function of time, when the mass is deflected by a small angle φ_0. We determine all the forces acting on the mass, because according to Newton's law of motion the acceleration force is given by the sum of all the forces.

Ignoring frictions, the only force acting on the mass is the gravitational force $F_G = mg$. Since the acceleration of the mass is perpendicular to the thread, the acceleration force acts as follows

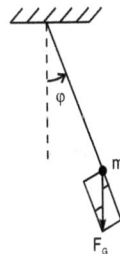

Figure 7.9.
Pendulum

$$F_B = -F_G \cdot \sin\varphi = -mg\sin\varphi.$$

In a circular motion, the velocity is $v(t) = l\omega = l \cdot \dot{\varphi}(t)$ and thus the acceleration is $a(t) = \dot{v}(t) = l\ddot{\varphi}(t)$. According to Newton's law

$$m\,l\,\ddot{\varphi}(t) = -m\,g\,\sin(\varphi(t)) \qquad \text{or} \qquad \boxed{\ddot{\varphi}(t) + \frac{g}{l}\sin\varphi(t) = 0.}$$

This equation is called a *differential equation*, because besides the function $\varphi(t)$ (= angular deflection at any time t), also its (second) derivative occurs. There is no closed solution $\varphi(t)$ for this equation. However, for small angles $\varphi(t) \approx 0$ we simplify this expression according to the Example 7.19 ②

$$\sin\varphi(t) \approx \varphi(t).$$

This simplifies the equation of motion to

$$\boxed{\ddot{\varphi}(t) + \frac{g}{l}\varphi(t) = 0.}$$

The solution of this differential equation can be written as

$$\varphi(t) = \varphi_0 \cos(\omega t)$$

where $\omega = \sqrt{\frac{g}{l}}$ and $\varphi_0 = \varphi(t=0)$ the initial deflection. This is checked by inserting the function $\varphi(t) = \varphi_0 \cos(\omega t)$ and $\ddot{\varphi}(t) = -\omega^2 \varphi_0 \cos(\omega t)$ directly into the differential equation:

$$\ddot{\varphi}(t) + \frac{g}{l}\varphi(t) = -\omega^2 \varphi_0 \cos(\omega t) + \frac{g}{l}\varphi_0 \cos(\omega t) = 0.$$

Summary: For small angular deflections, the angle $\varphi(t)$ of a pendulum is given by $\varphi(t) = \varphi_0 \cos(\omega t)$. This is a periodic motion with constant amplitude φ_0, fixed frequency $\omega = \sqrt{\frac{g}{l}}$ and a period $T = \frac{2\pi}{\omega} = 2\pi\sqrt{\frac{l}{g}}$.   □

7.4.2 Error Calculation

The differential of a function is a statement about how the value of a function changes for small changes in the argument. An application of the differential is the error calculus, since it gives a statement about the maximum change in the value of the function when the argument is known only up to $\pm\,dx$.

Error Calculation

Let f be a physical quantity calculated from the measurement of an independent quantity x associated with a measurement uncertainty dx. Then, the **Maximum Error** in linear approximation is defined by the absolute value of the differential

$$|df| = |f'(x)| \, |dx| \, .$$

Application Example 7.21 (Wheatstone Bridge).

Using the *Wheatstone bridge circuit,* an unknown ohmic resistance R can be determined by

$$R = R_0 \cdot \frac{x}{l - x}$$

Figure 7.10. Wheatstone bridge

when the sampling point D on the wire is moved so that the bridge is without current. Due to measurement inaccuracies, x is only known up to $\pm 1mm$. What is the maximum and relative error in the linear approximation when $R_0 = 100\Omega$, $l = 1m$ and $x = 0.530m$?

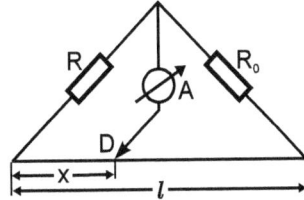

We look at R as a function of x:

$$R(x) = R_0 \frac{x}{l - x}$$

and find its differential

$$dR = R'(x) \cdot dx = R_0 \frac{1 \cdot (l - x) + x \cdot 1}{(l - x)^2} \, dx = \frac{R_0 \, l}{(l - x)^2} \, dx.$$

For $x_0 = 0.530m$ and $dx = 0.001m$, the maximum error in the linear approximation is given by

$$|dR| = 100 \frac{1}{0.47^2} \cdot 0.001\Omega = 0.452\Omega.$$

So the resistance is

$$R = R_0 \frac{x_0}{l - x_0} \pm |dR| = (112.76 \pm 0.45) \, \Omega$$

with a *relative error* of $\quad \dfrac{|dR|}{R} = 0.4\% \, .$ □

7.5 Applications in Mathematics

From the derivatives of a function we can identify special properties of a function. The first derivative tells us about monotony and the extreme values, and the second derivative tells us about the turning points. In particular, the characterization of the local extremes via the first derivative will be used to solve optimization problems.

7.5.1 Geometric Meaning of f' and f''

The derivative of a function is the slope of its tangent. If a function always has a positive slope, $f'(x) > 0$, then the increase is $df = f'(x) \cdot dx > 0$ and the function is strictly increasing (Fig. 7.11 (a)). On the other hand, if $f'(x) < 0$, then the function is strictly decreasing (Fig. 7.11 (b)).

Monotonic Behavior of Functions

Let f be a function differentiable in its domain $\mathbb{D}$.

(1) If $f'(x) > 0$ for all $x \in \mathbb{D}$ then f is strictly increasing.

(2) If $f'(x) < 0$ for all $x \in \mathbb{D}$ then f is strictly decreasing.

The second derivative of a function determines the curvature, since $f'' = (f')'$ gives the change in f', i.e. the *change* in the slope of the tangent. The following applies

---

Curvature Behavior of Functions

For a function that can be differentiated twice in its domain $\mathbb{D}$, the following holds for $x_0 \in \mathbb{D}$:

(1) If $f''(x_0) > 0$, then the slope of the tangent increases as it passes through the point $P(x_0, f(x_0))$. The tangent rotates in a positive direction (counter-clockwise). The curve has a *left curvature* at P (see Fig. 7.11 (b)).

(2) If $f''(x_0) < 0$, then the slope of the tangent decreases through the point $P(x_0, f(x_0))$. The tangent rotates in a negative (clockwise) direction. The curve has a *right curvature* at P (see Fig. 7.11 (a)).

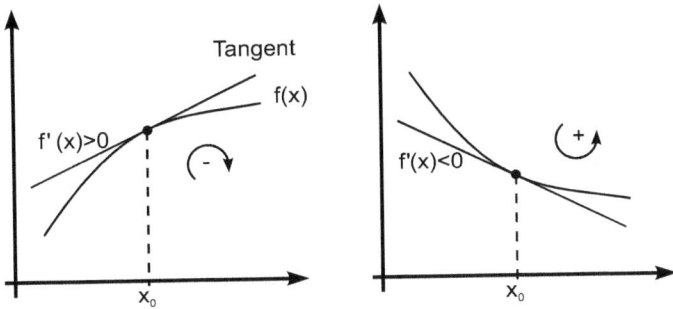

Figure 7.11. (a) Increasing with right curvature, (b) Decreasing with left curvature

7.5.2 Relative Extremes

The *relative extremes* indicate the points of a function where the function values have a relatively largest or smallest value.

> **Definition:** *A function f has at the location $x_0 \in \mathbb{D}$ a* **relative maximum** *or* **relative minimum**, *if in an environment of the point x_0 it applies*
>
> $$f(x_0) > f(x) \quad \text{for all } x \neq x_0 \qquad \text{(rel. Maximum)}$$
>
> $$f(x_0) < f(x) \quad \text{for all } x \neq x_0 \qquad \text{(rel. Minimum)}.$$

A differentiable function has a horizontal tangent at a local extreme. So a necessary condition for a relative extreme at x_0 is that

$$\boxed{f'(x_0) = 0.}$$

⚠ **Caution:** However, this condition is not sufficient. It may be that the curvature changes at such a point so that the second derivative is also zero

$$f''(x_0) = 0.$$

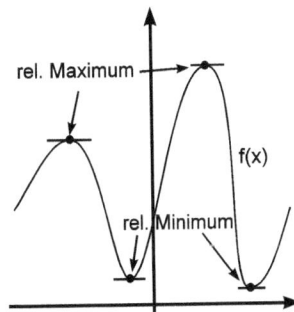

Figure 7.12. Relative extremes

Then the point is not necessarily an extreme, as the following example shows:

⚠ **Example 7.22.** For the function

$$f(x) = x^3$$

it is $f'(0) = 0$; but $x_0 = 0$ is not a local extreme. This function simultaneously changes its curvature at $x_0 = 0$ (from right to left curvature). Therefore, $f''(0) = 0$. So a horizontal tangent is sufficient for an extreme, if f at x_0 must not change the curvature: $f''(x_0)$ must not disappear.   □

Condition for a Relative Extreme Value

Let f be a twice continuously differentiable function in an environment of $x_0 \in \mathbb{D}$. If $f'(x_0) = 0$ and $f''(x_0) \neq 0$, then f has a relative extreme value at x_0.

In short:

$$f'(x_0) = 0, \ f''(x_0) < 0 \quad \Rightarrow \quad x_0 \text{ is a relative maximum.}$$
$$f'(x_0) = 0, \ f''(x_0) > 0 \quad \Rightarrow \quad x_0 \text{ is a relative minimum.}$$

Examples 7.23:

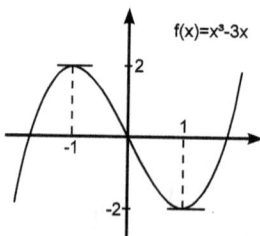

Figure 7.13. Extreme values

① We look for the relative extremes of the function

$$f(x) = x^3 - 3x.$$

The derivatives are

$$f'(x) = 3x^2 - 3, \quad f''(x) = 6x.$$

From $f'(x) = 0$ follows

$$3x^2 - 3 = 0 \hookrightarrow x_{1/2} = \pm 1.$$

At these points the function has a horizontal tangent. We insert these two expressions for a local extreme value in the second derivative:

i) $f''(x_1) = f''(1) > 0 \quad \Rightarrow \quad x_1 = 1$ is a local minimum,

ii) $f''(x_2) = f''(-1) < 0 \quad \Rightarrow \quad x_2 = -1$ is a local maximum.

② For the exponential function $f(x) = e^x$ it holds for all $x \in \mathbb{R}$: $f'(x) = e^x > 0$. So the exponential function has no extremes and is strictly increasing on $\mathbb{R}$.   □

7.5.3 Turning Points and Saddle Points

Definition:

(1) *Curve points at which the direction of rotation of the tangent changes are called* **Turning Points** (Fig. 7.14 (a)).

(2) *Turning points with a horizontal tangent are called* **Saddle points** (Fig. 7.14 (b)).

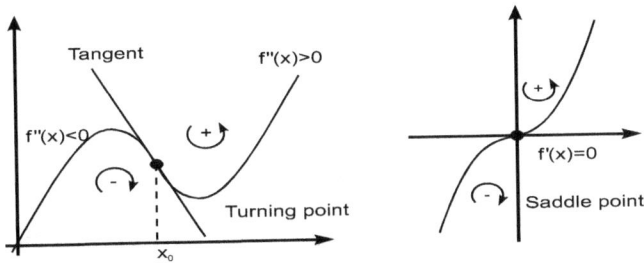

Figure 7.14. Turning and saddle points

A change in curvature occurs at the turning points. For a function that can be continuously differentiated twice, $f''(x_0) = 0$ holds at these points. This condition is not sufficient as the example $f(x) = x^4$ shows: At $x_0 = 0$ $f''(x_0) = 0$ is valid, but also $f'''(x_0) = 0$. The following conditions are sufficient:

Condition for Turning and for Saddle Points

Let the function f be continuously differentiable by 3 in its domain $\mathbb{D}$ and $x_0 \in \mathbb{D}$.

(1) Is $f''(x_0) = 0$ and $f'''(x_0) \neq 0$ $\Rightarrow x_0$ is a **turning point.**

(2) Is $f''(x_0) = 0$, $f'''(x_0) \neq 0$ and $f'(x_0) = 0$
$$\Rightarrow x_0 \text{ is a } \textbf{saddle point.}$$

Examples 7.24:

① The function $f(x) = x^3$ has a saddle point at $x_0 = 0$:

$$f'(x) = 3x^2, \quad f''(x) = 6x, \quad f'''(x) = 6.$$

Because $f'(0) = f''(0) = 0$ and $f'''(0) \neq 0$, $x_0 = 0$ is a saddle point.

② The function $f(x) = x^3 - 3x$ has a turning point at $x_0 = 0$:

$$f'(x) = 3x^2 - 3, \quad f''(x) = 6x, \quad f'''(x) = 6.$$

From $f''(x) = 6x = 0 \hookrightarrow x_0 = 0$. Since $f'''(0) = 6 \neq 0$, $x_0 = 0$ is a turning point.

③ The turning points of the sine function coincide with its zeroes:

$$f(x) = \sin x, \quad f'(x) = \cos x, \quad f''(x) = -\sin x, \quad f'''(x) = -\cos x.$$

From $f''(x) = -\sin x = 0$ follows $x_k = k \cdot \pi$. Because

$$f'''(x_k) = -\cos(k\pi) = -(-1)^k \neq 0$$

all zeroes are thus also turning points.   □

7.5.4 Curve Discussion

The characteristic properties of a function are recorded using differential calculus. Characteristic points are zero points and pole points but also the relative extremes, turning points and saddle points. A complete *Curve Discussion* is based on the following 10 points:

Curve Discussion

1. Domain

2. Symmetry Behavior

3. Zero Points

4. Poles

5. Asymptotes (function behavior for $x \to \pm\infty$)

6. Derivatives of the Function (up to order 3)

7. Relative Extremes

8. Turning Points and Saddle Points

9. Range of Values

10. Function Graph

Prime Example 7.25 .

Let's have a look at the rational function $y = \dfrac{-5x^2 + 5}{x^3}$.

Domain: The denominator becomes zero at $x_0 = 0 \Rightarrow \mathbb{D} = \mathbb{R} \setminus \{0\}$. $x_0 = 0$ is definition gap.

Symmetry Behavior: The function is point symmetric because it is the quotient of an even and an odd function. So it is odd in total.

Zeros: The numerator and denominator are divided into linear factors and the common factors are shortened:
$$y = \frac{-5x^2 + 5}{x^3} = \frac{-5\,(x+1)\,(x-1)}{x^3} \qquad \Rightarrow \qquad \text{Zeros are } x_{1/2} = \pm 1.$$

Poles: $x_3 = 0$.

Asymptotes: The degree of the numerator is less than the degree of the denominator. $\Rightarrow \lim\limits_{x \to \pm\infty} y = 0 \hookrightarrow x$-axis is an asymptote.

Derivatives:
$$y' = \frac{5\,(x^2 - 3)}{x^4},$$
$$y'' = \frac{-10\,(x^2 - 6)}{x^5},$$
$$y''' = \frac{30\,(x^2 - 10)}{x^6}.$$

Relative Extremes: $y' = 0 \hookrightarrow x^2 - 3 = 0 \Rightarrow x_{4/5} = \pm\sqrt{3}$.

$y''\left(\sqrt{3}\right) = \frac{10}{9}\sqrt{3} > 0 \Rightarrow \left(\sqrt{3}, -\frac{10}{9}\sqrt{3}\right)$ is a relative minimum.

$y''\left(-\sqrt{3}\right) = -\frac{10}{9}\sqrt{3} < 0 \Rightarrow \left(-\sqrt{3}, \frac{10}{9}\sqrt{3}\right)$ is a relative maximum.

Turning Points: $y'' = 0 \hookrightarrow x^2 - 6 = 0 \Rightarrow x_{6/7} = \pm\sqrt{6}$.
$y'''\left(\pm\sqrt{6}\right) \neq 0 \Rightarrow x_{6/7}$ are turning points.

Function Graph:

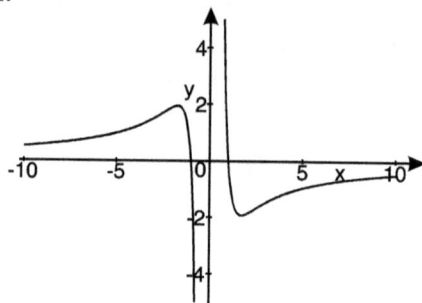

Figure 7.15. Function graph of $y = \frac{-5x^2+5}{x^3}$

Range of Values: $\mathbb{W} = \mathbb{R}$.   □

Application Example 7.26 (Damped Free Vibration).

The curve of a damped free vibration is given by

$$x(t) = 4\,e^{-0.1t} \cdot \cos(2t) \qquad t \geq 0.$$

Domain: This function has no restrictions. For physical reasons $\mathbb{D} = \mathbb{R}_{\geq 0}$.

Symmetry: none (negative t are not in the domain)

Zero Points: $x(t) = 0 \hookrightarrow e^{-0.1t} \cos(2t) = 0 \hookrightarrow \cos(2t) = 0$
$\Rightarrow 2t = \frac{\pi}{2} + k\pi \hookrightarrow t_k = \frac{\pi}{4} + k\frac{\pi}{2}, \qquad k = 0, 1, 2, 3, \ldots$

Poles: none

Asymptotes: For $t \to \infty$ is $x(t) \to 0$. The t-axis is an asymptote.

Derivatives:

$$\dot{x}(t) = (-0.4 \cos(2t) - 8 \sin(2t))\, e^{-0.1t},$$
$$\ddot{x}(t) = (-15.96 \cos(2t) + 1.6 \sin(2t))\, e^{-0.1t},$$
$$\dddot{x}(t) = (4.796 \cos(2t) + 31.76 \sin(2t))\, e^{-0.1t}.$$

Relative Extremes: $\dot{x}(t) = 0 \hookrightarrow -0.4 \cos(2t) - 8 \sin(2t) = 0$
$\hookrightarrow \tan(2t) = -\frac{0.4}{8} = -0.05$
$\hookrightarrow 2t = \arctan(-0.05) + k \cdot \pi$, because the arc tangent is ambiguous
$\hookrightarrow \boxed{2t_k = 3.091 + k \cdot \pi} \qquad k = 0, 1, 2, 3, \ldots$

(i) For even $k = 2n$, $\ddot{x}(t_k)$ is positive:
$$\ddot{x}(t_k) = \ddot{x}(t_{2n})$$
$$= (-15.96 \cos(3.091 + 2n\pi) + 1.6 \sin(3.091 + 2n\pi)) e^{-0.1 t_{2n}}$$
$$= 16.020 e^{-0.1 t_{2n}} > 0 \quad \Rightarrow \text{ relative minimum:}$$
$\text{Min}_1 = (1.545, -3.422); \text{Min}_2 = (4.687, -2.500);$
$\text{Min}_3 = (7.828, -1.826); \ldots$

(ii) For odd $k = 2n + 1$, $\ddot{x}(t_k)$ is negative:
$$\ddot{x}(t_k) = \ddot{x}(t_{2n+1})$$
$$= (-15.96 \cos(3.091 + \pi) + 1.6 \sin(3.091 + \pi)) e^{-0.1 t_{2n+1}}$$
$$= -16.020 e^{-0.1 t_{2n+1}} < 0 \quad \Rightarrow \text{ relative maximum:}$$
$\text{Max}_1 = (3.116, 2.925); \text{Max}_2 = (6.257, 2.136);$
$\text{Max}_3 = (9.399, 1.560); \ldots$

The distance between minimum and maximum is $\frac{\pi}{2}$.

Turning Points: $\ddot{x}(t) = 0 \hookrightarrow -15.96 \cos(2t) + 1.6 \sin(2t) = 0$
$$\Rightarrow \tan(2t) = \frac{15.96}{1.6} \Rightarrow \boxed{2t_k = 4.612 + k \cdot \pi.}$$
Since the third derivative is alternately positive and negative at these points, there are turning points.

Function Graph:

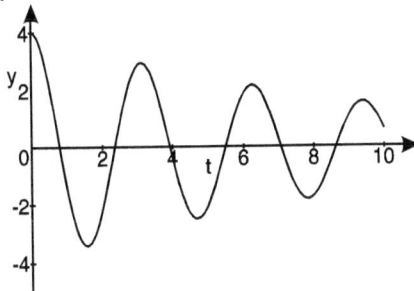

Figure 7.16. Function graph of $x(t) = 4 \exp(-0.1t) \cos(2t)$.

Range of Values: $\mathbb{W} = [-3.422, 4]$.   □

7.6 Extreme Value Problems (Optimization Problems)

Differential calculus is now used to solve optimization problems, where we search for the absolute maximum or minimum of a problem under constraints.

Usually, technical problems are transformed into a mathematical model (target function) and the solution of the problem is attributed to the determination of extreme values of this function. The absolute maximum (or minimum) can also be located at a boundary point of the interval. For simple extreme value problems, the following scheme can be chosen:

Extreme Value Problems

(1) Approach: The problem to be optimized is described mathematically. Due to boundary conditions, all variables except one are eliminated. The quantity to be optimized is then a function (*target function*) of this variable.

(2) The extremes of the target function are determined and their compatibility with the given constraints is checked.

(3) By comparing the extreme values with the function values at the boundary points, the **absolute** largest (or smallest) value of the function is obtained.

Application Example 7.27 (AC Circuit).

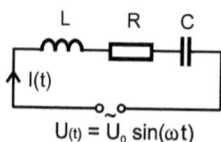

In a RCL circuit, an alternating voltage is applied

$$U(t) = U_0 \sin(\omega t)$$

leading to an alternating current

$$I(t) = I_0 \sin(\omega t + \varphi)$$

Figure 7.17. RLC circuit

$U(t) = U_0 \sin(\omega t)$

with a frequency-dependent amplitude

$$I_0(\omega) = U_0 \frac{1}{\sqrt{R^2 + \left(\omega L - \frac{1}{\omega C}\right)^2}}.$$

At what frequency ω does I_0 have its maximum value?

$I_0(\omega)$ is maximum when $\sqrt{R^2 + \left(\omega L - \frac{1}{\omega C}\right)^2}$ is minimum, or when the following function f becomes minimum:

$$f(\omega) = R^2 + \left(\omega L - \frac{1}{\omega C}\right)^2$$

$$f'(\omega) = 2(\omega L - \frac{1}{\omega C}) \cdot (L + \frac{1}{\omega^2 C}) \overset{!}{=} 0 \qquad \Rightarrow \omega_0 = \frac{1}{\sqrt{LC}}$$

$$f''(\omega) = 2(L + \frac{1}{\omega^2 C})^2 - \frac{4}{\omega^3 C}(\omega L - \frac{1}{\omega C}) \quad \Rightarrow f''\left(\frac{1}{\sqrt{LC}}\right) = 8L^2 > 0.$$

f has its relative minimum at $\boxed{\omega_0 = \frac{1}{\sqrt{LC}}}$ and therefore $I_0(\omega)$ has the maximum value $I_0(\omega_0) = \frac{U_0}{R}$.

Discussion: We check the values at the boundaries $\omega = 0$ and $\omega \to \infty$. Because of $I_0(\omega) \overset{\omega \to 0}{\longrightarrow} 0$ and $I_0(\omega) \overset{\omega \to \infty}{\longrightarrow} 0$, the relative maximum is also the absolute maximum. So I_0 has its maximum at the resonance frequency ω_0. The impedance is then equal to the ohmic resistance R. □

Application Example 7.28 (Power Adjustment at a Resistor).

An external resistance R (load resistance) is connected to a source with constant voltage U_0 and an internal resistance R_i. How large must the external resistance be to maximize the effective power?

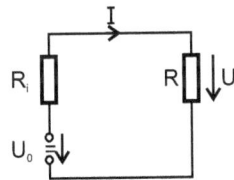

Figure 7.18. Electric circuit

The power at the resistor is $P = U \cdot I = R \cdot I \cdot I$. An additional condition is given by the mesh rule, which says that $U_0 = U_i + U = R_i I + R I = (R_i + R) I$. From this condition we obtain a constraint for the current

$$\Rightarrow I = \frac{U_0}{R_i + R}.$$

This gives the net power as a function of R:

$$P(R) = R \left(\frac{U_0}{R_i + R}\right)^2 = U_0^2 \frac{R}{(R_i + R)^2} \qquad \textbf{(Net Power)}$$

For the extreme value

$$P'(R) = U_0^2 \frac{1 \cdot (R_i + R)^2 - R \cdot 2 (R_i + R)}{(R_i + R)^4} = U_0^2 \frac{R_i - R}{(R_i + R)^3} \overset{!}{=} 0$$

$$\Rightarrow \boxed{R = R_i\,.}$$

For the second derivative of $P(R)$, the following applies

$$P''(R) = U_0^2 \frac{-1\,(R_i + R)^3 - (R_i - R)\,3\,(R_i + R)^2}{(R_i + R)^6}$$

$$= U_0^2 \frac{-(R_i + R) - 3\,(R_i - R)}{(R_i + R)^4}.$$

With this

$$P''(R_i) = U_0^2 \frac{-2\,R_i}{16\,R_i^4} = -\frac{1}{8} \frac{U_0^2}{R_i^3} < 0.$$

Discussion: For $\boxed{R = R_i}$ we obtain a relative extreme value. We check if the local maximum is also the global maximum: Since $P(R) \xrightarrow{R \to \infty} 0$ and $P(R) \xrightarrow{R \to 0} 0$, the relative maximum is the absolute maximum. The effective power becomes maximum when the load resistance is equal to the internal resistance. The maximum net power at this external resistance is

$$P_{\text{max}} = \frac{1}{4} \frac{U_0^2}{R_i}.$$

This is also known as the voltage divider rule: If the internal resistance R_i is equal to the external resistance R (=load resistance), then the voltage drop across R_i is equal to the voltage drop across R, i.e. $\frac{U_0}{2}$.   □

Application Example 7.29 (Beam bending).

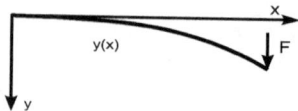

y(x)

F

y

Figure 7.19. Beam bending

The bending line of a beam of length L clamped on the left side deforms under the influence of a force F approximately as follows

$$\boxed{y(x) = \frac{F}{2\,E\,I}\,\left(L\,x^2 - \tfrac{1}{3}\,x^3\right),}$$

where E is the elasticity, I is the moment of inertia and $0 \le x \le L$. Where is the maximum deflection of the beam?

We find the relative extremes of the bending $y(x)$:

$$y'(x) = \frac{F}{2\,E\,I}\,(2\,L\,x - x^2) = 0 \Rightarrow x_1 = 0 \quad \text{or} \quad x_2 = 2\,L$$

$$y''(x) = \frac{F}{2\,E\,I}\,(2\,L - 2\,x)$$

and insert $x_1 = 0$, $x_2 = 2\,L$ into the 2nd derivative:

$$y''\,(0) = \frac{F\,L}{E\,I} > 0 \quad \Rightarrow \quad x_1 \text{ is a relative minimum;}$$

$$y''\,(2\,L) = -\frac{F\,L}{E\,I} < 0 \Rightarrow x_2 \text{ is a relative maximum.}$$

However, x_2 is outside the physical range $(0 \le x \le L)$.

⚠ The **maximum** deflection of the bend (absolute maximum), however, is at the free end $(x = L)$:

$$y_{\text{max}} = y\,(L) = \frac{F\,L^3}{3\,E\,I}\,. \qquad\qquad \square$$

7.7 Theorems on Differential Calculus

This section summarizes some important theorems of the differential calculus. The central theorem of differential calculus is the mean value theorem. We will use this theorem, for example, to prove the important rules of *l'Hospital* to be able to calculate zero divided by zero.

We start with a theorem which says that the differentiability of a function is a stronger property than its continuity:

> **Theorem:** Let $f : \mathbb{D} \to \mathbb{R}$ be a function that is differentiable at point $x_0 \in \mathbb{D}$, then f is also continuous at that point.

Proof: If f is differentiable at x_0, then the rules for computing limits apply:

$$\lim_{h \to 0} (f\,(x_0 + h) - f\,(x_0)) = \lim_{h \to 0} [h \cdot (f\,(x_0 + h) - f\,(x_0))\,/h]$$

$$= \lim_{h \to 0} h \cdot \lim_{h \to 0} \frac{1}{h}\,(f\,(x_0 + h) - f\,(x_0))$$

$$= 0 \cdot f'\,(x_0)\ =\ 0$$

$$\Rightarrow \quad \lim_{h \to 0} f\,(x_0 + h) = f\,(x_0)\,. \qquad\qquad \square$$

⚠ The inversion of this theorem does not hold, because for example the absolute value function $abs\,(x) = |x|$ at the position $x_0 = 0$ is continuous but not differentiable!

7.7.1 Theorem of the Exponential Function

An important property of the exponential function is that the derivative is again e^x: $(e^x)' = e^x$. It is the only function that has this property:

> **Property of the Exponential Function**
> _____
>
> Let a be a constant and $f : \mathbb{R} \to \mathbb{R}$ be a differentiable function with $f'(x) = a f(x)$ for all $x \in \mathbb{R}$.
> Then $f(x) = f(0) e^{ax}$ for all $x \in \mathbb{R}$.

Proof: Let's define the function $F(x) = f(x) \cdot e^{-ax}$. Then the derivative of $F(x)$ is calculated using the product rule:

$$\Rightarrow F'(x) = f'(x) e^{-ax} + f(x) e^{-ax} (-a)$$
$$= (f'(x) - a f(x)) e^{-ax} = 0 \quad \text{for all } x \in \mathbb{R}.$$

Since the derivative of F is zero, $F(x)$ must be the constant function with $F(0) = f(0) = const$ and so $F(x) = f(0)$ for all $x \in \mathbb{R}$.

$$\Rightarrow f(x) = F(x) \cdot e^{ax} = f(0) \cdot e^{ax} \quad \text{for all } x \in \mathbb{R}.$$
$$\Rightarrow f(x) = f(0) \cdot e^{ax}$$

$\square$

Consequences:

(1) $\exp : \mathbb{R} \to \mathbb{R}$ with $x \mapsto e^x$ is the only differentiable function with $f'(x) = f(x)$ and $f(0) = 1$.

(2) The solution of the differential equation

$$y'(x) = a y(x) \quad \text{with } y(0) = y_0$$

is given by $y(x) = y_0 e^{ax}$.

(3) The solution of the differential equation

$$y'(x) = a y(x) + f(x) \quad \text{with } y(0) = y_0$$

is unique.

7.7.2 Mean Value Theorem

An obvious statement is that every differentiable function $f : [a, b] \to \mathbb{R}$ with $f(a) = f(b)$ has an interior point with a horizontal tangent. This statement is called *Rolle's Theorem* (1652 - 1719).

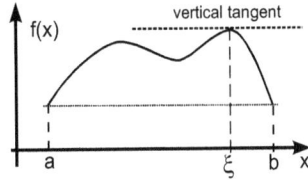

Figure 7.20. Rolle's Theorem

Theorem of Rolle

Let $f : [a, b] \to \mathbb{R}$ be a differentiable function with $f(a) = f(b)$. Then there exists an intermediate point $\xi \in (a, b)$ with

$$f'(\xi) = 0.$$

An extension of this statement gives the following mean value theorem, which we trace back to Rolle's theorem:

Mean Value Theorem

Let $f : [a, b] \to \mathbb{R}$ be a differentiable function. Then there exists an intermediate point $\xi \in (a, b)$ with

$$f'(\xi) = \frac{f(b) - f(a)}{b - a}.$$

Proof: We define the function $F : [a, b] \to \mathbb{R}$ according to

$$F(x) = f(x) - \frac{f(b) - f(a)}{b - a}(x - a).$$

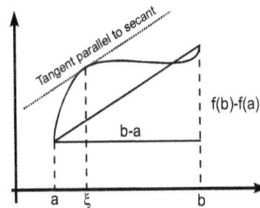

Figure 7.21. Mean value theorem

F is continuous and differentiable with $F(a) = f(a) = F(b)$. According to Rolle's theorem, there exists a $\xi \in (a, b)$ with

$F'(\xi) = f'(\xi) - \frac{f(b) - f(a)}{b - a} = 0.$

This checks the statement.   □

Geometrically the mean value theorem says, that there is at least one point where the graph of the function has a tangent parallel to the secant defined by the points $(a, f(a))$ and $(b, f(b))$.

7.7.3 l'Hospital's Rule

We use the mean value theorem to justify the rule of l'Hospital. The rule of l'Hospital (1661 - 1704) provides a method to calculate the limit $\lim\limits_{x \to x_0} \frac{f(x)}{g(x)}$ if $g(x_0) = f(x_0) = 0$. So to calculate expressions of the form zero by zero:

Rule of l'Hospital

If f and g are continuously differentiable in x_0 and $f(x_0) = g(x_0) = 0$, then we have

$$\lim_{x \to x_0} \frac{f(x)}{g(x)} = \lim_{x \to x_0} \frac{f'(x)}{g'(x)}.$$

Proof: According to the mean value theorem, we get for $a = x_0$ and $b = x_0 + h$

$$f(x_0 + h) = f(x_0) + h\,f'(\xi) \quad \text{when} \quad \xi \in (x_0, x_0 + h) .$$

With $\delta \in (0, 1)$ we can rewrite $\xi = x_0 + \delta\,h$. Hence, $f(x_0 + h) = f(x_0) + h\,f'(x_0 + \delta \cdot h)$. Applying this form of the mean value theorem to f and g, together with $f(x_0) = g(x_0) = 0$, we get

$$\frac{f(x_0 + h)}{g(x_0 + h)} = \frac{f(x_0) + h\,f'(x_0 + \delta_1 h)}{g(x_0) + h\,g'(x_0 + \delta_2 h)} = \frac{f'(x_0 + \delta_1\,h)}{g'(x_0 + \delta_2\,h)} .$$

The statement follows as $h \to 0$. $\qquad\qquad\qquad\qquad\qquad\qquad\qquad$ $\square$

Remarks:

(1) The rule of l'Hospital also applies to limits of the form "$\frac{\infty}{\infty}$":
If $\lim\limits_{x \to x_0} f(x) = \infty$ and $\lim\limits_{x \to x_0} g(x) = \infty$, then

$$\lim_{x \to x_0} \frac{f(x)}{g(x)} = \lim_{x \to x_0} \frac{f'(x)}{g'(x)} .$$

(2) The rules of l'Hospital also apply to the limits $x \to \infty$:

$$\lim_{x \to \infty} \frac{f(x)}{g(x)} = \lim_{x \to \infty} \frac{f'(x)}{g'(x)},$$

when $\left(\lim_{x \to \infty} f(x) = \lim_{x \to \infty} g(x) = 0 \right)$ or $\left(\lim_{x \to \infty} f(x) = \lim_{x \to \infty} g(x) = \infty \right)$.

(3) The l'Hospital rules always assume that the functions are differentiable in an environment of x_0.

(4) Sometimes l'Hospital rules have to be applied several times to reach the goal. But, there are also cases where the repeated application of the rules fails.

Examples 7.30:

① $\lim_{x \to 0} \dfrac{\sin x}{x} \overset{\frac{0}{0}}{=} \lim_{x \to 0} \dfrac{\cos x}{1} = 1.$

② $\lim_{x \to 0} \dfrac{e^x - 1}{2x} \overset{\frac{0}{0}}{=} \lim_{x \to 0} \dfrac{e^x}{2} = \dfrac{1}{2}.$

③ $\lim_{x \to 1} \dfrac{1 + \cos(\pi x)}{x^2 - 2x + 1} \overset{\frac{0}{0}}{=} \lim_{x \to 1} \dfrac{-\sin(\pi x)\pi}{2x - 2} \overset{\frac{0}{0}}{=} \lim_{x \to 1} \dfrac{-\cos(\pi x)\pi^2}{2} = \dfrac{\pi^2}{2}.$

④ $\lim_{x \to \infty} \dfrac{3x^2 - 5}{x + 4x^2} \overset{\frac{\infty}{\infty}}{=} \lim_{x \to \infty} \dfrac{6x}{8x + 1} \overset{\frac{\infty}{\infty}}{=} \lim_{x \to \infty} \dfrac{6}{8} = \dfrac{3}{4}.$

⑤ $\lim_{x \to \infty} \dfrac{\ln(2x - 1)}{e^x} \overset{\frac{\infty}{\infty}}{=} \lim_{x \to \infty} \dfrac{\frac{2}{2x-1}}{e^x} = \lim_{x \to \infty} \dfrac{2}{(2x - 1)\, e^x} = 0.$ □

⊘ **Expressions of the form** $0 \cdot \infty$, $\infty - \infty$, 0^0, ∞^0, 1^∞

The rules of l'Hospital apply only to indefinite expressions of the form "$\frac{0}{0}$" or "$\frac{\infty}{\infty}$". Other indefinite expressions such as $0 \cdot \infty$, $\infty - \infty$, 0^0, ∞^0, 1^∞ can be reduced to one of the above cases by the following elementary transformations:

Calculating $0 \cdot \infty$, $\infty - \infty$, 0^0, ∞^0, 1^∞

	Expression	$\lim\limits_{x \to x_0} \varphi(x)$	Transformation
(A)	$u(x) \cdot v(x)$	$0 \cdot \infty$	$\dfrac{u(x)}{1/v(x)}$ or $\dfrac{v(x)}{1/u(x)}$
(B)	$u(x) - v(x)$	$\infty - \infty$	$\dfrac{1/v(x) - 1/u(x)}{1/(u(x) \cdot v(x))}$
(C)	$u(x)^{v(x)}$	0^0, 0^∞, ∞^0, 1^∞	$\exp(v(x)\ln(u(x)))$

Examples 7.31:

① $\lim\limits_{x \to \infty} x \cdot \ln\left(1 + \frac{1}{x}\right) \overset{(A)}{=} \lim\limits_{x \to \infty} \dfrac{\ln\left(1 + \frac{1}{x}\right)}{\frac{1}{x}} \overset{\frac{0}{0}}{=} \lim\limits_{x \to \infty} \dfrac{\frac{1}{1+1/x} \cdot \left(-\frac{1}{x^2}\right)}{\left(-\frac{1}{x^2}\right)}$

$$= \lim_{x \to \infty} \dfrac{1}{1 + \frac{1}{x}} = 1.$$

② $\lim\limits_{x \to 0} \left(\frac{1}{x} - \frac{1}{\sin x}\right) \overset{(B)}{=} \lim\limits_{x \to 0} \dfrac{\sin x - x}{x \cdot \sin x} \overset{\frac{0}{0}}{=} \lim\limits_{x \to 0} \dfrac{\cos x - 1}{\sin x + x \cdot \cos x}$

$$\overset{\frac{0}{0}}{=} \lim_{x \to 0} \dfrac{-\sin x}{2\cos x - x \cdot \sin x} = \dfrac{0}{2} = 0.$$

③ $\lim\limits_{x \to \infty} \left(1 + \frac{1}{x}\right)^x \overset{(C)}{=} \lim\limits_{x \to \infty} e^{x \ln\left(1 + \frac{1}{x}\right)}$.

After ① is $\lim\limits_{x \to \infty} x \ln\left(1 + \frac{1}{x}\right) = 1$ and therefore

$$\lim_{x \to \infty} \left(1 + \dfrac{1}{x}\right)^x = \exp\left(\lim_{x \to \infty} x \ln\left(1 + \dfrac{1}{x}\right)\right) = e^1 = e.$$

④ $\lim\limits_{x \to 0} (1 + tx)^{\frac{1}{x}} \overset{(C)}{=} \lim\limits_{x \to 0} \exp(\frac{1}{x} \ln(1 + tx))$.

Because $\lim\limits_{x \to 0} \dfrac{\ln(1 + tx)}{x} \overset{\frac{0}{0}}{=} \lim\limits_{x \to 0} \dfrac{\frac{1}{1+tx} \cdot t}{1} = t$ holds

$$\lim_{x \to 0} (1 + tx)^{\frac{1}{x}} = \exp\left(\lim_{x \to 0} \dfrac{1}{x} \ln(1 + tx)\right) = \exp(t) = e^t. \qquad \square$$

7.8 Newton's Method

The *Newton method* is a fast, numerical method for approximating the zeros of functions. So, we are introducing a method to approximately calculate a zero of a function f

$$f(x) = 0$$

in the interval $[a, b]$.

The method: The Newton method is an iterative method, i.e. an initial estimate x_0 for the zero point is used and this value is improved in further iteration steps: We calculate the tangent of f at the point x_0 and determine the intersection of the tangent with the x-axis. This value is x_1. x_1 is usually closer to the zero point than x_0. Now the tangent of the function is calculated at x_1 and the intersection of the axes is calculated at x_2. Continuing the procedure, the zero point is approached (see Fig. 7.22).

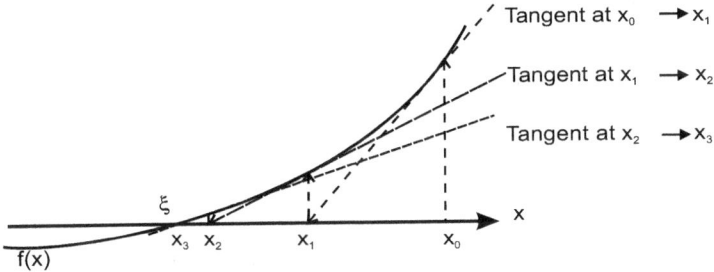

Figure 7.22. Geometric interpretation of Newton's method

Setting up the formulas: The tangent equation at x_0 has the form

$$y = f(x_0) + f'(x_0)(x - x_0)$$

and the intersection x_1 with the x-axis is defined by $y = 0$:

$$0 = f(x_0) + f'(x_0)(x_1 - x_0).$$

This means

$$x_1 = x_0 - \frac{f(x_0)}{f'(x_0)}.$$

The following procedure is obtained by iteration:

Algorithm (Newton's Method)

(1) **Initialization:** Choose initial value x_0; $\quad \delta := 10^{-5}$.

(2) **Iteration:** $\quad x_{n+1} = x_n - \dfrac{f(x_n)}{f'(x_n)} \qquad n = 0, 1, 2, 3, \ldots$

(3) **Stop condition:**

If $|x_{n+1} - x_n| < \delta$, then $\xi = x_{n+1}$. Stop.
If $|x_{n+1} - x_n| \geq \delta$, then continue with (2).

Features of the Method

(1) Very fast convergence (only a few iterations are needed).

(2) ⚠ The procedure may diverge if the initial value is not close enough to the zero.

(3) For the Newton method, the error estimate is

$$|x_{n+1} - x^*| \leq L^n \, |x_{n+1} - x_1| \leq L^n \, (b - a) \,,$$

when $|f'(x)| \leq L$ for all $x \in I = [a, b]$.

(4) There must be an explicit formula for the derivative.

(5) The Newton method always converges for a convex or concave function that has a zero with $f'(x_0) \neq 0$.

(6) The convergence of Newton's method can also be guaranteed if f is a function that can be continuously differentiated 3 times with the property

$$f'(x) \neq 0 \text{ for all } x \in I \quad \text{and} \quad \left| \frac{f(x) \, f''(x)}{f'(x)^2} \right| \leq K < 1 \text{ for all } x \in I.$$

The smaller the constant K, the better the convergence.

Hint: If there is a bad estimate for the initial value, first use the bisection method ($\rightarrow$ 6.4) to get a better initial approximation, and then use Newton's method. Often a good initial value x_0 can be found by drawing and examining the function graph.

Example 7.32 (With MAPLE-Worksheet). Find the solution to the equation $1 - \frac{1}{5}z = e^{-z}$, i.e. find the zero of the function

$$f(z) = 1 - \frac{1}{5}z - e^{-z}.$$

Using the derivative

$$f'(z) = -\frac{1}{5} + e^{-z},$$

the corresponding Newton sequence

$$x_{n+1} = x_n - \frac{1 - \frac{1}{5}x_n - e^{-x_n}}{-\frac{1}{5} + e^{-x_n}} = x_n - \frac{(5 - x_n)\,e^{x_n} - 5}{-e^{x_n} + 5}$$

is set up. With an initial value of $x_0 = 2$, the result is

n	x_n	$f(x_n)$
0	2	0.4646
1	9.1857	$-.8372$
2	4.9973	$-.00622$
3	4.9651	$-0.3 \cdot 10^{-5}$

After 3 iterations, an approximation of the equation $z \approx 4.9651$ is obtained with an accuracy of 4 decimal places. □

Animation: The algorithm of Newton's method is written directly into the procedure **Newton newton**. The animation visualizes the convergence process of the Newton sequence to zero.

⊘ **Applying Newton's Method to Roots**

We want to find the square root $\sqrt{a}$ of a positive number a. We interpret $\sqrt{a}$ as the positive zero of the function

$$f(x) = x^2 - a.$$

For the numerical calculation, we apply Newton's method to this function. Using

$$f'(x) = 2x$$

we obtain the Newton sequence

$$x_{n+1} = x_n - \frac{f(x_n)}{f'(x_n)} = x_n - \frac{x_n^2 - a}{2\,x_n} = \frac{2\,x_n^2 - x_n^2 + a}{2\,x_n} = \frac{1}{2}\left(x_n + \frac{a}{x_n}\right).$$

If $x_0 := a$ is set and iterated according to

$$x_{n+1} := \frac{1}{2}\left(x_n + \frac{a}{x_n}\right)$$

for $n = 0, 1, 2, \ldots$, there is a fast converging sequence (see Example 6.4: Babylonian root extraction).

Example 7.33. Calculation of $\sqrt{3}$ with the above scheme

n	0	1	2	3	4	5
x_n	3.00000	2.00000	1.75000	1.73214	1.73205	1.73205

The given method is one of the best for calculating square roots. Most computer programs are based on it.   □

⊘ Calculation of k-th Roots

This method is not limited to calculating square roots, but can also be used to calculate k-th roots. Since $\sqrt[k]{a}$ is the positive zero of

$$f(x) = x^k - a.$$

With $f'(x) = k\,x^{k-1}$ the Newton sequence is

$$x_{n+1} = x_n - \frac{f(x_n)}{f'(x_n)} = x_n - \frac{x_n^k - a}{k x_n^{k-1}} = \frac{k-1}{k} x_n + \frac{a}{k x_n^{k-1}}$$

or

$$x_{n+1} = \frac{1}{k}\left[(k-1)\,x_n + \frac{a}{x_n^{k-1}}\right] \qquad n = 0, 1, 2, 3, \ldots.$$

Example 7.34. Calculation of $\sqrt[3]{8}$ with the above scheme. For this we set $a = 8$ and $k = 3$ and choose 6 for the number of iterations.

n	0	1	2	3	4	5	6
x_n	8.00000	5.37500	3.67563	2.64780	2.145565	2.00965	2.00004

7.9 Problems on Differential Calculus

7.1 Determine the first derivative of the functions using the power rule:

a) $y = 8x^7 - 10x^3 + \frac{10}{x^3} - \frac{8}{x^7}$ b) $y = 12\sqrt[4]{x^3} - 7\sqrt[7]{x^4} + 11x - \frac{8}{\sqrt{x^3}}$

c) $y(l) = 2\sqrt[4]{\sqrt[15]{l}} + 3\sqrt[5]{\sqrt[12]{l}} - 3\sqrt[3]{\sqrt[20]{l}} - 3\sqrt[6]{\sqrt[10]{l}}$ d) $y(a) = \frac{9}{a^5\sqrt{a^3}}$

e) $y(x) = \left(a + bx^2\right)(c + ex)^3$ f) $y(x) = \left(x^3 + x^2\right)\sqrt{x}$

g) $y(x) = x^\alpha x^\beta$

7.2 Determine the derivative of the functions using product and quotient rule:

a) $\sin(x) \cdot \dfrac{10}{x^3}$ b) $\sin(x) \cdot \cos(x)$ c) $x^n e^x$

d) $\dfrac{x^2 - 5x + 6}{x^2 - 12x + 20}$ e) $\dfrac{\sin(\varphi)}{(1 - \cos(\varphi))}$ f) $\dfrac{4t}{(t^2 - 1)(t + 1)}$

g) $\dfrac{x}{x^2 + 2}$ h) $\dfrac{x^2 e^x}{e^x - 1}$ i) $\dfrac{x \cdot \ln(x)}{(x - 1)^2}$

7.3 Determine the first derivative of the functions applying the chain rule

a) $y(x) = \cos(3x + 2)$ b) $y(x) = (3x - 2)^3$

c) $y(x) = 3 \cdot \sin(5x)$ d) $y(x) = e^{4x^2 - 3x + 2}$

e) $y(x) = 10 \cdot \ln(1 + x^2)$ f) $x(t) = A \cdot \sin(\omega t + \varphi)$

g) $y(x) = \ln(\sin(2x - 3))$ h) $y(x) = \sqrt{\ln(x^2 - 1)}$

7.4 Calculate by logarithmic differentiation the derivative of

a) $y(x) = x^x$ b) $y(x) = x^{\sin x}$

7.5 What is the derivative of

a) $f_1(x) = x^{(x^x)}$ b) $f_2(x) = (x^x)^x$ c) $f_3(x) = x^{(x^a)}$

d) $f_4(x) = x^{(a^x)}$ e) $f_5(x) = a^{(x^x)}$

7.6 Determine the first derivative of

a) $y(t) = \ln\sqrt{a^2 - t^2}$ b) $y = \ln\sqrt{\frac{1 - x^2}{1 + x^2}}$ c) $y = \ln\frac{(x-5)^3}{(x+1)^2}$

d) $y(x) = a^{\ln(x-3)}$ e) $y = e^x \cdot \sqrt{\frac{1+x}{1-x}}$ f) $y = e^{\ln x}$

7.7 Given are the functions

$\sinh : \mathbb{R} \to \mathbb{R}$ with $\sinh(x) := \frac{1}{2}\left(e^x - e^{-x}\right)$ (Hyperbolic sinus)

$\cosh : \mathbb{R} \to \mathbb{R}$ with $\cosh(x) := \frac{1}{2}\left(e^x + e^{-x}\right)$ (Hyperbolic cosine)

$\tanh : \mathbb{R} \to \mathbb{R}$ with $\tanh(x) := \dfrac{\sinh(x)}{\cosh(x)}$ (Hyperbolic tangent)

i) Draw the graph of the 3 hyperbolic functions.

ii) Calculate the derivative of the functions.

iii) Show that $\cosh^2(x) - \sinh^2(x) = 1$.

7.8 Calculate the derivatives of the arc functions

$$\arcsin(x),\ \arccos(x),\ \arctan(x),\ \operatorname{arccot}(x)$$

as the derivative of the inverse function of the trigonometric functions.

7.9 Calculate the derivative of the area functions

$$\text{arsinh}(x) \quad \text{and} \quad \text{arcosh}(x)$$

as derivative of the inverse function of sinh and cosh.

7.10 Prove the power rule $y(x) = x^n \Rightarrow y'(x) = n x^{n-1}$ applying logarithmic differentiation.

7.11 Calculate the first derivative of the implicitly given functions
a) $e^{x \cdot y(x)} + y^3(x) \ln x = \cos(2x)$

b) $y^{\left(e^{-x\, y(x)}\right)} = \left(\sqrt[x]{ y(x) \sqrt{a^2 x + 2\, a^2 x} } \right)^{-\frac{x}{y(x)}}$

c) $\ln y(x) - \sqrt{y(x)} - x = 0$
d) $\sin y(x) = y(x) \cdot x^2$

7.12 Determine by implicit differentiation the slope of the circular rod at the point $P_0 = (4,\ y_0 > 0)$ of the circle $(x - 2)^2 + (y - 1)^2 = 25$.

7.13 Given are the functions
a) $f_1(x) = \sqrt{1 + x^4}$; $x_0 = 1$ b) $f_2(x) = 3 \ln (1 + 3 x^5)$; $x_0 = 3$
c) $y(x) = 2 \cos x$; $x_0 = \frac{\pi}{4}$.
Calculate for the functions   i) the total differential   ii) the total differential at point x_0 iii) the tangent at point x_0 and   iv) the linearization at the point x_0.   v) Express an approximate value for $f(x_0 + 0.01)$ and compare it with the exact value.

7.14 A steamed spring-mass-system has the space-time-law

$$x(t) = A e^{-\gamma t} \cos(\omega t).$$

i) Calculate the velocity and acceleration at any time.
ii) Give a condition for the secondary maximum.

7.15 The potential energy for an ion in a crystal lattice is approximated by

$$V(r) = -D \left(\frac{2a}{r} - \frac{a^2}{r^2} \right) \quad (D > 0).$$

Show that $V(r)$ has a relative minimum at $r_0 = a$.

7.16 For the mirror dimension with scale and telescope, the deflection x is measured at a fixed scale distance s. How does a small measurement error of x influence the value of the result α, if $\alpha = \arctan \frac{x}{s}$? ($s = 2\,m$, $x = 250\,mm$, $dx = 1\,mm$.) What is the relative error?

7.17 Where do the following functions have relative extremes?
a) $y(x) = -8 x^3 + 12 x^2 + 18 x$ b) $z(t) = t^4 - 8 t^2 + 16$
c) $u(z) = \sqrt{1 + z} + \sqrt{1 - z}$ d) $y(x) = x e^{-x}$
e) $y(x) = \sin x \cdot \cos x$ f) $y(x) = \frac{2x - 2x^2}{x^2 - x - 6}$

7.18   Discuss the slope of the following functions:

a) $y = \dfrac{x^2 + 1}{x - 3}$
b) $y = \dfrac{(x - 1)^2}{x + 1}$
c) $y = \dfrac{\ln x}{x}$
d) $y = \sin^2 x$

7.19   Determine the function values applying the rules of l'Hospital

a) $\displaystyle\lim_{x \to a} \dfrac{x^2 - a^2}{x - a}$
b) $\displaystyle\lim_{x \to 0} \dfrac{\sin(2x)}{\sin(x)}$
c) $\displaystyle\lim_{x \to 0} \dfrac{\sin^2 x}{1 - \cos x}$

d) $\displaystyle\lim_{x \to 0} \dfrac{x^2 - 2 + 2\cos x}{x^4}$
e) $\displaystyle\lim_{x \to 0} \left(\dfrac{1}{x} - \dfrac{1}{\sin x}\right)$
f) $\displaystyle\lim_{x \to 1} \dfrac{\ln x - x + 1}{(x - 1)^2}$

g) $\displaystyle\lim_{x \to 0} x^x$
h) $\displaystyle\lim_{x \to \infty} \left(1 + \dfrac{a}{x}\right)^x$

Index

Homepage of the Book: Additional Material

iMath: iMath is an interactive maths application: In this pedagogically appealing app, easy-to-understand exercises from this book are solved in detail. The app can also be used for exam preparation. It can be launched directly from

http://www.imathhome.de/iMatheWeb

YouTube videos: On the Westermann YouTube channel, many of the topics covered in this book are explained in detail in short videos. In the form of summaries, the most important aspects are summarised in a short and easily understandable way. The corresponding links to the videos can be found at

https://www.youtube.com/channel/UChzktnND8kk9pmwQmybSx-w

Additional material is available on the homepage of the book. All information, MAPLE procedures, MAPLE worksheets and additional chapters can be downloaded free of charge from

http://www.imathhome.de/books/mathe/start.htm

MAPLE-**Worksheets:** All the MAPLE worksheets for the problems and examples mentioned in the text. In particular, the worksheets for all visualizations can be found here.

Animations: All animations shown or described in the text are available on the homepage as Animated-Gif, so that they can be started directly in the browser.

http://www.imathhome.de/animations

Additional chapters, which are not available in printed form, such as
Numerical solution of equations;
Numerical differentiation and integration;
Numerical solution of differential equations.

Solutions to the exercises: Complete solutions are given for all problems.